Advances in Experimental Medicine and Biology

Volume 1351

Advances in Experimental Medicine and Biology provides a platform for scientific contributions in the main disciplines of the biomedicine and the life sciences. This series publishes thematic volumes on contemporary research in the areas of microbiology, immunology, neurosciences, biochemistry, biomedical engineering, genetics, physiology, and cancer research. Covering emerging topics and techniques in basic and clinical science, it brings together clinicians and researchers from various fields.

Advances in Experimental Medicine and Biology has been publishing exceptional works in the field for over 40 years, and is indexed in SCOPUS, Medline (PubMed), Journal Citation Reports/Science Edition, Science Citation Index Expanded (SciSearch, Web of Science), EMBASE, BIOSIS, Reaxys, EMBiology, the Chemical Abstracts Service (CAS), and Pathway Studio.

2020 Impact Factor: 2.622

More information about this series at https://link.springer.com/bookseries/5584

Dong-Wook Han • Suck Won Hong
Editors

Multifaceted Biomedical Applications of Graphene

Springer

Editors
Dong-Wook Han
Department of Cogno-Mechatronics Engineering
College of Nanoscience and Nanotechnology, Pusan National University
Busan, South Korea

Suck Won Hong
Department of Cogno-Mechatronics Engineering
College of Nanoscience and Nanotechnology, Pusan National University
Busan, South Korea

ISSN 0065-2598   ISSN 2214-8019 (electronic)
Advances in Experimental Medicine and Biology
ISBN 978-981-16-4922-6   ISBN 978-981-16-4923-3 (eBook)
https://doi.org/10.1007/978-981-16-4923-3

This Springer imprint is published by the registered company Springer Nature Singapore Pte Ltd.
The registered company address is: 152 Beach Road, #21-01/04 Gateway East, Singapore 189721, Singapore

Preface

During the past few decades, two-dimensional (2D) nanomaterials have attracted enormous interests due to their outstanding performance in various areas ranging from electronics, photonics, biomedicine, and energy storage/harvesting. Likewise, graphene family nanomaterials have emerged since 2010 as functional 2D materials to offer a wide spectrum of technological opportunities.

Owing to their exceptional physicochemical and thermomechanical characteristics as well as biological and optoelectronical properties, multifunctional graphene-based nanomaterials (MFGNs) are promising for many applications, especially in the biomedical fields such as drug delivery, bioimaging, biosensor, theranostics, therapeutics, and tissue engineering. Recently, much attention has been paid to the potential of MFGNs to stimulate the differentiation of stem cells into specific lineages and a range of strategies for the development of novel biomaterials based on MFGNs to serve as tissue regeneration scaffolds.

Firstly, this book provides an overview of the chemical composition and physical properties of graphene and its derivatives as well as their potential toxicity and biosafety. The book also covers the potential of some graphene-based multidimensional formations including particulates, coatings, matrices, patterns, foams, and 3D structures as new innovative options for stem cell fate modulation and tissue regeneration. Together with these potentials, the book subsequently focuses on the capability of MFGNs as agents for drug delivery, bioimaging, theranostics, and therapeutics as well as their usefulness as potential platforms for nanobiosensors, biochips, and antibacterial agents.

The book is divided into 12 in-depth chapters, each of which is organized to enrich readers' understanding of MFGNs' biofunctionality and their varied biomedical applications. The pivotal roles of this book make it essential reading for scientists and engineers of all the biomedical research fields seeking to fully utilize novel MFGN-incorporated composite materials and develop various functional devices based on them. Besides its value as a fundamental book for graduate students, it offers a well-organized reference guide for all individuals working in biomedical science and engineering.

Busan, South Korea

Dong-Wook Han
Suck Won Hong

Contents

Part I

Basic Features and Potential Toxicity of Graphene Family Nanomaterials

Principles and Biomedical Application of Graphene Family Nanomaterials

Iruthayapandi Selestin Raja, Saifullah Lone, Dong-Wook Han, and Suck Won Hong

Abstract

Two-dimensional graphene family nanomaterials (GFNs) are extensively studied by the researchers for their quantum size effect, large surface area, numerous reactive functional sites, and biocompatibility. The hybrid materials of GFNs exhibit an increased level of mechanical strength, optical, electronic, and catalytic activity due to their incorporation. The application of GFNs in the energy, environment, electric and electronic, personal care, and health sectors is abundant, which is not only by their unique physicochemical properties but also by their ease and large production by various synthetic approaches and economically inexpensiveness. Their general biomedical applications include bioimaging, biosensing, drug delivery, tissue engineering, killing the microbes, and demolishing the cancer tumor. The first chapter of this book describes definitions, synthetic methods, unique properties, and biomedical applications of GFNs, including graphene, graphene oxide, and reduced graphene oxide.

Keywords

Graphene family nanomaterials · Principles · Synthetic approaches · Biomedical applications

Dong-Wook Han and Suck Won Hong equally contributed to this work.

I. S. Raja · S. Lone
BIO-IT Fusion Technology Research Institute, Pusan National University, Busan, South Korea

D.-W. Han · S. W. Hong (✉)
Department of Cogno-Mechatronics Engineering, College of Nanoscience and Nanotechnology, Pusan National University, Busan, South Korea
e-mail: nanohan@pusan.ac.kr; swhong@pusan.ac.kr

D.-W. Han, S. W. Hong (eds.), *Multifaceted Biomedical Applications of Graphene*, Advances in Experimental Medicine and Biology 1351,
https://doi.org/10.1007/978-981-16-4923-3_1

1 Introduction

The wonder material called "graphene" is two-dimensional (2D) atomic-scale graphite with sp^2 hybridized carbon atoms flawlessly amassed in a honeycombed alignment (Lee et al. 2015a; Yoo et al. 2015). Since the ground-breaking discovery of graphene in 2004, the disruptive material and the corresponding compounds, i.e., graphene-family nanomaterials (GFNs) such as pristine graphene, graphene oxide (GO), and reduced graphene oxide (rGO), have received an overwhelming response from chemistry, materials, physics, and biomedical fields (Shin et al. 2017; Shin et al. 2018a; Kang et al. 2020). Their unique physicochemical properties include unparallel electrical conductivity (1738 siemens/m), highest known mechanical strength (about 1100 GPa in perfect crystalline form), high surface area (2630 m^2/g), and exceptional thermal conductivity (5000 W/m/K) (Shin et al. 2015, 2018b; Hu et al. 2017). The family of GFNs and their definitions are provided in Table 1.

Single-layer graphene is thermodynamically unstable in the free state compared to any curved structures like fullerenes and nanotubes (Novoselov et al. 2004). Though pristine graphite is hydrophobic, its derived material, graphite oxide, is hydrophilic due to intercalating water molecules between the layers. The intrinsic oxygen-doped surface of GO has more binding energy towards the metal atoms compared to pristine graphene. Further, GO is economically inexpensive and more readily available than graphene (Fan et al. 2015). Several studies have been carried out to understand the development of GFMs in recent years. The estimated value of graphene in the global market was around US$12 million in 2013 and US$195 million by 2018. It is expected that the value would reach US$1.3 billion by 2023 with a 47.1% of annual growth rate from 2018 to 2023 (Zurutuza and Marinelli 2014).

2 Synthetic Approaches of Graphene Family Nanomaterials

The synthesis routes of GFNs are generally categorized into top-down and bottom-up approaches. The top-down methods include exfoliation of graphite oxide and micromechanical cleavage, whereas bottom-up methods comprise chemical vapor deposition and other methods (Hu et al. 2017). The top-down and bottom-up synthetic routes to prepare GFNs are shown schematically in Fig. 1.

The micromechanical cleavage of highly oriented pyrolytic graphite yields single-layer graphene (SLG) nanosheets by the repeated peeling of small mesas with tape (Tyurnina et al. 2013). Direct observation of graphene nanosheet formation is the advantage of this simple method. However, the micromechanical cleavage method bears some disadvantages as the process requires a high purity condition. The SLG produced deposits randomly on the substrate limiting its application. Further, the method is not apt for large-scale production due to its low reproducibility (Machado and Serp 2011). Mechanical exfoliation of graphite oxide followed by post-treatment using reducing agents is the most advanced method to obtain a high yield of GFNs. Owing to low-cost raw material (graphite) and high production yield,

Table 1 List of graphene family nanomaterials (GFNs) and their definitions

GFMs	Definition	References
Graphene	Graphene is a single-atom-thick sheet with sp^2-bonded carbon atoms that are suspended or adhered to a foreign substrate. The lateral dimensions of graphene can vary from nanometers to the macroscale.	Bianco et al. (2013)
Graphene layer	The graphene layer is a single-atom-thick sheet with sp^2-bonded carbon atoms occurring within a carbon material structure. The "graphene layer" is a structural unit that has been used for many years to describe the structure and texture of 3D carbon materials with primary sp^2-hybridized bonding.	Bianco et al. (2013) and Syama and Mohanan (2019)
Multi-layer graphene	A 2D (sheet-like) material consisting of a number between 2 and about 10 stacked graphene layers of extended lateral dimension. Carbon films contain small lateral graphene layers called "carbon thin films" or "multi-layer graphene."	Bianco et al. (2013)
Graphene nanosheet	A single-atom-thick sheet with sp^2-bonded carbon atoms that is not an integral part of carbon material but is suspended or adhered to a foreign substrate and has a lateral dimension less than 100 nm	Bianco et al. (2013) and McCallion et al. (2016)
Graphene oxide (GO)	GO is characterized by a distinctive structure of sp^2 carbon, oxygen, and hydrogen in variable ratios. In GO, the carbon atoms covalently bonded to oxygen functional groups such as hydroxyl, epoxy, and carboxy containing sp^3 hybrids groups which are displaced above or below the graphene plane.	Marcano et al. (2010), Mas-Ballesté et al. (2011), and Bianco et al. (2013)
Reduced graphene oxide (rGO)	The reduced GO (rGO) sheets are usually considered a kind of chemically derived graphene. The GO can be reduced to graphene-like sheets by removing the oxygen-containing groups with the recovery of a conjugated structure by thermal, chemical, or electrical treatments	Pei and Cheng (2012)

Reproduced from De Marchi et al. 2018, Copyright (2018) Elsevier

this method is the most feasible procedure adopted in industries (Hu et al. 2015). Initially, graphite is converted to graphite oxide introducing many functional groups such as hydroxyl, alkoxy, epoxy, and carboxyl on its surface and within the layers. Subsequently, the graphite oxide is subjected to exfoliation to produce graphene oxide by ultrasonic, mechanical, chemical, or thermal procedures. The interlayer distance of GO and graphite oxide is increased, and the planar sp^2 configuration is

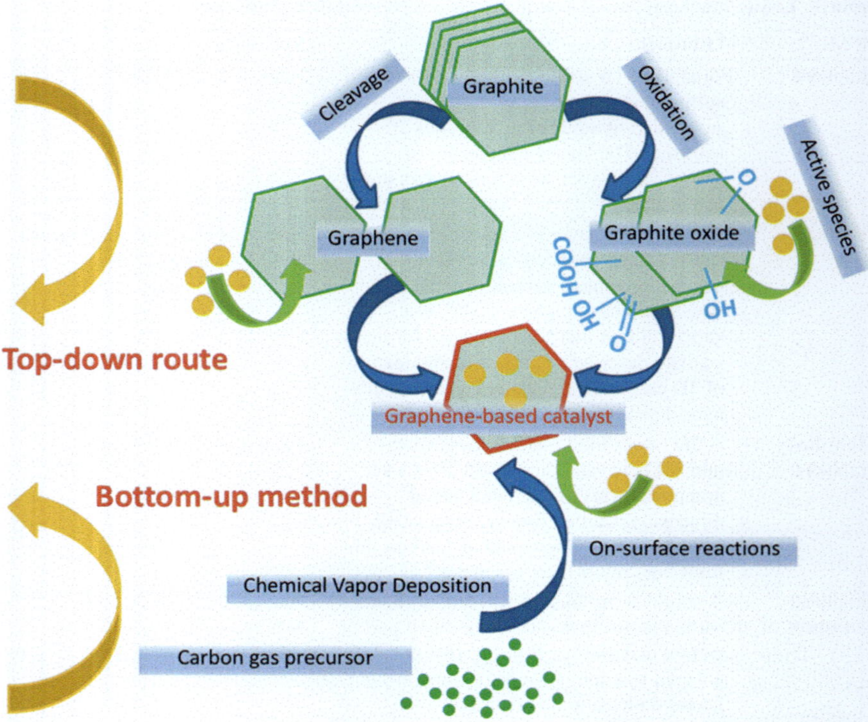

Fig. 1 Top-down and bottom-up synthetic routes for the preparation of graphene-family nanomaterials (GFNs), including graphite, graphite oxide, and graphene (Reproduced from Hu et al. 2017, Copyright (2017) American Chemical Society)

changed to tetrahedral sp^3 by the presence of oxygen-containing functional groups (Hua et al. 2011; Wang and Hu 2011).

Liquid phase exfoliation of graphite has been an emerging method nowadays to produce GFNs as it bypasses the redox reactions, which is reasonable for the materials' poor conductivity (Hernandez et al. 2008). The surface energy of the solvent is the important factor for the high quality and large quantity of the yield. However, the solvents, including N-methyl pyrrolidone, used for the reaction are expensive and need more attention during handling. As the surface energy of water is too high, a water-surfactant solution is used for the process (Lotya et al. 2009). It has been reported that Coulombic repulsion forces stabilize the dispersed graphitic flakes due to the adsorbed surfactant, sodium dodecylbenzene sulfonate (SDBS). Then, FLG sheets are obtained by the ultrasonication of graphite sample in water-SDBS solution (Mao et al. 2013). Carbon nanotubes can be described as a rolled-up cylindrical form of graphene sheets. The longitudinal unzipping of carbon nanotubes can represent a strategy to create graphene nanoribbons (GNRs) with precise dimensions (Kosynkin et al. 2009). An extremely high yield of GNRs (nearly

100%) was produced by unzipping multiwalled carbon nanotubes using potassium permanganate oxidation (Cano-Márquez et al. 2009). Argon plasma etching has also been used to unzip multiwalled carbon nanotubes partially embedded in a poly (methyl methacrylate) film to obtain GNRs with controllable widths and smooth edges (Jiao et al. 2009). When the etching time is controlled, the high quality of GNRs results in a width distribution of 10–20 nm. The decrease in width of GNRs, sized less than 10 nm, causes a transformation from semimetallic to semiconducting property (Akhavan et al. 2012; Hu et al. 2017).

Epitaxial chemical vapor deposition (CVD) growth is the most developed bottom-up method to produce many GFNs due to its low cost and high quality. CVD has been considered a commercially viable technique to manufacture continuous graphene films, which are often used to prepare high-performance solid semiconductors (Berger et al. 2004; Emtsev et al. 2009). Plasma enhanced and microwave-assisted chemical vapor deposition procedures are the different techniques of CVD (Jacob et al. 2015). Heteroatom-doped graphene can be synthesized using CVD by depositing carbon atoms and heteroatom simultaneously on the substrates such as Ni, Cu, Pt, Co, and Ru. Ito et al. synthesized N and S co-doped graphene using CVD with enhanced electrical conductivity, environmental stability, and photocatalytic properties (Ito et al. 2015). This method's advantage is the obtained large area epitaxial graphene films can be transferred to other substrates for practical device application after etching of the metallic support (Mao et al. 2013). There are many other methods for the synthesis of GFNs apart from the physicochemical approaches discussed above. Monodispersed graphene sheets in polar aprotic solvents have been produced by the one-step electrochemical exfoliation of graphite (Liu et al. 2008). Graphene discs with carbon atoms up to 222 have been synthesized through a bottom-up organic synthesis approach (Simpson et al. 2002). The electrical arc discharge is an inexpensive and less complicated technique for synthesizing graphene between two graphitic electrodes. The research group of Subrahmanyam et al. synthesized 2–4 layered graphene flakes using the arc discharge method under a mixture of different proportional hydrogen and helium without using any catalyst (Subrahmanyam et al. 2009). Wang et al. produced graphene nanosheets using the same air pressure and exposed that high pressure of air favors the formation of graphene whereas low pressure produces only non-graphene structures (Wang et al. 2010b). The sonochemical approach is another method for the synthesis of graphene. Recently Xu et al. produced polystyrene functionalized graphene using a sonochemical approach. They discovered that styrene could act as a solvent in graphite exfoliation and as a monomer to form the polymeric free radicals (Xu and Suslick 2011).

3 Physicochemical Properties and Characterizations of GFNs

The general physicochemical characteristics of GFNs are a high level of thermal conductivity, electrical conductivity, elastic modulus, intrinsic flexibility, and aspect ratio of the flakes. Their attributed functional properties are semiconductivity, low

Table 2 Properties and general application of GFNs

Property	Application	References
Optical	Anti-reflection coatings and tailored refractive index of surfaces	Li et al. (2013) and Lin et al. (2014)
Magnetic	Increased density storage media and improved magnetic resonance images	Singh (2016) and Reina et al. (2017)
Thermal	Improve the efficiency of coolants	Cabaleiro et al. (2017)
Mechanical	Improve wear resistance, new anti-corrosion properties, and new structural materials and composites	Nine et al. (2015) and Tapasztó et al. (2017)
Electronic	High performance and smaller components (e.g., mobile phones) and high conductivity materials	Asadi et al. (2015) and Secor et al. (2015)
Energy	High energy density and durable batteries, electrocatalysts for high-efficiency fuel cells, ultra-high-performance solar cells, and catalysts for combustion engines to improve efficiency	Long et al. (2013) and Acik and Darling (2016)
Biomedical	Sensor for diseases detection (quantum dots) and programmed release drug delivery systems	Dubey et al. (2015) and Elkhenany et al. (2015)
Environmental	Clean up soil contamination and pollution (e.g., oil, biodegradable polymers), treatment of industrial emissions, and water filtration	Konatham et al. (2013), Cohen-Tanugi and Grossman (2014), and Nicolaï et al. (2014)
Personal care	Inorganic sunscreens and dye protection	Qiu et al. (2014) and Benítez-Martínez et al. (2016)

Reproduced from De Marchi et al. 2018, Copyright (2018) Elsevier

permeability, optical transportation, anisotropic transport, and fluorescence quenching (Wang et al. 2016). Table 2 represents the general characteristics of GFNs and their related applications. The chemical functionality of GFNs involves the incorporation of various oxygen-containing functional groups, which allows the growth of organic moieties and metallic nanoparticles. The combination of graphene with many metal oxides and complex oxides and their derivatives acts as photocatalysts and adsorbents in many applications. The insulating or semiconducting behavior of GO can be controlled by tuning its redox parameters. The photoluminescence property of GO enables the materials to apply fluorescence tags, biosensing, and optoelectronics (Li et al. 2020).

More than two techniques are coupled to provide information of GFNs' intrinsic properties accurately. Density functional theory (DFT) describes the physicochemical properties of functionalized materials by analyzing the electronic structure. The DFT computational study results determine the active sites and rate-limiting steps examining activation energy, catalytic pathway, and adsorption and activation mechanisms (Nørskov et al. 2011; Boukhvalov 2013). Raman spectroscopy is a robust analytical technique for the qualitative and quantitative analyses of GFNs. In a Raman spectrum, the predominant features are D, G, and 2D bands at 1320–1350 cm^{-1}, 1570–1585 cm^{-1}, and 2640–2680 cm^{-1}, respectively, to characterize graphene materials. The D to G band's intensity ratio is helpful to compare the defects and disorder qualitatively among various GFNs. The 2D bands evaluate the number of layers related to the surface area (Begliarbekov et al. 2010; Balachandran and Jose 2011; Deng et al. 2011). X-ray Photoelectron Spectroscopy (XPS) characterizes the dopants' presence in a graphite structure and reveals the types of different binding energies using the deconvoluted curve fitting. Moreover, XPS helps detect active sites and study the related reaction mechanisms (Wang et al. 2012; Susi et al. 2015). Scanning Tunneling Microscopy (STM) is a powerful technique to probe the materials' charge density at the Fermi level. STM images are helpful to establish the defects, morphology, and electronic properties of the doped graphene (Deng et al. 2011). Diffuse Reflectance Fourier Transform Infrared Spectroscopy is an effective tool to distinguish differential vibrational bands of the adsorbed basic probe moieties in a catalyst (Wang et al. 2013a). Atomic force microscopy (AFM) analyzes the sample's topography using a tip to scan across the surface. The differential height measurements at the edge would count the number of layers on the material's surface (Shearer et al. 2016). X-ray diffraction (XRD) pattern shows characteristic 2θ peaks at 26° and 11° for pristine graphite and graphene oxide, respectively (Rao et al. 2009). The high-resolution transmission electron microscopy (HRTEM) also offers information on the crystalline character of graphene by its electron diffraction pattern (Priya et al. 2016). The nitrogen adsorption–desorption analyses (at 77 K) by a Brunauer-Emmett-Teller (BET) method provides information to identify the number of layers qualitatively (Hu et al. 2014). Besides, scanning electron microscope, energy dispersive spectroscope, selected area electron diffraction, inductively coupled plasma mass spectrometry, Fourier transform infrared spectroscope, and UV–vis diffuse reflectance spectroscope analyses were also used to characterize GFNs (Hu et al. 2017).

3.1 Biomedical Applications of GFNs

Graphene can be chemically functionalized both in dry and liquid states. The graphene's ability to adsorb various aromatic biomolecules via π–π stacking /or electrostatic interaction makes it ideal for biosensing and drug loading applications (Dong et al. 2010; Hong et al. 2014; Shin et al. 2017). The number of articles published under the exact word of "graphene" and "biomedical application of graphene" in the title for the last 10 years is provided using a scientific community, Scopus database (date of search: 18 March 2021) (Fig. 2). We can observe a steady growth rate in both terms, which indicates that the scientific community is showing immense interest in the field of graphene family nanomaterials.

In biomedical science, innovative therapies with effective and site-specific treatment are of tremendous importance. The materials currently employed in commercially available biomedical technologies are far from satisfactory, given their less-efficient intrinsic properties. For instance, metal and silicon are still

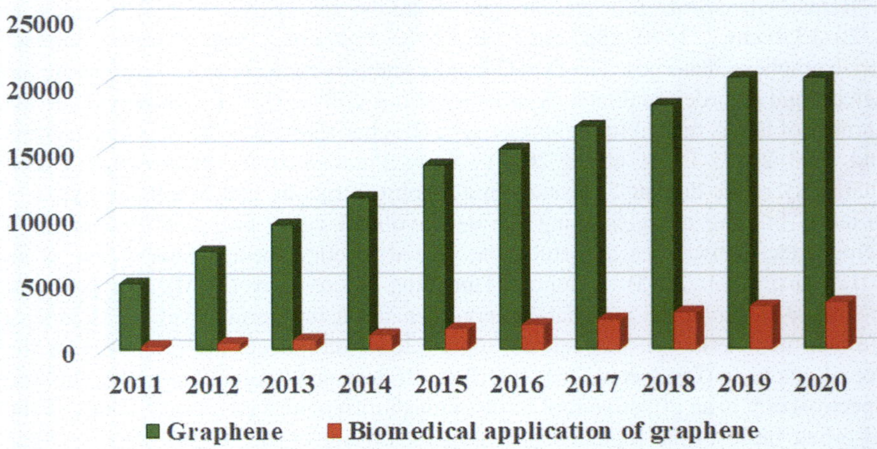

Fig. 2 The number of papers published under the article title/abstract/keywords "graphene" and the search within the result "biomedical application" in 2011–2020 (Using Scopus database; date of search: 18 March 2021)

among the most used materials to fabricate affordable and conventional implant devices (Molina-Fernández et al. 2021). However, their limitations are still of serious concern due to their long-term stability in physiological environments, rigid mechanical properties, and high inflammatory potential. In contrast, the applications guided by the intriguing features of GFNs are unbound in the areas of energy technology, sensors, catalysis, biotechnology, and biomedicine due to incomparable mechanical properties, accompanied by biocompatibility, transparency, and electrical conductivity (Fig. 3) (Paul and Sharma 2011; Gerasimova et al. 2019; Kokkilagadda et al. 2020). This chapter will summarize some of the notable advances of GFNs in biomedical applications, including bioimaging, 3D bioprinting, tissue engineering, drug delivery, photothermal therapy, 3D bioprinting, biosensors, gene therapy, and antibacterial activity.

3.1.1 GFNs in Bioimaging Application

Optical imaging is a superior, inexpensive, widely available, highly sensitive, non-invasive, and real-time visualization technique that has revolutionized how we study the human body's inner workings (Weissleder 2006; Cai and Chen 2007). The technology relies upon the visible light and photons' unique properties to visualize organs, tissues, smaller structures, cells, and molecules in diagnostics, therapeutics, drug discovery, and development to understand the nanoscale reactions, including enzymatic conversions and protein-protein interactions (James and Gambhir 2012; Nolting et al. 2012). The bioimaging applications of GFNs are emerging rapidly, especially at the interface of nanotechnology, biochemical, physical, and material science. The integration of GFNs with organic dyes, polymers, and nanoparticles (NPs) promoted by the presence of abundant oxygen functionalities on GO and rGO

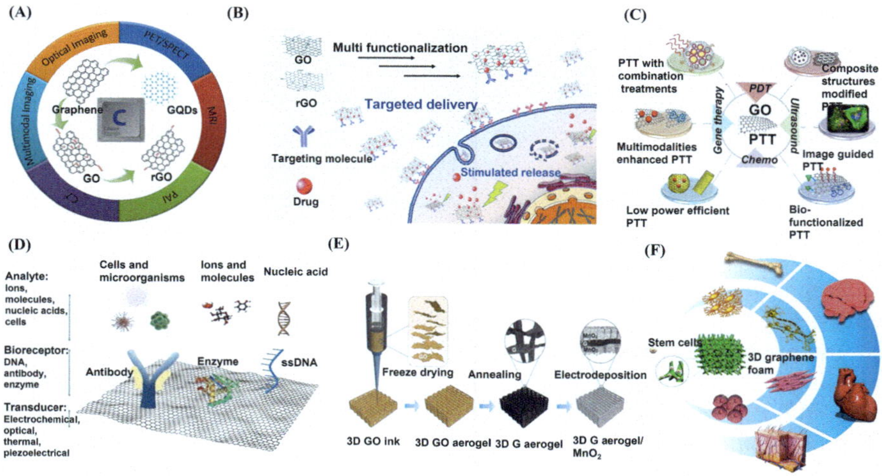

Fig. 3 Schematic overview of the wide-spectrum biomedical applications of graphene-based nanomaterials (GFNs). (**a**) Different molecular imaging modalities (Reproduced from Lin et al. 2016, Copyright (2016) Elsevier). (**b**) GO and rGO in drug delivery (Reproduced from Reina et al. 2017, Copyright (2017) Royal Society of Chemistry). (**c**) Photothermal therapy (Reproduced from Chen et al. 2016, Copyright (2016) Elsevier). (**d**) The attachment of bioreceptors, such as an antibody, DNA, and enzymes onto the graphene surfaces (Reproduced from Peña-Bahamonde et al. 2018, Copyright (2018) Springer Nature). (**e**) 3D printed graphene aerogel for electrode fabrication (Reproduced from Yao et al. 2019, Copyright (2019) Elsevier). (**f**) 3D graphene foam for tissue engineering (Reproduced from Amani et al. 2019, Copyright (2019) American Chemical Society)

surfaces has shown great potential for the detection of biomolecules such as thrombin, ATP, oligonucleotide, amino acid, and dopamine (Chang et al. 2010; Wang et al. 2010a). Previously, cellular visualization was carried out by organic dyes and inorganic semiconductor-based fluorophores. Nevertheless, such materials' practical applications are hampered, given the drawbacks of photobleaching, low extinction coefficient, low water solubility, and intrinsic toxicity (Jiang and Tian 2018; Bisquert 2020). Graphene and its derivatives are extensively explored for bioimaging applications to play a vital role in developing specific molecular probes/or contrast agents to detect and characterize early-stage diseases, hence providing a quick approach to outline the treatment response. More importantly, due to their surface functionalization and ultrahigh surface area, GFNs can be easily modified by small molecular dyes, polymers, nanoparticles, and drugs to obtain graphene-based nanomaterials for various bioimaging applications at a microscopic scale. In general, GFNs have been applied for different molecular imaging modalities such as (a) fluorescence imaging (Baker and Baker 2010; Janib et al. 2010), (b) two-photon fluorescence imaging (Li et al. 2012a; Yoo et al. 2015), (c) positron emission tomography/or single-photon emission computed tomography (Park et al. 2009; Hong et al. 2012), (d) magnetic resonance imaging (Song et al. 2015; Yu et al. 2015), (e) Raman imaging (Bergholt et al. 2013; Kang et al. 2015),

(f) photoacoustic imaging (Cheng et al. 2014; Mei et al. 2014), and (g) multimodal imaging (Lee et al. 2011a).

3.1.2 Significance of GFNs in 3D Bioprinting

Graphene materials can be altered from monolayers to ordered hierarchical structures, i.e., 2D graphene sheets into 3D composite materials using supramolecular chemistry. 3D bioprinting is a form of additive engineering that uses cells/and other biomolecules as "inks" (also called bioinks) to print living structures layer-by-layer, which could mimic the behavior of natural living systems improving the bio-integrated tissue substitutes' efficiency. The research of graphene inks is increased by dispersing graphene nanoplatelets into the existing material to fabricate 3D systems. Relying upon the remarkable mechanical strength over metals/or metal composites, GFNs are emerging as notable replacements (Zhang et al. 2012; Zhong et al. 2017; Palmieri et al. 2020). The GFNs with accelerated mesenchymal stem cell adhesion, proliferation, and differentiation have exhibited osteogenic and adipogenic inductive properties to prove they are crucial in rehabilitation medicine. Mainly, GO and rGO possess adjustable lateral size and oxygen content, given their honeycomb packing of carbon atoms with the surface functionalities available to interact with the biomolecules (for instance, proteins, peptides, DNA, or RNA). This peculiar surface chemistry and biomolecule interaction designate graphene as ideal for biomedical engineering (Lee et al. 2011b; Li et al. 2020). The precision of 3D printed graphene substitutes has been found advantageous to fit the surgical need. Besides the scalability and mechanical tenability, graphene-based scaffolds are also biocompatible, bioactive, and biodegradable (Jakus et al. 2015; Vlăsceanu et al. 2019). The cytocompatibility of graphene 3D scaffolds has been successfully shown to endow bone regeneration and cartilage reconstruction (Lee et al. 2011b).

3.1.3 Role of GFNs in Antimicrobial Activities

Graphene is an effective antimicrobial agent showing better cytocompatibility. Graphene's unique selective toxicity has attracted substantial interest in its use as antimicrobial agents against microbes such as *Escherichia coli*, *Staphylococcus aureus*, *Streptococcus mutans*, *Pseudomonas aeruginosa*, *Candida albicans*, and *Salmonella typhimurium* (Ji et al. 2016; Szunerits and Boukherroub 2016). The interaction of active surface functional groups on graphene, and various microbial interfaces like proteins and peptides have given rise to a thorough investigation of the antibacterial application of GFNs (Zou et al. 2017). Though the exact mechanism for the antimicrobial activity of graphene is not properly elucidated, several plausible models, including electron transfer at the microbial membrane, reactive oxygen species generation at the membrane interface yet, membrane stress that in effect disrupts the microbial membranes, cell entrapment, and photothermal effect have been proposed (Liu et al. 2011; Shi et al. 2016). The particle size and sharp edges of the nanomaterials significantly affect their adsorption and desorption into the microorganisms. It was reported that the larger the lateral size of the GFNs, the higher the adsorption capacity. The number of layers also plays a pivotal role in determining antimicrobial activity. As the number of GFNs increases, the resulting thickness causes a nano knife effect and decreases their interaction with microorganisms (Wang et al. 2013b; Ansari 2020). However, Mangadlao et al. reported that an increased amount of GO layers caused a severe antimicrobial effect against *E. coli*. A research finding demonstrates that GO and rGO nanowalls' sharp edges significantly reduced the population of *S. aureus* and *E. coli* (Mangadlao et al. 2015). Zou et al. demonstrated that the rough-bottom-side graphene was effective against only rod-shaped *P. aeruginosa* while the smooth-top-side graphene presented its efficiency against round shape as well as *P. aeruginosa* (Zou et al. 2016).

3.1.4 GFNs in Biosensors

The applications of pristine graphene and its derivatives as scaffolds in sensor-based technologies are limitless in biosensing for molecules, including thrombin (Chang et al. 2010), oligonucleotides (Tang et al. 2011), ATP (Shao et al. 2010), amino corrosives (Dong et al. 2010), and dopamine (Wang et al. 2009). In general, sensors are made up of two components: (i) receptor, an organic or inorganic material that interacts specifically with the target molecule (organic, inorganic, or even whole cells), and (ii) transducer, which converts chemical information into a measurable signal. Given the large surface area, electrical conductivity, high electron transfer rate, and capacity to immobilize different molecules, GFNs have been deployed for designing biosensors with different transduction modes; for example, the electron transfer between the bioreceptor and transducer can be facilitated by the conjugated structure of graphene to generate high signal sensitivity for electrochemical sensors (Park et al. 2016; Chauhan et al. 2017). GFNs can serve as a quencher in the transducer to generate fluorescent biosensors (Kasry et al. 2012). Furthermore, self-assembling and controllable graphene biomolecules permit assembling ultrasensitive biosensors to recognize DNA and different atoms (Premkumar and Geckeler 2012). Graphene-based devices could increase traditional batteries' life span by charging faster and retaining power for a longer duration. With the unique features of weightlessness and flexibility, GFNs are used nowadays as a coating to improve touch sensitivity in smartphones and as an integral part of portable devices attached to clothing.

3.1.5 Applications of GFNs in Drug Delivery and Gene Therapy

The advantage of GFNs for drug delivery is their larger surface area to provide numerous anchoring sites for various drug molecules and their permeability to the cell membranes readily. They can interact with the drugs through hydrogen bonding or hydrophobic forces (Zhao et al. 2017). Jin et al. developed a hematin-dextran-GO hybrid via π-π interactions to release doxorubicin for destroying multidrug-resistant MCF-7/ADR cells. The hybrid's drug loading capacity was 3.4 mg/mg nGO, and the drug delivery was pH-dependent, releasing more drugs at lower pH. Besides, the drug-loaded hybrids accumulated abundantly in both cytosol and nucleus to kill the cells (Jin et al. 2013). Encapsulation of GFNs by a surfactant or polymer improves their stability against aggregation in the respective medium. Xu et al. stabilized GO with PEG functionalization in a physiological buffer and loaded paclitaxel onto the composite with a high loading capacity of 11.2 wt% (Xu et al. 2014).

One of graphene's promising applications is gene therapy, which involves GFNs as a nanocarrier for delivering therapeutic agents to the target cells. Gene therapy needs a safe and efficient gene carrier to deliver a specific cell to overcome the intra- and extracellular obstacles (Findeis 2001). Nonetheless, the genetic structure's negative charge prevents it from passing through the cell barrier itself; thus, the complexes formed between the carrier and genetic material require electrostatic interactions (Jin and Ye 2007). GO modification could provide an efficient cationic surface to facilitate electrostatic interaction with the anionic genes. The exterior features (i.e., shape and size) play a crucial role in the interaction of graphene with the cell membrane and its intracellular uptake and the functionalization of graphene (Li et al. 2013). Previously, graphene's functionalization has been exploited to deliver the drugs in the cytosolic compartment, unlike other nanocarriers that enter the cell by endocytosis (Zhang et al. 2011). Different graphene-modifiers have been used for gene delivery; for example, functionalization of GO with polyethyleneimine has been an excellent gene delivering agent for short interfering RNA into the cells and has resulted in a significant decrease in the cytotoxicity (Kim et al. 2011). The complex (polyethyleneimine-GO) can condense DNA at a low mass ratio (Paul et al. 2014). Chitosan-modified GO has shown reduced cytotoxicity and enhanced cellular uptake (Shim et al. 2016). Due to its large surface area and conjugated structure, GO is a dual-mode carrier for simultaneous gene and drug carriers. One such example is the chitosan functionalized GO used to deliver the anticancer drug and the nucleotides (Bao et al. 2011).

3.1.6 Graphene-Based Nanomaterials in Photothermal Therapy

Based on the optical absorption in the near-infrared region (NIR), graphene has been an excellent material for novel cancer therapies (Ma et al. 2012; An et al. 2015). The primary mechanism involved is the photo-ablation of cancer cells through heating using near-infrared rays. Without damaging the healthy cells, graphene should absorb maximum in the NIR region (700–1100 nm), as living cells do not show absorption in this range. In recent years, many research works have been dedicated to exploring the in vitro and in vivo behavior of GFNs in living cells as they offer excellent stabilization in the living system (An et al. 2015). Optical/or thermal properties of graphene open many new biomedicine directions to treat cancer and other diseases, including Alzheimer's disease (Li et al. 2012b). Studies show that photothermal damage to tumor cells appears at 41 °C and significant damage is observed at 50 °C (Xie et al. 2011; Yang et al. 2011). Salaheldin et al. revealed that the use of graphene-based nanocomposites with transition metals like iron (Fe) enhanced the photothermal-therapeutic effect on human liver cancer cells (Salaheldin et al. 2019). Li et al. covalently linked GO with thioflavin-S to selectively attach amyloid β aggregates, which remotely heated and dislocated the aggregate with NIR irradiation (Li et al. 2012b).

3.1.7 Importance of GFNs in Tissue Engineering

Due to the superior physicochemical and biological properties, the GFNs have been identified and deployed for tissue engineering and regenerative medicine applications. Besides regulating the cardiomyogenic, neurogenic, osteogenic, and cartilaginous stem cells' abilities, GFNs provide an impeccable selection for enhancing the mechanical and surface properties of biomaterials (Lee et al. 2015b; Yu et al. 2016; Shang et al. 2019; Maleki et al. 2020). Furthermore, they significantly stimulate the growth and other cellular activities of electrically excitable cells, including cardiomyocytes, neurons, and improved cellular signals recording. Coupling the tunable chemical and mechanical properties enables the graphene surface for 3D printing to create the complex graphene architectures for their enhanced application spectrum. GFNs combined with other materials have been reported as innovative multifunctional materials with variable wettability/or flexibility for diverse applications, such as stimuli-responsive and shape-memory/or self-folding properties essential for biomedical applications. Xie et al. prepared a highly porous 3D GO-hydroxyapatite hydrogel for effective bone tissue engineering. The hydrogel had a robust mechanical property and a high electrical conductivity due to the incorporation of GO nanosheets (Xie et al. 2015). Though GFNs show tremendous biomedical applications, the cytotoxic or genotoxic effects of these nanomaterials are yet to be solved. Various modifications, such as protein or cytokine functionalization, have been attempted to resolve graphene's biocompatibility concerns. Given the drawbacks, more in-depth investigation is required before graphene-based derivatives are applied to the clinics.

4 Conclusions

The present chapter encompasses the definition, properties, synthetic approaches, and general biomedical application of graphene family nanomaterials. The list of publications based on GFNs shows how important the materials are in the field of biomedicine. The primary reasons for the high exploration of graphene and its derivatives in the scientific community are that they possess a large surface area to volume, hydrophobic, hydrophilic interactions, economically inexpensive, and biocompatible. The synthetic procedure of nanomaterials includes top-down and bottom-up methods. Epitaxial growth and mechanical exfoliation are the conventional methods to produce GFNs abundantly. The biomedical applications of GFNs include a biosensor, drug loading and release, antibacterial activity, bioimaging, photothermal therapy, and tissue regeneration. Though GFNs alone exhibit biomedical properties, they are sometimes doped with other nanomaterials or bioactive molecules to obtain hierarchical structures to meet the desired properties and enhanced functionality. The incorporation of GFNs into different substrates/ scaffolds improves their physicochemical properties such as mechanical strength, magnetic, thermal, optical, and electronic. We hope that the successive chapters will provide a comprehensive review of each biomedical application with their future directions.

Acknowledgments This research was supported by 2-Year Research grant from the Pusan National University.

References

Acik M, Darling SB (2016) Graphene in perovskite solar cells: device design, characterization and implementation. J Mater Chem A 4(17):6185–6235

Akhavan O, Ghaderi E, Emamy H (2012) Nontoxic concentrations of PEGylated graphene nanoribbons for selective cancer cell imaging and photothermal therapy. J Mater Chem 22(38):20626–20633

Amani H, Mostafavi E, Arzaghi H, Davaran S, Akbarzadeh A, Akhavan O, Pazoki-Toroudi H, Webster TJ (2019) Three-dimensional graphene foams: synthesis, properties, biocompatibility, biodegradability, and applications in tissue engineering. ACS Biomater Sci Eng 5(1):193–214

An S, Wu S-Y, Hulme J (2015) Current applications of graphene oxide in nanomedicine. Int J Nanomedicine 10:9–24

Ansari M (2020) Antimicrobial mechanisms and effectiveness of graphene and graphene-functionalized biomaterials. A scope review. Front Bioeng Biotechnol 8:465

Asadi K, Timmering EC, Geuns TCT, Pesquera A, Centeno A, Zurutuza A, Klootwijk JH, Blom PWM, de Leeuw DM (2015) Up-scaling graphene electronics by reproducible metal-graphene contacts. ACS Appl Mater Interfaces 7(18):9429–9435

Baker S, Baker G (2010) Luminescent carbon nanodots: emergent nanolights. Angew Chem 49(38):6726–6744

Balachandran M, Jose A (2011) Raman spectrum of graphite layers in Indian coal. AIP Conf Proc 1391:140

Bao H, Pan Y, Ping Y, Sahoo N, Wu T, Li L, Li J, Gan LH (2011) Chitosan-functionalized graphene oxide as a nanocarrier for drug and gene delivery. Small 7(11):1569–1578

Begliarbekov M, Sul O, Kalliakos S, Yang E-H, Strauf S (2010) Determination of edge purity in bilayer graphene using micro-Raman spectroscopy. Appl Phys Lett 97(3):031908

Benítez-Martínez S, López-Lorente ÁI, Valcárcel M (2016) Determination of TiO_2 nanoparticles in sunscreen using N-doped graphene quantum dots as a fluorescent probe. Microchim Acta 183(2):781–789

Berger C, Song Z, Li T, Li X, Ogbazghi AY, Feng R, Dai Z, Marchenkov AN, Conrad EH, First PN, de Heer WA (2004) Ultrathin epitaxial graphite: 2D electron gas properties and a route toward graphene-based nanoelectronics. J Phys Chem B 108(52):19912–19916

Bergholt M, Zheng W, Huang Z (2013) Development of a multiplexing fingerprint and high wavenumber Raman spectroscopy technique for real-time in vivo tissue Raman measurements at endoscopy. J Biomed Opt 18(3):30502

Bianco A, Cheng H-M, Enoki T, Gogotsi Y, Hurt RH, Koratkar N, Kyotani T, Monthioux M, Park CR, Tascon JMD, Zhang J (2013) All in the graphene family-a recommended nomenclature for two-dimensional carbon materials. Carbon 65:1–6

Bisquert J (2020) Chapter 3 Optical transitions in organic and inorganic semiconductors. In: Bisquert J (ed) The physics of solar cells: perovskites, organics, and photovoltaic fundamentals. CRC Press, Boca Raton, pp 43–70. https://doi.org/10.1201/b22380-3

Boukhvalov D (2013) DFT modeling of the covalent functionalization of graphene: from ideal to realistic models. RSC Adv 3(2):7150–7159

Cabaleiro D, Colla L, Barison S, Lugo L, Fedele L, Bobbo S (2017) Heat transfer capability of (ethylene glycol+water)-based nanofluids containing graphene nanoplatelets: design and thermophysical profile. Nanoscale Res Lett 12(1):53

Cai W, Chen X (2007) Nanoplatforms for targeted molecular imaging in living subjects. Small 3(11):1840–1854

Cano-Márquez AG, Rodríguez-Macías FJ, Campos-Delgado J, Espinosa-González CG, Tristán-López F, Ramírez-González D, Cullen DA, Smith DJ, Terrones M, Vega-Cantú YI (2009)

Ex-MWNTs: graphene sheets and ribbons produced by lithium intercalation and exfoliation of carbon nanotubes. Nano Lett 9(4):1527–1533
Chang H, Tang L, Wang Y, Jiang J, Li J (2010) Graphene fluorescence resonance energy transfer aptasensor for the thrombin detection. Anal Chem 82(6):2341–2346
Chauhan N, Maekawa T, Kumar DNS (2017) Graphene based biosensors-accelerating medical diagnostics to new-dimensions. J Mater Res 32(15):2860–2882
Chen Y-W, Su Y-L, Hu S-H, Chen S-Y (2016) Functionalized graphene nanocomposites for enhancing photothermal therapy in tumor treatment. Adv Drug Deliv Rev 105(PtB):190–204
Cheng L, Liu J, Gu X, Gong H, Shi X, Liu T, Wang C, Wang X, Liu G, Xing H, Bu W, Sun B, Liu Z (2014) Imaging: PEGylated WS_2 Nanosheets as a multifunctional theranostic agent for in vivo dual-modal CT/photoacoustic imaging guided photothermal therapy. Adv Mater 26(12): 1886–1893
Cohen-Tanugi D, Grossman JC (2014) Mechanical strength of nanoporous graphene as a desalination membrane. Nano Lett 14(11):6171–6178
De Marchi L, Pretti C, Gabriel B, Marques PAAP, Freitas R, Neto V (2018) An overview of graphene materials: properties, applications and toxicity on aquatic environments. Sci Total Environ 631-632:1440–1456
Deng D, Pan X, Yu L, Cui Y, Jiang Y, Qi J, Li W-X, Fu Q, Ma X, Xue Q, Sun G, Bao X (2011) Toward N-doped graphene via solvothermal synthesis. Chem Mater 23(5):1188–1193
Dong X, Cheng J, Li J, Wang Y (2010) Graphene as a novel matrix for the analysis of small molecules by MALDI-TOF MS. Anal Chem 82(14):6208–6214
Dubey N, Bentini R, Islam I, Cao T, Castro Neto AH, Rosa V (2015) Graphene: a versatile carbon-based material for bone tissue engineering. Stem Cells Int 2015:804213
Elkhenany H, Amelse L, Lafont A, Bourdo S, Caldwell M, Neilsen N, Dervishi E, Derek O, Biris AS, Anderson D, Dhar M (2015) Graphene supports in vitro proliferation and osteogenic differentiation of goat adult mesenchymal stem cells: potential for bone tissue engineering. J Appl Toxicol 35(4):367–374
Emtsev KV, Bostwick A, Horn K, Jobst J, Kellogg GL, Ley L, McChesney JL, Ohta T, Reshanov SA, Rohrl J, Rotenberg E, Schmid AK, Waldmann D, Weber HB, Seyller T (2009) Towards wafer-size graphene layers by atmospheric pressure graphitization of silicon carbide. Nat Mater 8(3):203–207
Fan X, Zhang G, Zhang F (2015) ChemInform abstract: multiple roles of graphene in heterogeneous catalysis. Chem Soc Rev 44(10):3023–3035
Findeis MA (2001) Nonviral vectors for gene therapy: methods and protocols. Humana Press, Totowa. https://doi.org/10.1385/1592591396
Gerasimova A, Smolsky G, Melezhik A, Galunin E, Memetov N, Tkachev A (2019) Synthesis, study and applications of graphene materials. In: Proceedings of the 4th world congress on recent advances in nanotechnology. https://doi.org/10.11159/icnnfc19.103
Hernandez Y, Nicolosi V, Lotya M, Blighe FM, Sun Z, De S, McGovern IT, Holland B, Byrne M, Gun'Ko YK, Boland JJ, Niraj P, Duesberg G, Krishnamurthy S, Goodhue R, Hutchison J, Scardaci V, Ferrari AC, Coleman JN (2008) High-yield production of graphene by liquid-phase exfoliation of graphite. Nat Nanotechnol 3(9):563–568
Hong H, Zhang Y, Engle JW, Nayak TR, Theuer CP, Nickles RJ, Barnhart TE, Cai W (2012) In vivo targeting and positron emission tomography imaging of tumor vasculature with Ga-66-labeled nano-graphene. Biomaterials 33(16):4147–4156
Hong SW, Lee JH, Kang SH, Hwang EY, Hwang Y-S, Lee MH, Han D-W, Park J-C (2014) Enhanced neural cell adhesion and neurite outgrowth on graphene-based biomimetic substrates. Biomed Res Int 2014:212149
Hu M, Hui KS, Hui K (2014) Role of graphene in MnO_2/graphene composite for catalytic ozonation of gaseous toluene. Chem Eng J 254:237–244
Hu F, Kou L, Frauenheim T (2015) Controllable magnetic correlation between two impurities by spin-orbit coupling in graphene. Sci Rep 5:8943

Hu M, Yao Z, Wang X (2017) Graphene-based nanomaterials for catalysis. Ind Eng Chem Res 56(13):3477–3502
Hua B, Li C, Shi G (2011) Functional composite materials based on chemically converted graphene. Adv Mater 23(9):1089–1115
Ito Y, Cong W, Fujita T, Tang Z, Chen M (2015) High catalytic activity of nitrogen and sulfur co-doped nanoporous graphene in the hydrogen evolution reaction. Angew Chem 54(7): 2131–2136
Jacob M, Rawat RS, Ouyang B, Bazaka K, Kumar S, Taguchi D, Iwamoto M, Neupane R, Varghese OK (2015) Catalyst-free plasma enhanced growth of graphene from sustainable sources. Nano Lett 15(9):5702–5708
Jakus A, Secor E, Rutz A, Jordan S, Hersam M, Shah R (2015) Three-dimensional printing of high-content graphene scaffolds for electronic and biomedical applications. ACS Nano 9(4): 4636–4648
James ML, Gambhir SS (2012) A molecular imaging primer: modalities, imaging agents, and applications. Physiol Rev 92(2):897–965
Janib S, Moses A, Mackay J (2010) Imaging and drug delivery using theranostic nanoparticles. Adv Drug Deliv Rev 62(11):1052–1063
Ji H, Sun H, Qu X (2016) Antibacterial applications of graphene-based nanomaterials: recent achievements and challenges. Adv Drug Deliv Rev 105(Pt B):176–189
Jiang Y, Tian B (2018) Inorganic semiconductor biointerfaces. Nat Rev Mater 3(12):473–490
Jiao L, Zhang L, Wang X, Diankov G, Dai H (2009) Narrow graphene nanoribbons from carbon nanotubes. Nature 458(7240):877–880
Jin S, Ye K (2007) Nanoparticle-mediated drug delivery and gene therapy. Biotechnol Prog 23(1): 32–41
Jin R, Ji X, Yang Y, Wang H, Cao A (2013) Self-assembled graphene-dextran nanohybrid for killing drug-resistant cancer cells. ACS Appl Mater Interfaces 5(15):7181–7189
Kang JW, So P, Dasari R, Lim D-K (2015) High resolution live cell Raman imaging using subcellular organelle-targeting SERS-sensitive gold nanoparticles with highly narrow intra-nanogap. Nano Lett 15(3):1766–1772
Kang MS, Lee JH, Song SJ, Shin DM, Jang JH, Hyon SH, Hong SW, Lee JH, Han DW (2020) Graphene oxide-functionalized nanofibre composite matrices to enhance differentiation of hippocampal neuronal cells. Mater Adv 1(9):3496–3506
Kasry A, Afzali A, Tulevski G, Menges B, Copel M, Vyklicky L (2012) Highly efficient fluorescence quenching with graphene. J Phys Chem C 116(4):2858
Kim H, Namgung R, Singha K, Oh I-K, Kim W (2011) Graphene oxide-polyethylenimine nanoconstruct as a gene delivery vector and bioimaging tool. Bioconjug Chem 22(12): 2558–2567
Kokkilagadda V, Kumar T, Ramesh K, Chandra S, Varma M, Hanumanaik B, Maddiboyina HM, Maddiboyina B (2020) Pharmaceutical and biomedical applications of graphene based nanoparticles. Int J Pharm Ind Res 10(1):35–44
Konatham D, Yu J, Ho TA, Striolo A (2013) Simulation insights for graphene-based water desalination membranes. Langmuir 29(38):11884–11897
Kosynkin DV, Higginbotham AL, Sinitskii A, Lomeda JR, Dimiev A, Price BK, Tour JM (2009) Longitudinal unzipping of carbon nanotubes to form graphene nanoribbons. Nature 458(7240): 872–876
Lee D-E, Koo H, Sun I-C, Ryu J, Kim K, Kwon I (2011a) ChemInform abstract: multifunctional nanoparticles for multimodal imaging and theragnosis. Chem Soc Rev 41:2656–2672
Lee WC, Lim YX, Shi H, Tang LAL, Wang Y, Lim CT, Loh K (2011b) Origin of enhanced stem cell growth and differentiation on graphene and graphene oxide. ACS Nano 5(9):7334–7341
Lee JH, Shin YC, Jin OS, Kang SH, Hwang Y-S, Park J-C, Hong SW, Han D-W (2015a) Reduced graphene oxide-coated hydroxyapatite composites stimulate spontaneous osteogenic differentiation of human mesenchymal stem cells. Nanoscale 7(27):11642–11651

Lee JH, Shin YC, Lee S-M, Jin OS, Kang SH, Hong SW, Jeong C-M, Huh JB, Han D-W (2015b) Enhanced osteogenesis by reduced graphene oxide/hydroxyapatite nanocomposites. Sci Rep 5: 18833

Li J-L, Bao H, Hou X-L, Sun L, Wang X-G, Gu M (2012a) Graphene oxide nanoparticles as a nonbleaching optical probe for two-photon luminescence imaging and cell therapy. Angew Chem Int Ed Engl 51(8):1830–1834

Li M, Yang X, Ren J, Qu K, Qu X (2012b) Using graphene oxide high near-infrared absorbance for photothermal treatment of Alzheimer's disease. Adv Mater 24(13):1722–1728

Li Y, Yuan H, von dem Bussche A, Creighton M, Hurt RH, Kane AB, Gao H (2013) Graphene microsheets enter cells through spontaneous membrane penetration at edge asperities and corner sites. Proc Natl Acad Sci 110(30):12295–12300

Li J, Liu X, Crook J, Wallace G (2020) 3D printing of cytocompatible graphene/alginate scaffolds for mimetic tissue constructs. Front Bioeng Biotechnol 8:824

Lin P-C, Lin S, Wang PC, Sridhar R (2014) Techniques for physicochemical characterization of nanomaterials. Biotechnol Adv 32(4):711–726

Lin J, Chen X, Huang P (2016) Graphene-based nanomaterials for bioimaging. Adv Drug Deliv Rev 105(Pt B):242–254

Liu N, Luo F, Wu H, Liu Y, Zhang C, Chen J (2008) One-step ionic-liquid-assisted electrochemical synthesis of ionic-liquid-functionalized graphene sheets directly from graphite. Adv Funct Mater 18(10):1518–1525

Liu S, Zeng T, Hofmann M, Burcombe E, Wei J, Jiang R, Kong J, Chen Y (2011) Antibacterial activity of graphite, graphite oxide, graphene oxide, and reduced graphene oxide: membrane and oxidative stress. ACS Nano 5(9):6971–6980

Long CM, Nascarella MA, Valberg PA (2013) Carbon black vs. black carbon and other airborne materials containing elemental carbon: physical and chemical distinctions. Environ Pollut 181: 271–286

Lotya M, Hernandez Y, King PJ, Smith RJ, Nicolosi V, Karlsson LS, Blighe FM, De S, Wang Z, McGovern IT, Duesberg GS, Coleman JN (2009) Liquid phase production of graphene by exfoliation of graphite in surfactant/water solutions. J Am Chem Soc 131(10):3611–3620

Ma X, Tao H, Yang K, Feng L, Cheng L, Shi X, Li Y, Guo L, Liu Z (2012) A functionalized graphene oxide-iron oxide nanocomposite for magnetically targeted drug delivery, photothermal therapy, and magnetic resonance imaging. Nano Res 5(3):199–212

Machado B, Serp P (2011) Graphene-based materials for catalysis. Cat Sci Technol 2(1):54–75

Maleki M, Zarezadeh R, Nouri M, Raei Sadigh A, Pouremamali F, Asemi Z, Kafil HS, Alemi F, Yousefi B (2020) Graphene oxide: a promising material for regenerative medicine and tissue engineering. Biomol Concepts 11(1):182–200

Mangadlao J, Santos C, Felipe M, Leon A, Rodrigues D, Advincula R (2015) On the antibacterial mechanism of graphene oxide (GO) Langmuir-Blodgett films. Chem Commun 51(14): 2886–2889

Mao H, Laurent S, Chen W, Akhavan O, Imani M, Ashkarran AA, Mahmoudi M (2013) Graphene: promises, facts, opportunities, and challenges in nanomedicine. Chem Rev 113(5):3407–3424

Marcano DC, Kosynkin DV, Berlin JM, Sinitskii A, Sun Z, Slesarev A, Alemany LB, Lu W, Tour JM (2010) Improved synthesis of graphene oxide. ACS Nano 4(8):4806–4814

Mas-Ballesté R, Gómez-Navarro C, Gómez-Herrero J, Zamora F (2011) 2D materials: to graphene and beyond. Nanoscale 3(1):20–30

McCallion C, Burthem J, Rees-Unwin K, Golovanov A, Pluen A (2016) Graphene in therapeutics delivery: problems, solutions and future opportunities. Eur J Pharm Biopharm 104:235–250

Mei C, Tang S, Guo Z, Wang X, Mo S, Huang X, Liu G, Zheng N (2014) Core-shell Pd@Au nanoplates as theranostic agents for in-vivo photoacoustic imaging, CT imaging, and photothermal therapy. Adv Mater 26(48):8210–8216

Molina-Fernández I, Hadij-Elhouati A, Luque-González JM, Pereira D, Postigo AS, Wangüenmert-Perez G, Ortega-Moñux A, Halir R, Rubio JDO, Schmid JH, Cheben P, Ctyroky J (2021)

Subwavelength grating silicon photonic devices. In: Proceedings volume 11691, Silicon potonics XVI, p 1169109. https://doi.org/10.1117/12.2577455

Nicolaï A, Sumpter BG, Meunier V (2014) Tunable water desalination across graphene oxide framework membranes. Phys Chem Chem Phys 16(18):8646–8654

Nine MJ, Cole MA, Tran DNH, Losic D (2015) Graphene: a multipurpose material for protective coatings. J Mater Chem A 3(24):12580–12602

Nolting DD, Nickels ML, Guo N, Pham W (2012) Molecular imaging probe development: a chemistry perspective. Am J Nucl Med Mol Imaging 2(3):273–306

Nørskov JK, Abild-Pedersen F, Studt F, Bligaard T (2011) Density functional theory in surface chemistry and catalysis. Proc Natl Acad Sci 108(3):937–943

Novoselov KS, Geim AK, Morozov S, Jiang D, Zhang Y, Dubonos SV, Grigorieva IV, Firsov AA (2004) Electric field effect in atomically thin carbon films. Science 306(5696):666–669

Palmieri V, Lattanzi W, Perini G, Augello A, Papi M, De Spirito M (2020) 3D-printed graphene for bone reconstruction. 2D Mater 7(2):022004

Park K, Lee S, Kang E, Kim K, Choi K, Kwon I (2009) New generation of multifunctional nanoparticles for cancer imaging and therapy. Adv Funct Mater 19(10):1553–1566

Park C, Yoon H, Kwon OS (2016) Graphene-based nanoelectronic biosensors. J Ind Eng Chem 38: 13–22

Paul W, Sharma C (2011) Blood compatibility and biomedical applications of graphene. Trends Biomater Artif Organs 25:91–94

Paul A, Hasan A, Al Kindi H, Gaharwar A, Rao V, Nikkhah M, Shin SR, Krafft D, Dokmeci MR, Shum-Tim D, Khademhosseini A (2014) Injectable graphene oxide/hydrogel-based angiogenic gene delivery system for vasculogenesis and cardiac repair. ACS Nano 8(8):8050–8062

Pei S, Cheng H-M (2012) The reduction of graphene oxide. Carbon 50(9):3210–3228

Peña-Bahamonde J, Nguyen HN, Fanourakis SK, Rodrigues DF (2018) Recent advances in graphene-based biosensor technology with applications in life sciences. J Nanobiotechnol 16(1):75

Premkumar T, Geckeler K (2012) Graphene-DNA hybrid materials: assembly, applications, and prospects. Prog Polym Sci 37(4):515–529

Priya B, Shandilya P, Raizada P, Thakur P, Singh N, Singh P (2016) Photocatalytic mineralization and degradation kinetics of ampicillin and oxytetracycline antibiotics using graphene sand composite and chitosan supported BiOCl. J Mol Catal A Chem 423:400–413

Qiu Y, Wang Z, Owens ACE, Kulaots I, Chen Y, Kane AB, Hurt RH (2014) Antioxidant chemistry of graphene-based materials and its role in oxidation protection technology. Nanoscale 6(20): 11744–11755

Rao CNR, Biswas K, Subrahmanyam KS, Govindaraj A (2009) Graphene, the new nanocarbon. J Mater Chem 19(17):2457–2469

Reina G, González-Domínguez JM, Criado A, Vázquez E, Bianco A, Prato M (2017) Promises, facts and challenges for graphene in biomedical applications. Chem Soc Rev 46(15):4400–4416

Salaheldin T, Loutfy S, Ramadan M, Youssef T, Mousa S (2019) IR-enhanced photothermal therapeutic effect of graphene magnetite nanocomposite on human liver cancer HepG2 cell model. Int J Nanomedicine 14:4397–4412

Secor EB, Ahn BY, Gao TZ, Lewis JA, Hersam MC (2015) Rapid and versatile photonic annealing of graphene inks for flexible printed electronics. Adv Mater 27(42):6683–6688

Shang L, Qi Y, Lu H, Pei H, Li Y, Qu L, Wu Z, Zhang W (2019) Chapter 7 Graphene and graphene oxide for tissue engineering and regeneration. In: Cui W, Zhao X (eds) Theranostic bionanomaterials. Elsevier, Amsterdam, pp 165–185. https://doi.org/10.1016/B978-0-12-815341-3.00007-9

Shao Y, Wang J, Wu H-X, Liu J, Aksay IA, Lin Y (2010) Graphene based electrochemical sensors and biosensors: a review. Electroanalysis 22(10):1027–1036

Shearer CJ, Slattery AD, Stapleton AJ, Shapter JG, Gibson CT (2016) Accurate thickness measurement of graphene. Nanotechnology 27(12):125704

Shi L, Chen J, Teng L, Wang L, Guanglin Z, Liu S, Luo Z, Shi X, Wang Y, Ren L (2016) The antibacterial applications of graphene and its derivatives. Small 12(31):4165–4184
Shim G, Kim M-G, Park J, Oh Y-K (2016) Graphene-based nanosheets for delivery of chemotherapeutics and biological drugs. Adv Drug Deliv Rev 105:205–227
Shin YC, Lee J, Kim MINJ, Hong SW, Kim B, Hyun JK, Choi YS, Park J-C, Han D-W (2015) Stimulating effect of graphene oxide on myogenesis of C_2C_{12} myoblasts on RGD peptide-decorated PLGA nanofiber matrices. J Biol Eng 9(1):22
Shin YC, Song SJ, Hong SW, Jeong SJ, Chrzanowski W, Lee JC, Han DW (2017) Multifaceted biomedical applications of functional graphene nanomaterials to coated substrates, patterned arrays and hybrid scaffolds. Nanomaterials 7(11):369
Shin YC, Song SJ, Hong SW, Oh JW, Hwang YS, Choi YS, Han DW (2018a) Graphene-functionalized biomimetic scaffolds for tissue regeneration. Adv Exp Med Biol 1064:73–89
Shin YC, Song SJ, Jeong SJ, Kim B, Kwon IK, Hong SW, Oh JW, Han DW (2018b) Graphene-based nanocomposites as promising options for hard tissue regeneration. Adv Exp Med Biol 1078:103–117
Simpson CD, Brand JD, Berresheim AJ, Przybilla L, Räder HJ, Müllen K (2002) Synthesis of a giant 222 carbon graphite sheet. Chemistry 8(6):1424–1429
Singh Z (2016) Applications and toxicity of graphene family nanomaterials and their composites. Nanotechnol Sci Appl 9:15–28
Song X-R, Wang X, Yu S-X, Jianbo C, Li S-H, Li J, Liu G, Yang H-H, Chen X (2015) Co_9Se_8 nanoplates as a new theranostic platform for photoacoustic/magnetic resonance dual-modal-imaging-guided chemo-photothermal combination therapy. Adv Mater 27(21):3285–3291
Subrahmanyam KS, Panchakarla LS, Govindaraj A, Rao CNR (2009) Simple method of preparing graphene flakes by an arc-discharge method. J Phys Chem C 113(11):4257–4259
Susi T, Pichler T, Ayala P (2015) X-ray photoelectron spectroscopy of graphitic carbon nanomaterials doped with heteroatoms. Beilstein J Nanotechnol 6:177–192
Syama S, Mohanan PV (2019) Comprehensive application of graphene: emphasis on biomedical concerns. Nano-Micro Lett 11(1):6
Szunerits S, Boukherroub R (2016) Antibacterial activity of graphene-based materials. J Mater Chem B 4(43):6892–6912
Tang L, Wang Y, Liu Y, Li J (2011) DNA-directed self-assembly of graphene oxide with applications to ultrasensitive oligonucleotide assay. ACS Nano 5(5):3817–3822
Tapasztó O, Balko J, Puchy V, Kun P, Dobrik G, Fogarassy Z, Horvath ZE, Dusza J, Balazsi K, Balazsi C, Tapaszto L (2017) Highly wear-resistant and low-friction Si_3N_4 composites by addition of graphene nanoplatelets approaching the 2D limit. Sci Rep 7(1):10087
Tyurnina A, Tsukagoshi K, Hiura H, Obraztsov A (2013) Structural and charge transport characteristics of graphene layers obtained from CVD thin film and bulk graphite materials. Carbon 52:49–55
Vlăsceanu G, Iovu H, Ionita M (2019) Graphene inks for the 3D printing of cell culture scaffolds and related molecular arrays. Compos Part B Eng 162(3):712–723
Wang H, Hu Y (2011) Effect of oxygen content on structures of graphite oxides. Ind Eng Chem Res 50(10):6132–6137
Wang Y, Li Y, Tang L, Lu J, Li J (2009) Application of graphene-modified electrode for selective detection of dopamine. Electrochem Commun 11(4):889–892
Wang Y, Li Z, Hu D, Lin C-T, Li J, Lin Y (2010a) Aptamer/graphene oxide nanocomplex for in situ molecular probing in living cells. J Am Chem Soc 132(27):9274–9276
Wang Z, Li N, Shi Z, Gu Z (2010b) Low-cost and large-scale synthesis of graphene nanosheets by arc discharge in air. Nanotechnology 21(17):175602
Wang H, Maiyalagan T, Wang X (2012) Review on recent progress in nitrogen-doped graphene: synthesis, characterization, and its potential applications. ACS Catal 2(5):781–794
Wang H, Deng T, Cui X, Qi Y, Mu X, Hou X, Zhu Y (2013a) Graphene oxide as a facile acid catalyst for the one-pot conversion of carbohydrates into 5-ethoxymethylfurfural. Green Chem 15(9):2379–2383

Wang J, Wei Y, Shi X, Gao H (2013b) Cellular entry of graphene nanosheets: the role of thickness, oxidation and surface adsorption. RSC Adv 3(36):15776–15782

Wang K, Dong J, Sun L, Chen H, Wang Y, Wang C, Dong L (2016) Effects of elemental doping on the photoluminescence properties of graphene quantum dots. RSC Adv 6(94):91225–91232

Weissleder R (2006) Molecular imaging in cancer. Science 312(5777):1168–1171

Xie X, Shao X, Gao F, Jin H, Zhou J, Lehui D, Zhang Y, Ouyang W, Wang X, Zhao L, Zhang X, Tang J (2011) Effect of hyperthermia on invasion ability and TGF-β1 expression of breast carcinoma MCF-7 cells. Oncol Rep 25(6):1573–1579

Xie X, Hu K, Fang D, Shang L, Tran S, Cerruti M (2015) Graphene and hydroxyapatite self-assemble into homogenous, free standing nanocomposite hydrogels for bone tissue engineering. Nanoscale 7(17):7992–8002

Xu H, Suslick KS (2011) Sonochemical preparation of functionalized graphenes. J Am Chem Soc 133(24):9148–9151

Xu Z, Wang S, Li Y, Wang M, Shi P, Huang X (2014) Covalent functionalization of graphene oxide with biocompatible poly(ethylene glycol) for delivery of paclitaxel. ACS Appl Mater Interfaces 6(19):17268–17276

Yang K, Wan J, Zhang S, Tian B, Zhang Y, Liu Z (2011) The influence of surface chemistry and size of nanoscale graphene oxide on photothermal therapy of cancer using ultra-low laser power. Biomaterials 33(7):2206–2214

Yao B, Chandrasekaran S, Zhang J, Xiao W, Qian F, Zhu C, Duoss EB, Spadaccini CM, Worsley MA, Li Y (2019) Efficient 3D printed pseudocapacitive electrodes with ultrahigh MnO_2 loading. Joule 3(2):459–470

Yoo JM, Kang JH, Hong BH (2015) Graphene-based nanomaterials for versatile imaging studies. Chem Soc Rev 44(14):4835–4852

Yu J, Yin W, Zheng X, Tian G, Tao B, Dong X, Wang Z, Gu Z, Ma X, Zhao Y (2015) Smart MoS_2/Fe_3O_4 nanotheranostic for magnetically targeted photothermal therapy guided by magnetic resonance/photoacoustic imaging. Theranostics 5(9):931–945

Yu P, Bao R-Y, Shi X-J, Yang W, Yang M-B (2016) Self-assembled high-strength hydroxyapatite/graphene oxide/chitosan composite hydrogel for bone tissue engineering. Carbohydr Polym 155:507–515

Zhang L, Lu Z, Zhao Q, Huang J, Shen H, Zhang Z (2011) Enhanced chemotherapy efficacy by sequential delivery of siRNA and anticancer drugs using PEI-grafted graphene oxide. Small 7(4):460–464

Zhang Y, Nayak T, Hong H, Cai W (2012) Graphene: a versatile nanoplatform for biomedical applications. Nanoscale 4(13):3833–3842

Zhao H, Ding R, Zhao X, Li Y, Qu L, Pei H, Yildirimer L, Wu Z, Zhang W (2017) Graphene-based nanomaterials for drug and/or gene delivery, bioimaging, and tissue engineering. Drug Discov Today 22(9):1302–1317

Zhong J, Zhou G-X, He P, Yang Z-H, Jia D-C (2017) 3D printing strong and conductive geo-polymer nanocomposite structures modified by graphene oxide. Carbon 117:421–426

Zou X, Zhang L, Wang Z, Luo Y (2016) Mechanisms of the antimicrobial activities of graphene materials. J Am Chem Soc 138(7):2064–2077

Zou X, Wei S, Jasensky J, Xiao M, Wang Q, Brooks C, Chen Z (2017) Molecular interactions between graphene and biological molecules. J Am Chem Soc 139(5):1928–1936

Zurutuza A, Marinelli C (2014) Challenges and opportunities in graphene commercialization. Nat Nanotechnol 9(10):730–734

Differential Toxicity of Graphene Family Nanomaterials Concerning Morphology

Iruthayapandi Selestin Raja, Anara Molkenova, Moon Sung Kang, Seok Hyun Lee, Ji Eun Lee, Bongju Kim, Dong-Wook Han, and Timur Sh. Atabaev

Abstract

Graphene family nanomaterials (GFNs) are well-known carbonaceous materials, which find application in several fields like optoelectronics, photocatalysis, nanomedicine, and tissue regeneration. Despite possessing many advantages in biomedical applications, GFNs exhibited toxicity depending on various parameters including dosage, size, exposure time, and kinds of administration. GFNS are majorly classified into nanosheets, quantum dots, nanoplatelets, and nanoribbons based on morphology. Understanding the toxic effects of GFNs would provide new suggestions as to how the materials can be utilized

Iruthayapandi Selestin Raja and Anara Molkenova equally contributed to this work.

I. S. Raja · A. Molkenova
BIO-IT Fusion Technology Research Institute, Pusan National University, Busan, South Korea

M. S. Kang · S. H. Lee · D.-W. Han (✉)
Department of Cogno-Mechatronics Engineering, College of Nanoscience and Nanotechnology, Pusan National University, Busan, South Korea
e-mail: nanohan@pusan.ac.kr

J. E. Lee
Department of Optics and Mechatronics Engineering, College of Nanoscience and Nanotechnology, Pusan National University, Busan, South Korea

B. Kim
Dental Life Science Research Institute/Innovation Research and Support Center for Dental Science, Seoul National University Dental Hospital, Seoul, South Korea

T. S. Atabaev
Department of Chemistry, School of Sciences and Humanities, Nazarbayev University, Nur-Sultan, Kazakhstan
e-mail: timur.atabaev@nu.edu.kz

D.-W. Han, S. W. Hong (eds.), *Multifaceted Biomedical Applications of Graphene*, Advances in Experimental Medicine and Biology 1351,
https://doi.org/10.1007/978-981-16-4923-3_2

effectively. Hence, we are summarizing here some of the recent findings in cellular and animal level toxicity studies of GFNs on the perspective of their different morphologies. Notwithstanding, we highlight progress, challenges, and new toxicological approaches to ensure biosafety of GFNs for future directions.

Keywords

Graphene family nanomaterials · Morphology · Toxicity · Biosafety

1 Introduction

Graphene family nanomaterials (GFNs) are the abundantly used carbonaceous materials in the modern society, which expose potentially toxic materials to the environment at an increasing rate (Seabra et al. 2014). It is essential to assess biosafety of GFNs to control their damaging effects (Shin et al. 2018). It was established that chemical structure, lateral dimension, layer number, surface functionalization, cell type, dosage, exposure time, and administration route of the nanoparticles are the factors to determine their toxicity level (Lee et al. 2012; Ou et al. 2016; Gies and Zou 2018; Han and Chrzanowski 2018; Raja et al. 2021). In graphene, carbon atoms are arranged into a layer assuming a honeycomb-like structure (Georgakilas et al. 2015). When graphene is chemically treated, oxygen functional groups are cleaved from its carbon lattice. The physicochemical properties of graphene can be tuned to various preparation procedures. Graphene-based nanomaterials are graphene, graphene oxide, reduced graphene oxide, and ultrafine graphite, which can be prepared in different forms like nanolayers, nanosheets, nanoribbons, nanoplatelets (above 10 graphene sheets but <100 nm in thickness), and quantum dots (Guo and Mei 2014). The layered graphene may be single, few (2–10 layers), or multi-typed (above 10 layers) nanoparticle (Goh and Pumera 2010). Identification of multiple graphene forms and their concerned toxic effects will provide a clear insight into the biosafety measurements. The formation of different graphene-based nanomaterials from their parent compounds has been shown schematically in Fig. 1.

Two-dimensional graphene nanosheets (GNSs) have been widely used for the development of artificial tissue engineering scaffolds due to their remarkable properties such as electrical conductivity and mechanical strength (Castro Neto et al. 2009; Lee et al. 2008). With the combined properties of high in-plane stiffness, strength, and flexibility, nanosheets of graphene and graphene oxide can be easily transferred and coated onto different substrates (Kenry et al. 2018). The surface of graphene can be easily modified by incorporating functional groups through either covalent or non-covalent interactions. In comparison to other nanomaterials, GFNs act as highly attractive scaffolds for directing stem cell growth and differentiation (Lee et al. 2011). Graphene quantum dots (GQDs), composed of sp^2-hybridized

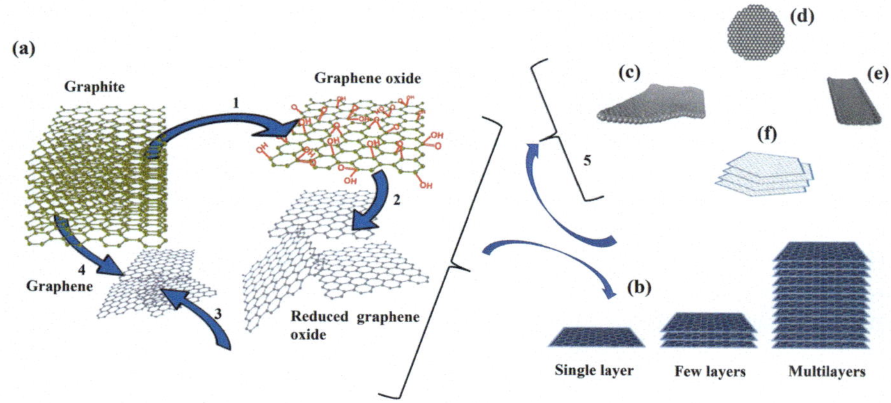

Fig. 1 (**a**) Formation of graphene oxide, reduced graphene oxide and graphene from graphite using different techniques (Reproduced from Filip and Tkac 2014, Copyright (2014) Elsevier). (**b**) Graphene family nanomaterials containing single layer, few layers, and multilayers (Reproduced from Goh and Pumera 2010, Copyright (2010) Elsevier). Different forms of graphene family nanomaterials: (**c**) graphene sheet, (**d**) graphene quantum dot, (**e**) graphene nanoribbon (Reproduced from Georgakilas et al. 2015, Copyright (2015) American Chemical Society), and (**f**) graphene nanoplatelet (Reproduced from Saeed et al. 2018, Copyright (2018) Elsevier). The preparation methods are (1) oxidation and exfoliation, (2) reduction by chemical, thermal, or electrochemical methods, (3) chemical vapor deposition, (4) mechanical cleavage, and (5) top-down methods: oxidizing, cutting, reduction, and surface passivation

carbon atoms, are obtained when a graphene monolayer was cut into small disks with 2–20 nm (Georgakilas et al. 2015). The quantum confinement and edge effects of GQDs are more pronounced than graphene nanosheet precursors, which are sized less than 100 nm (Ponomarenko et al. 2008). GQDs have non-zero band gaps, which are attributed to their strong photoluminescence emission (Ritter and Lyding 2009; Neubeck et al. 2010). The literature reports reveal that the bandgap of GQDs can be altered by changing their geometry, size, and shape (Lu et al. 2011). GQDs have traced their steps profoundly in bioimaging, photocatalysis, phototransistors, and photovoltaic devices (Yan et al. 2010).

GQDs can be produced either by top-down or bottom-up approaches. Conventionally, GQDs are prepared from graphitic structures by the mechanism of the cutting process as a top-down approach (Dong et al. 2012, 2013). Carving graphene into GQDs has been achieved using a high-resolution electron-beam lithography technique, which is a highly specialized one providing low yield of product (Ponomarenko et al. 2008). The hydrothermal treatment oxidized reduced graphene oxide (rGO) in the presence of strong acid at 200 °C to produce flat GQDs containing 1–3 graphene layers with a diameter of about 10 nm (Pan et al. 2010). The literature reports reveal that hydrothermal cutting of GO assisted by poly-(ethylenimine) formed monodispersed GQDs with a diameter of 5 nm. Poly-(ethylenimine) has been known to act as a chemical scissor, which not only cut

GO nanosheet into small disks but also prevent stacking of produced GQDs (Xue et al. 2013). The electrolysis process followed by reduction was utilized to achieve synthesis of GQDs. In such a method, a graphite rod immersed in alkaline solution acted as anode in electrolysis and hydrazine solution involved in the reduction process to produce light brown GQDs with a diameter range of 5–10 nm (Zhang et al. 2012). It has been well known that the electrochemical unzipping of MWNTs generates highly qualified graphene nanoribbons in aqueous medium. Contrarily, such reaction proceeds to produce GQDs if the medium is organic solvent (Georgakilas et al. 2015). The bottom-up approach produced GQDs from the aromatic precursors, which eliminated the drawback of the top-down approaches that yields nanoparticles with a broad size distribution (Liu et al. 2011). A large-scale synthesis of GQDs was achieved using a thermal plasma, which exhibited a strong and tunable photoluminescence upon irradiation like carbon dots jet (Kim and Suh 2014). GQDs emit difference photoluminescence like green, yellow, or blue depending on size, distribution, and functionalization (Shinde and Pillai 2012; Lee et al. 2013). The PEG-functionalized single-layer GQDs were reported to exhibit strong photoluminescence intensity than as-prepared GQDs (Dong et al. 2012, 2013).

Graphene nanoribbons are one-dimensional nanomaterial composed of sp^2 hybridized carbon atoms. They can be classified into three types, viz. zigzag, armchair, and chiral nanoribbons, depending on their defined edge termination (Terrones et al. 2010; Georgakilas et al. 2015). According to Raman spectral analysis, graphene nanoribbons prepared from MWCNTs exhibited a strong D band because numerous defect sites were produced in particles with high edge-to-surface ratios (Higginbotham et al. 2010). The graphene nanoribbons produced by chemical treatment of MWCNTs using sulfuric acid and potassium permanganate possessed carbonyl, hydroxyl, and carboxyl groups at the edges. The disruption of aromatic chains at the oxidized sites caused poor electrical conductivity in graphene nanoribbons (Kosynkin et al. 2009). GNRs were also prepared by calcination in air followed by sonication. Apart from oxidation, GNRs could be synthesized in different techniques such as electrical unwrapping, lithographic methods, chemical vapor deposition, and high-pressure treatment (Georgakilas et al. 2015). The condensation of polycyclic aromatic hydrocarbons including polyphenylene precursors and hexaphenylbenzene-type polymers yielded GNRs in a bottom-up approach. GNRs were prepared with different architectures using the combined procedures of surface covalent synthesis and polymerization of aromatic precursors with peripheral halogen groups (Chen et al. 2012).

Raman spectra of graphene nanoplatelets exhibited a prominent G band, D band, and 2D band at 1580 cm^{-1}, 1350 cm^{-1}, and 2700 cm^{-1}, respectively. With a higher I_D/I_G ratio, graphene nanoplatelets had a larger number of oxygen sites and defects (Wang et al. 2008; Georgakilas et al. 2015). Due to the presence of strong covalent bonds (C-C) between adjoining sp^2 carbon atoms, GNPs are the strong graphene nanomaterials with high flexibility (Akhavan et al. 2012). When conductive polymeric composite materials were prepared by dispersing GNPs, the electrical conductivity of such materials was generally determined by the degree of connections among the nanomaterials (Shao et al. 2010). Apart from graphene sheets, quantum dots, ribbons, and platelets, there are still different forms of graphene materials being studied by the researchers. They include graphene nanowalls (Akhavan and Ghaderi 2010), flakes (Mao et al. 2016), and foams (Amani et al. 2019), very few literatures of which are available to describe their potential biomedical applications and possible toxicity studies (Han and Atabaev 2020).

2 Toxicity Studies of Graphene Family Nanomaterials

2.1 Nanosheets

The DNA damaging potential of few-layer graphene (FLG) nanosheets towards human primary umbilical vein endothelial cells was confirmed from the experiments such as combined single-cell gel electrophoresis and comet assay. The average lateral dimension and layer thickness of the FLG structure was 160 ± 48.5 nm and of 0.8 ± 0.42 nm, respectively. The cDNA microarray analysis showed that the adverse effects of FLG could be triggered by the downregulation of key genes associated with DNA damage responses. The interaction of FLG with cell surface affected cytoskeletal integrity elevating ROS level, and reduced cell viability significantly. Further, they caused glutathione oxidation and lipid peroxidation in endothelial cells. They stimulated the overexpression of intracellular calcium ions in a dose-dependent manner (0–10 μg/mL) (Sasidharan et al. 2016). Ivask et al. reported toxicity profile of single-layer GO sheets when T- and B-lymphocytes and primary lymphocyte cells were treated with nanoparticles at a concentration of above 50 μg/mL. GO sheets elicited genotoxicity by increasing the number of micronuclei in the lymphocyte cells. It was described that the physical interaction of GO with cell membrane and cytoskeleton disrupted chromosome segregation during mitosis (Ivask et al. 2015).

When primary human monocytes were exposed to 50 μg/mL of FLG sheets (lateral size 500–750 nm) at different periods 1, 4, 12, and 24 h, there was a dramatic decrease in monocyte population as evidenced by the morphological dot plots. There was a significant increase in necrotic cell population from 6.2% (control) to 29% (1 h) followed by 36.5% (4 h). Eventually, the percentage of cell death reached 71.5% at 24 h of post-exposure of FLG. The induction of FLG on monocytes upregulated numerous chemokines and cytokines with proportional to incubation time (Russier et al. 2017). Size- and dose-dependent toxicity of single-layer graphene oxide (SLGO) and multilayer graphene oxide (MLGO) nanosheets were examined on THP-1 human leukemia monocytic cells at different concentrations 10, 20, 40, 80, and 100 μg/mL of GOs. AFM analyses exposed that the topographic heights of SLGO and MLGO were 0.7–1.2 nm and 2.8–3.3 nm, respectively. GOs were reported to cause severe cytotoxicity and stimulate immune responses, which were evidenced by phagocytosis and cytokine secretion. Comparatively, SLGO showed significant protein adsorption than MLGO, confirmed by bicinchoninic acid assay. Acute and chronic in vivo toxicity studies of GOs were observed in lung and kidney following their intravenous administration in mice. Both GOs induced a high level of inflammation and size-dependent fibrosis in mice. Further, MLGO produced reactive oxygen species along with induced apoptosis and necrosis to a higher extent than SLGO (Cho et al. 2016).

The toxicity of FLG nanosheets with an average hydrodynamic diameter of 500 nm was investigated in different tissues of shrimp *Litopenaeus vannamei* following a diet exposure for 28 days. The results revealed that the nanoparticle produced more reactive oxygen species, reduced glutathione level, and altered antioxidant enzymes including glutathione-S-transferase and glutamate-cysteine ligase. Further, FLG treatment led to morphological changes in hepatopancreas tissues (Fernandes et al. 2017). The toxicity of three different graphene family nanoparticles including graphene oxide (GO), hydrazine reduced GO (H.rGO), and ascorbic acid reduced GO (AA.rGO) were studied against A549 human lung epithelial carcinoma cells. When GOs interacted with A549 cells at a final concentration of 20, 50, 100, 200, and 400 ppm for 24 h, they induced dose-dependent reduction in cell toxicity and caused alveolar toxicity at high concentrations. TEM measurements revealed that the size of GOs varied from 500 to 600 nm. AA.rGO was reported to affect cell membrane integrity and increase the release of lactate dehydrogenase enzyme significantly than H.rGO (Dervin et al. 2018).

The reproductive toxicity of GO nanosheets was investigated in adult male and female ICR-strain mice following the tail vein and abdominal injections (25 mg/kg mouse) to male mice, which were then allowed to mate with female mice. The results discovered that GO nanosheets (54.9 ± 23.1 nm) were accumulated in the liver, spleen, and lung of mice at large quantity. However, there was no trace of nanoparticles in testes and epididymides even at 60 days of post-injection. The nanoparticles treated mice had the same male to female ratio, growth of offspring and survival rate with those of controls. Besides, GO nanosheets did not influence in altering sex-linked enzymatical activities and hormones, which indicated that nanoparticles had insignificant male reproductive toxicity (Liang et al. 2015). Single-layer GO with an average size of 5.1 nm was described to induce hemolysis of RBCs by releasing ATP and destroy the membrane integrity by extracting the lipid content (Wang et al. 2015). TEM image of a few-layered graphene nanosheet has been shown in Fig. 2a.

Graphene oxide nanosheets of large (l-GO) and small lateral dimensions (s-GO) were investigated for their toxicity effects in enzymatically dissociated hippocampal neurons at a concentration of 1 or 10 μg/mL. The s-GO had the potential to alter synapses and stimulate glial cell reaction whereas l-GO induced unequivocal hippocampal cell loss preventing membrane-nanosheet interaction at 6 days of post-exposure. The cellular composition of control and sGO treated hippocampal cultures was determined by observing astrocytes, neurons, and nuclei of the cells by immunostaining with GFAP, β-tubulin III, and DAPI, respectively. As shown in Fig. 2b, sGO treated cells exhibited lesser density than control, showing the influence of nanoparticles in reducing synapse formation without changes in cell survival. The activity of small clusters of neurons was measured to demonstrate the distribution of neuronal excitation. The characterization of cell ability to generate

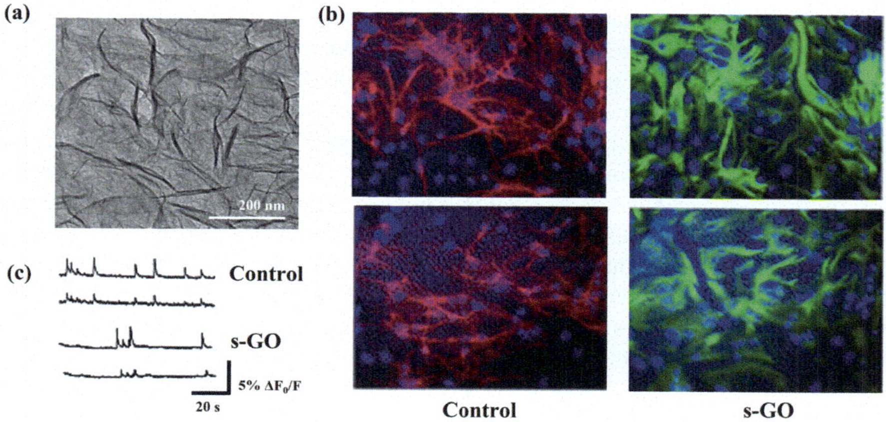

Fig. 2 (**a**) TEM image of few-layered crumpled graphene nanosheets (FLGs) with scale bar 200 nm (Reproduced from Yadav et al. 2017, Copyright (2017) Elsevier) (**b**) Immunofluorescence-stained images of cells shown to visualize neurons (anti-β-tubulin III, in red), glial cells (anti-GFAP, in green), and nuclei (DAPI, in blue) in hippocampal neuron cultures incubated with or without (control) single-layer graphene oxide (s-GO) nanosheets. Sample concentration 10 μg/mL; scale bar 50 μm. (**c**) Repetitive spontaneous Ca^{2+} oscillations recorded at 8–10 DIV. Recording of two cells was selected from each field sample (Reproduced from Rauti et al. 2016, Copyright (2016) American Chemical Society)

Ca^{2+} repetitive oscillations was compared between control and sGO (10 μg/mL) treated cultures (Fig. 2c). The control displayed active cells in all the recorded fields, whereas s-GO treated cells displayed only 44% active cells (n = 8 out of 18) (Rauti et al. 2016).

The potential toxicological impacts of rGO nanosheets towards lung cancer cells A549 and SKMES-1 were studied to determine oxidative stress and apoptosis as a function of concentration. The cells were treated with 5, 50, 250, 500, and 1000 μg/mL of rGO for 24 h. Even at a low concentration (>5 μg/mL), rGO nanosheets were found to disintegrate cellular membrane producing late apoptosis and necrosis significantly (Tabish et al. 2017). Single-layer GO nanosheets sized around 120 nm were treated with primary human corneal epithelial cells (hCorECs) and human conjunctiva epithelium cells (hConECs) at a concentration range of 12.5–100 μg/mL. The outcome of the results revealed that acute exposure of GOs did not cause cytotoxicity to hCorECs; however a short-term exposure of GO to cells triggered cytotoxicity with increased production of ROS. GOs did not exert any adverse effects such as conjunctival redness, abnormality of iris, corneal opacity, and chemosis in albino rabbits at a concentration of 100 μg/mL. However, the repeated exposure of GO (50 and 100 μg/mL) for 5 days induced reversible damages to corneal epithelium in Sprague-Dawley (SD) rats (Wu et al. 2016).

The influence of rGO nanosheet exposure was demonstrated on the reproductive ability of female mice and the development of offspring at three different concentrations 6.25, 12.5, and 25 mg/kg per mouse. A high dose of rGO (25 mg/

kg) by intravenous injection did not cause any significant changes in sex hormone level of female mice when compared to control. However, further investigations showed that rGO (12.5 mg/kg) treatment induced abortions at late gestational stage. Many pregnant mice were reported to die at a higher concentration of rGO (Xu et al. 2015). The potential effect of orally administered rGO nanosheets on animal behaviors was studied in male C57 black/6 mice. The buffer dispersed rGO (60 mg/kg body weight) was injected into mice at every 24 h for 5 days. The rGO treated mice exerted little changes in spatial learning, memory behavior, general locomotor activity, and neuromuscular coordination compared to control groups. However, the activities of hippocampal acetylcholinesterase and choline acetyltransferase enzymes and the morphology of hippocampus and neuroglia cells were not affected in rGO treated animal groups (Zhang et al. 2015).

2.2 Quantum Dots

A comparative size-dependent toxicity study comparing graphene quantum dots (GQDs) with graphene oxide sheets (GOSs) was investigated at a concentration range of 0–400 μg/mL in human gastric cancer cells (MGC-803) and breast cancer cells (MCF-7). The average particle size of GQDs was about 20 nm and the thickness was 1 nm. The lateral dimension of GO sheets was up to 1 μm and thickness was 1 nm. The TEM image of GQDs has been shown in Fig. 3a. Though GQDs and GOSs were internalized into the cells, their localized regions in cells were quite different. GOSs preferred to reside in the cytoplasm after a long-term incubation in cell culture, while GQDs accumulated in the nucleus and endoplasmic reticulum. Caveolae-mediated endocytosis was the major pathway for the effective cellular uptake of GQDs. From the results of cell viability analysis, GQDs were found to be more toxic against cancerous cells causing severe damages to

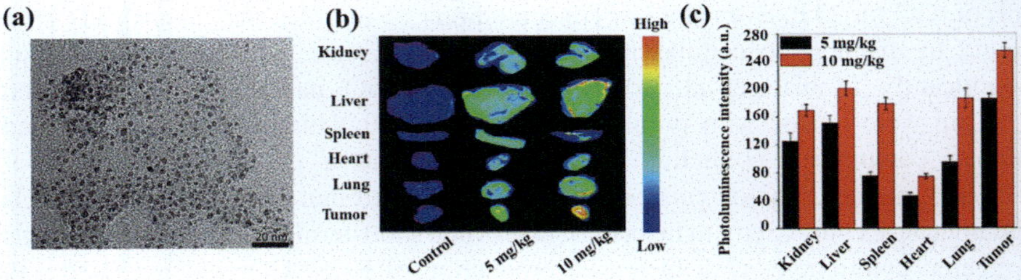

Fig. 3 (**a**) TEM image of graphene quantum dots (GQDs) with scale bar 20 nm (Reproduced from Wang et al. 2015, Copyright (2015) Elsevier). (**b**) Ex vivo images of different isolated organs from mice after 24 h of post intravenous administration of carboxylated GQDs. The tumor created by injecting epidermal carcinoma KB cells on the back of mice. (**c**) Photoluminescence intensities of carboxylated GQDs, which were extracted from isolated organs of mice following the treatment of different concentrated nanoparticles, 5 and 10 mg/kg. The data was represented with standard deviation ($n = 3$) ($p < 0.05$, one-way ANOVA) (Reproduced from Nurunnabi et al. 2013, Copyright (2013) American Chemical Society)

mitochondrial membrane potential, compared to microsized GOSs (Wu et al. 2013). The cytotoxic effects of hydroxyl-modified GQDs (OH-GQDs) were investigated on lung carcinoma A549 (wild type p53) and H1299 (p53-null) cells at different concentrations 0, 20, 40, 60, 80, and 100 μg/mL. The OH-GQDs significantly decreased the cell viability of both types of cells and increased the production of reactive oxygen species at a concentration of 50 μg/mL. Further, OH-GQDs treatment was reported to trigger cell senescence and G0–G1 arrest inducing overexpression of p21, a cyclin-dependent kinase inhibitor. TEM measurement showed that the average diameter of monodispersed OH-GQDs was 5.6 ± 1.1 nm, and dynamic light scattering analysis revealed that the nanoparticles had a hydrodynamic diameter of 10.3 ± 1.9 nm and were negatively charged with -5.0 mV zeta potential (Tian et al. 2016).

The potential toxicity of carboxylated photoluminescent GQDs (5 nm) was evaluated in four different cell line models such as epidermal carcinoma cells (KB), triple-negative breast cancer cells (MDA-MB231), adenocarcinoma human alveolar basal epithelial cells (A549), and Madin-Darby Canine Kidney epithelial cells (MDCK). It was discovered that GQDs were ingested effectively by all the investigated cells but GQDs triggered neither acute toxicity nor morphological changes at different concentrations 50, 100, 250, and 500 μg/mL. According to in vivo and ex vivo studies (Fig. 3b, c), the intravenously administered GQDs (5 and 10 mg/kg) were distributed in different organs such as liver, lung, spleen, kidney, and tumor sites of SD rats after 24 h of exposure. Also, there were no reports of obvious organ damage and lesions for the GQDs treated animals (Nurunnabi et al. 2013). When HeLa cells (A549) were treated with PEG-GQDs at concentrations 10, 20, 40, 80, and 160 μg/mL for 24 h, it was found that over 95% of HeLa cells survived up to 160 μg/mL. According to TEM micrograph results, GQD contained 1–2 graphene layers, and the size and height was measured to be 3–5 nm and 0.5–1 nm, respectively. The biosafety of PEG-GQDs was investigated by injecting the solution of nanoparticles to Balb/c mice intravenously or intraperitoneally for 14 days. The nanotoxicity of PEG-GO was also examined for the comparison. The analyses revealed that the PEG-GQDs treated mice survived without any obvious difference in body weight compared to control groups. Meanwhile, 25% of PEG-GQDs treated mice were reported to die after fourth administration because the nanoparticles were aggregated in liver and spleen without the sign of body loss (Chong et al. 2014).

When zebrafish embryos were co-cultured with GQDs sized 2.3–6.4 nm for in vivo bioimaging, it was observed that the uptake of few numbers of nanoparticles triggered little toxicity to the cardiovascular system of zebrafish. When cell viability assay was carried out introducing GQDs to HeLa cells at various concentrations 0, 12.5, 25, 50, 100, and 200 μg/mL for 24 h, the cytotoxicity was observed to decrease in a dose-dependent manner. Though there were no dramatic changes among the studied concentrations, GQDs exhibited 80% cell viability showing a low toxicity to cells at a higher concentration of 200 μg/mL (Jiang et al. 2015).

2.3 Nanoplatelets

A concentration-dependent cell viability study was carried out using human mesenchymal stem cells (hMSCs), which was treated with rGO nanoplatelets (rGONPs) with an average lateral dimension of 11 ± 4 nm. Besides, the result of cytotoxicity was compared with that of rGO nanosheets (rGONSs), lateral dimension of which was about 3.8 ± 0.4 μm. The final concentration of GOs was 0.01, 0.1, 1.0, 10, and 100 μg/mL. Oxidative stress and associated membrane damage were observed only in rGONSs treated cells but not in rGONPs treated cells at the concentration of 1.0 μg/mL. Meanwhile, rGONPs were reported to exhibit little genotoxicity such as DNA fragmentation and chromosomal aberrations (Akhavan et al. 2012). Toxic effects of two different polyvinylpyrrolidone (PVP) solubilized nanoplatelets of graphene oxide (GO-PVP) and reduced graphene oxide (rGO-PVP) were assessed in marine mussel (*Mytilus galloprovincialis*) hemocytes at various concentrations from 0.001 to 100 mg/L. The outcome of in vitro results revealed that rGO-PVP was found to be more toxic than GO-PVP and their LC50 values were 33.94 mg/L and 43.72 mg/L, respectively. Though functionalized graphene oxides were not much toxic reportedly, they caused ROS mediated oxidative stress a little followed by membrane rupture and lysosomal damage in mussel hemocytes (Katsumiti et al. 2017).

An inhalation toxicology study of graphene nanoplatelets (GNPs) was performed in male SD rats for 28 days (5 days/week). The experimental rats were exposed to airborne graphene by nose-only inhalation system at concentrations of 0.12, 0.47, and 1.88 mg/m^3 for 6 h per day. GNPs had an average lateral dimension larger than 2 μm with surface area 750 m^2/g. Figure 4a shows a TEM image of stacked GNPs. No significant changes were observed in body weight, bronchoalveolar lavage fluid, and blood biochemical markers at 1 day and 28 days of post-exposure. The results revealed that the inhaled graphene nanoplatelets were mostly translocated to lung lymph nodes and ingested by macrophages. The GNPs treated rats were reported to have a low toxicity even at a high dose (1.88 mg/m^3) and 28 days of post-exposure (Kim et al. 2016). The toxic response of commercially available GNPs was explored in cellular and animal level studies. The specific surface area and thickness of GNPs was 735 m^2/g and 3–4 nm, respectively. The concentration of GNPs tested for cell viability assay using human bronchial epithelial cell line (BEAS-2B cells) was 2.5, 5, 10, and 20 μg/mL. The nanoplatelets were engulfed by autophagosome-like vacuoles after 24 h of treatment, evidenced by TEM micrograph (Fig. 4d). A dose-dependent decrease in cell viability was observed along with the downregulation of ROS production, mitochondrial damage, and suppression of ATP production. The in vivo analysis revealed that GNPs localized in the lung of ICR mice at the concentrations of 2.5 and 5 mg/kg until 28 days of a single intratracheal instillation (Park et al. 2015).

When female C57BL/6 strain mice were exposed to 50 μg of GNPs by pharyngeal aspiration, GNPs were found to accumulate at a large quantity in the lungs. The specific surface area and projected diameter of the investigated GNPs was 100 m^2/g and 5.64 ± 4.56 nm, respectively. The mice aspirated with GNPs had expressed a

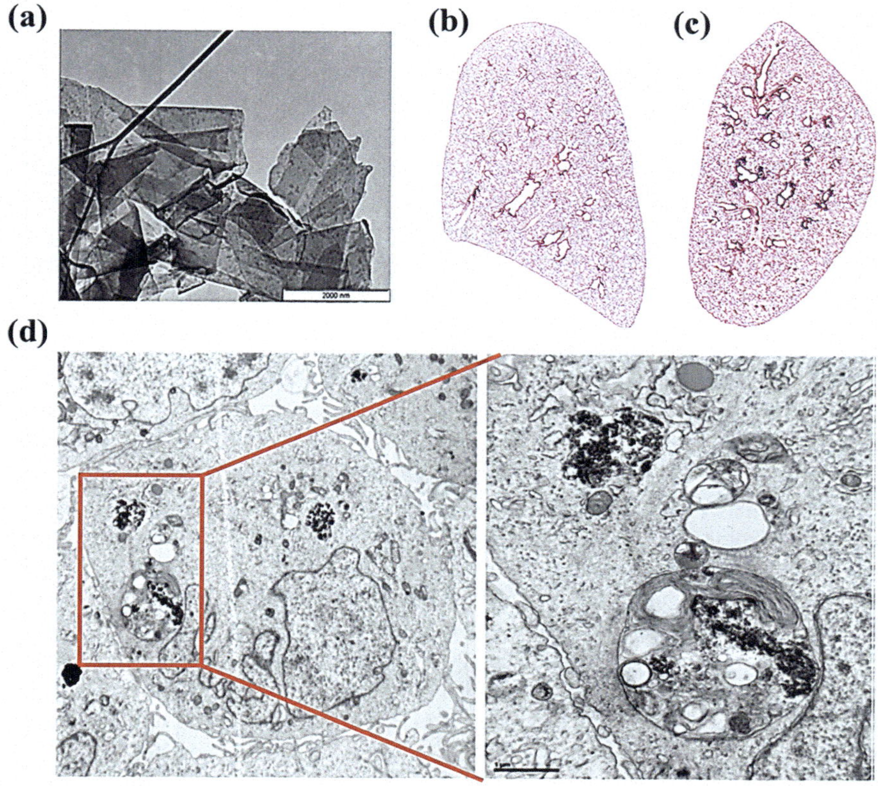

Fig. 4 (**a**) TEM image of graphene nanoplatelets (GNPs) with scale bar 2 μm (Reproduced from Fatima et al. 2017, Copyright (2017) Royal Society of Chemistry). (**b**) and (**c**) denote Haematoxylin & Eosin (H&E) staining of entire lung section from vehicle control and GNPs aspirated mice at 24 h. Granulomatous areas were appeared in the lung section of GNPs treated mouse (Reproduced from Schinwald et al. 2012, Copyright (2012) American Chemical Society). (**d**) TEM image showing autophagosome formation in BEAS-2B cells due to the exposure of GNPs (20 μg/mL) for 24 h. Enlarged view of nucleus section at the right panel with scale bar 1 μm (Reproduced from Park et al. 2015, Copyright (2015) Springer Nature)

significant level of toxicity in bronchoalveolar lavage (BAL) cells and polymorphonuclear leucocytes as compared to control after 24 h exposure. The histological analyses (Fig. 4b, c) of lung tissue harvested from GNPs exposed mice exhibited granulomatous lesions in the bronchiole lumen, which was not observed in the lung tissue of control group (Schinwald et al. 2012). Lammel T et al. reported that carboxyl modified GO nanoplatelets with 0.8–2.4 nm and 2–3 layers had the potential to permeate the plasma membrane of human hepatoma cells (Hep G2), reach the cytosol, and concentrate in intracellular vesicles. The exposure of cells to nanoplatelets (1–32 μg/mL) elevated intracellular ROS level, increased number of autophagosomes, and perturbed mitochondrial function (Lammel et al. 2013).

2.4 Nanoribbons

The shape-dependent cyto- and genotoxic effects of the single-layer rGO nanoribbons (rGONRs) were studied in human mesenchymal stem cells compared with those of rGO nanosheets (rGONSs). It was found that the lateral thickness of nanoribbons and nanosheets was about 1 nm and 0.8 nm, respectively. Cell viability assays showed that rGONRs exhibited cytotoxicity significantly at 10 μg/mL and 1 h post-exposure while the rGONSs displayed the same level of cytotoxicity at 100 μg/mL and 96 h post-exposure. The rGONSs damaged cell membrane to some extent owing to induction of intracellular oxidative stress. The rGONRs permeated cell membrane at 1 μg/mL and induced DNA fragmentations, and chromosomal aberrations after 1 h of exposure. In contrast to rGONSs, the rGONRs did not involve in the generation of reactive oxygen species (Akhavan et al. 2013). A comparative study in cytotoxicity was examined between graphene-oxide nanoribbons (GONRs) and graphene-oxide nanoplatelets (GONPs) in human lung carcinoma epithelial cell line (A549 cells). The oxidative treatment of multi-walled carbon nanotubes yielded GONRs and the lateral unzipping of stacked graphene nanofibers produced GONPs. The lateral size of GONRs and GONPs was 22.2 nm and 19.1 nm, respectively. The in vitro results revealed that the GONRs induced more cytotoxicity than GONPs at the investigated concentration range of nanoparticles, 0–400 μg/mL (Khim Chng et al. 2014).

The structural disruption of graphene oxide nanoribbons (GONRs) prepared by two different techniques, low-energy bath and high-energy probe sonication, was investigated through in vitro and in vivo biological systems. The outcome of the results informed that high-energy probe sonicated GONRs, at 20 μg/mL, had decreased metabolic rate in human breast cancer cells (MCF-7) and increased mortality in Japanese medaka Fish embryos when compared to low-energy bath sonicated GONRs (Mullick Chowdhury et al. 2014). AFM image of graphene oxide nanoribbons has been shown in Fig. 5a. The partial unzipping of graphene nanoribbons from the carbon nanotubes can be observed in Fig. 5b. The graphene oxide nanoribbons were functionalized using PEG-DSPE (1,2-distearoyl-sn-glycero-3-phosphoethanolamine-N-[amino(polyethylene glycol)]) to obtain water-solubilized O-GNR-PEG-DSPEs, and its cytotoxicity was evaluated in four different cell lines such as Henrietta Lacks (HeLa) cervical cancer cells, Michigan cancer foundation-7 (MCF-7) breast cancer cells, Sloan Kettering breast cancer cells (SKBR3), and National Institute of Health 3T3 (NIH-3T3) mouse fibroblast cells. The average width and length of the nanomaterial was about 125–220 nm and 500–2500 nm, respectively. The final concentration of the nanoribbons was 10, 50, 100, 200, 300, and 400 μg/mL. A concentration- and time-dependent decrease in cell viability was observed in all the investigated cells. However, HeLa cells witnessed higher cytotoxicity than MCF-7 and SKBR3 cells. As shown in Fig. 5c, d, TEM micrographs revealed that O-GNR-PEG-DSPEs were ingested into HeLa cells at a large quantity as compared to other cell types. The functionalized nanoribbons induced necrosis and subsequently caused cell membrane damages (Mullick Chowdhury et al. 2013).

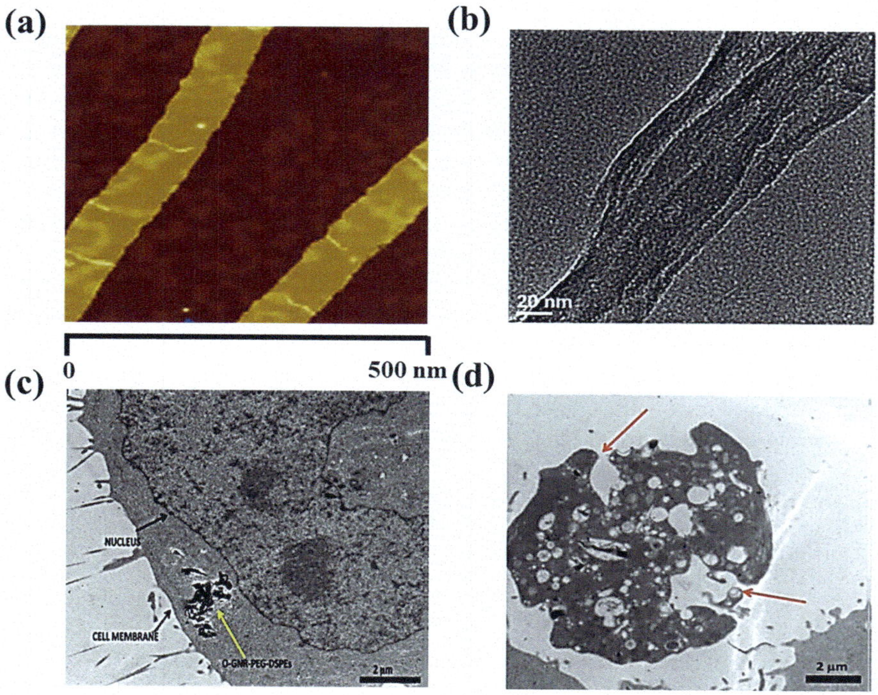

Fig. 5 (**a**) AFM image of graphene oxide nanoribbons aligned by a paint-brushing method (Reproduced from Akhavan et al. 2013, Copyright (2013) Elsevier). (**b**) TEM image of partially unzipped graphene nanoribbons from the carbon nanotubes longitudinally with scale bar 20 nm. These nanoribbons were still connected with their tips linked (Reproduced from Long et al. 2011, Copyright (2011) Royal Society of Chemistry). (**c**) TEM images of water-solubilized O-GNR-PEG-DSPEs (denoted by yellow arrow) accumulated inside the HeLa cell and (**d**) the plasma membrane ruptured (red arrows) due to necrosis induced by functionalized nanoribbons. The scale bar represents 2 μm (Reproduced from Mullick Chowdhury et al. 2013, Copyright (2013) Elsevier)

3 Conclusions and Perspectives

We have discussed the toxicity effects of graphene family nanomaterials for different morphologies such as nanosheet, quantum dots, nanoplatelets, and nanoribbons. The health risks associated with GFNs have been explored through collective pieces of evidence from in vitro and in vivo toxicological measurements. Besides, the current chapter demonstrates variation in toxicity of GFNs depending on physicochemical parameters such as particle size, shape, lateral dimension, and layer thickness. Still, there are some issues to be addressed. (1) Though in vitro toxicity of GFNs with different morphologies are extensively studied, comprehensive studies are required to investigate whether the same reflect at in vivo animal studies for different species. (2) When encapsulating agents are used for increasing solubility and biosafety of

GFNs, the surface properties like zeta potential, functional groups, and polydispersion index are not yet focused individually. (3) Biological interaction is of paramount importance to determine the toxicity of nanoparticle exposure. A systematic study is required while studying the interaction of nanoparticles with biological macromolecules. The following are few suggestions for the future of this field: (1) Determining differential toxicity of GFNs against normal and cancerous cells of the same source to conclude whether the studied nanoparticles are specific about their toxicity. (2) Selecting encapsulating agents with the same molecular weight and preparing the nanoparticles observing similar methodology to achieve homogeneity in the surface properties. (3) Elucidating the interaction of graphene-based nanoparticles with biological macromolecules concerning different functional groups on the surface of the nanoparticles to provide insight into the mechanism of triggering toxicity in the human body. We hope that the researchers with interdisciplinary backgrounds will advance the application of graphene family nanomaterials considering the problems and the concerned suggestions.

Acknowledgments This research was supported by National Research Foundation of Korea (NRF) funded by the Ministry of Science (NRF-2021R1A2C2006013) and by Korea Evaluation Institute of Industrial Technology (KEIT) grant funded by the Ministry of Trade, Industry and Energy (MOTIE, Korea) (No. 20014399).

References

Akhavan O, Ghaderi E (2010) Toxicity of graphene and graphene oxide nanowalls against bacteria. ACS Nano 4(10):5731–5736

Akhavan O, Ghaderi E, Akhavan A (2012) Size-dependent genotoxicity of graphene nanoplatelets in human stem cells. Biomaterials 33(32):8017–8025

Akhavan O, Ghaderi E, Emamy H, Akhavan F (2013) Genotoxicity of graphene nanoribbons in human mesenchymal stem cells. Carbon 54:419–431

Amani H, Mostafavi E, Arzaghi H, Davaran S, Akbarzadeh A, Akhavan O, Pazoki-Toroudi H, Webster TJ (2019) Three-dimensional graphene foams: synthesis, properties, biocompatibility, biodegradability, and applications in tissue engineering. ACS Biomater Sci Eng 5(1):193–214

Castro Neto AH, Guinea F, Peres NMR, Novoselov KS, Geim AK (2009) The electronic properties of graphene. Rev Mod Phys 81(1):109–162

Chen L, Hernandez Y, Feng X, Mullen K (2012) From nanographene and graphene nanoribbons to graphene sheets: chemical synthesis. Angew Chem Int Ed Engl 51(31):7640–7654

Cho YC, Pak PJ, Joo YH, Lee H-S, Chung N (2016) In vitro and in vivo comparison of the immunotoxicity of single- and multi-layered graphene oxides with or without pluronic F-127. Sci Rep 6:38884

Chong Y, Ma Y, Shen H, Tu X, Zhou X, Xu J, Dai J, Fan S, Zhang Z (2014) The in vitro and in vivo toxicity of graphene quantum dots. Biomaterials 35(19):5041–5048

Dervin S, Murphy J, Aviles R, Pillai SC, Garvey M (2018) An in vitro cytotoxicity assessment of graphene nanosheets on alveolar cells. Appl Surf Sci 434:1274–1284

Dong Y, Chen C, Zheng X, Gao L, Cui Z, Yang H, Guo C, Chi Y, Li CM (2012) One-step and high yield simultaneous preparation of single- and multi-layer graphene quantum dots from CX-72 carbon black. J Mater Chem 22(18):8764–8766

Dong Y, Pang H, Ren S, Chen C, Chi Y, Yu T (2013) Etching single-wall carbon nanotubes into green and yellow single-layer graphene quantum dots. Carbon 64:245–251

Fatima S, Ali SI, Iqbal MZ, Rizwan S (2017) The high photocatalytic activity and reduced band gap energy of La and Mn co-doped BiFeO3/graphene nanoplatelet (GNP) nanohybrids. RSC Adv 7(57):35928–35937

Fernandes AL, Josende ME, Nascimento JP, Santos AP, Sahoo SK, da Silva FMR, Romano LA, Furtado CA, Wasielesky W, Monserrat JM, Ventura-Lima J (2017) Exposure to few-layer graphene through diet induces oxidative stress and histological changes in the marine shrimp Litopenaeus vannamei. Toxicol Res 6(2):205–214

Filip J, Tkac J (2014) Is graphene worth using in biofuel cells? Electrochim Acta 136:340–354

Georgakilas V, Perman JA, Tucek J, Zboril R (2015) Broad family of carbon nanoallotropes: classification, chemistry, and applications of fullerenes, carbon dots, nanotubes, graphene, nanodiamonds, and combined superstructures. Chem Rev 115(11):4744–4822

Gies V, Zou S (2018) Systematic toxicity investigation of graphene oxide: evaluation of assay selection, cell type, exposure period and flake size. Toxicol Res 7(1):93–101

Goh MS, Pumera M (2010) Multilayer graphene nanoribbons exhibit larger capacitance than their few-layer and single-layer graphene counterparts. Electrochem Commun 12(10):1375–1377

Guo X, Mei N (2014) Assessment of the toxic potential of graphene family nanomaterials. J Food Drug Anal 22(1):105–115

Han DW, Atabaev TS (2020) The two faces of nanomaterials: toxicity and bioactivity. MDPI, Basel. https://doi.org/10.3390/books978-3-03928-984-4

Han DW, Chrzanowski W (2018) Frontiers in toxicity and functionalization of nanomaterials. MDPI, Basel. https://doi.org/10.3390/books978-3-03842-735-3

Higginbotham AL, Kosynkin DV, Sinitskii A, Sun Z, Tour JM (2010) Lower-defect graphene oxide nanoribbons from multiwalled carbon nanotubes. ACS Nano 4(4):2059–2069

Ivask A, Voelcker NH, Seabrook SA, Hor M, Kirby JK, Fenech M, Davis TP, Ke PC (2015) DNA melting and genotoxicity induced by silver nanoparticles and graphene. Chem Res Toxicol 28(5):1023–1035

Jiang D, Chen Y, Li N, Li W, Wang Z, Zhu J, Zhang H, Liu B, Xu S (2015) Synthesis of luminescent graphene quantum dots with high quantum yield and their toxicity study. PLoS One 10(12):e0144906–e0144906

Katsumiti A, Tomovska R, Cajaraville MP (2017) Intracellular localization and toxicity of graphene oxide and reduced graphene oxide nanoplatelets to mussel hemocytes in vitro. Aquat Toxicol 188:138–147

Kenry LWC, Loh KP, Lim CT (2018) When stem cells meet graphene: opportunities and challenges in regenerative medicine. Biomaterials 155:236–250

Khim Chng EL, Chua CK, Pumera M (2014) Graphene oxide nanoribbons exhibit significantly greater toxicity than graphene oxide nanoplatelets. Nanoscale 6(18):10792–10797

Kim J, Suh JS (2014) Size-controllable and low-cost fabrication of graphene quantum dots using thermal plasma jet. ACS Nano 8(5):4190–4196

Kim JK, Shin JH, Lee JS, Hwang JH, Lee JH, Baek JE, Kim TG, Kim BW, Kim JS, Lee GH, Ahn K, Han SG, Bello D, Yu IJ (2016) 28-Day inhalation toxicity of graphene nanoplatelets in Sprague-Dawley rats. Nanotoxicology 10(7):891–901

Kosynkin DV, Higginbotham AL, Sinitskii A, Lomeda JR, Dimiev A, Price BK, Tour JM (2009) Longitudinal unzipping of carbon nanotubes to form graphene nanoribbons. Nature 458(7240): 872–876

Lammel T, Boisseaux P, Fernandez-Cruz ML, Navas JM (2013) Internalization and cytotoxicity of graphene oxide and carboxyl graphene nanoplatelets in the human hepatocellular carcinoma cell line Hep G2. Part Fibre Toxicol 10:27

Lee C, Wei X, Kysar JW, Hone J (2008) Measurement of the elastic properties and intrinsic strength of monolayer graphene. Science 321(5887):385–388

Lee WC, Lim CHYX, Shi H, Tang LAL, Wang Y, Lim CT, Loh KP (2011) Origin of enhanced stem cell growth and differentiation on graphene and graphene oxide. ACS Nano 5(9): 7334–7341

Lee JH, Shin YC, Jin OS, Lee EJ, Han DW, Kang SH, Hong SW, Ahn JY, Kim SH (2012) Cytotoxicity evaluations of pristine graphene and carbon nanotubes in fibroblastic cells. J Kor Phys Soc 61(6):873–877

Lee E, Ryu J, Jang J (2013) Fabrication of graphene quantum dots via size-selective precipitation and their application in upconversion-based DSSCs. Chem Commun 49(85):9995–9997

Liang S, Xu S, Zhang D, He J, Chu M (2015) Reproductive toxicity of nanoscale graphene oxide in male mice. Nanotoxicology 9(1):92–105

Liu R, Wu D, Feng X, Mullen K (2011) Bottom-up fabrication of photoluminescent graphene quantum dots with uniform morphology. J Am Chem Soc 133(39):15221–15223

Long D, Li W, Qiao W, Miyawaki J, Yoon S-H, Mochida I, Ling L (2011) Partially unzipped carbon nanotubes as a superior catalyst support for PEM fuel cells. Chem Commun 47 (33):9429–9431

Lu J, Yeo PS, Gan CK, Wu P, Loh KP (2011) Transforming C60 molecules into graphene quantum dots. Nat Nanotechnol 6(4):247–252

Mao L, Hu M, Pan B, Xie Y, Petersen EJ (2016) Biodistribution and toxicity of radio-labeled few layer graphene in mice after intratracheal instillation. Part Fibre Toxicol 13(1):7

Mullick Chowdhury S, Lalwani G, Zhang K, Yang JY, Neville K, Sitharaman B (2013) Cell specific cytotoxicity and uptake of graphene nanoribbons. Biomaterials 34(1):283–293

Mullick Chowdhury S, Dasgupta S, McElroy AE, Sitharaman B (2014) Structural disruption increases toxicity of graphene nanoribbons. J Appl Toxicol 34(11):1235–1246

Neubeck S, Ponomarenko LA, Freitag F, Giesbers AJM, Zeitler U, Morozov SV, Blake P, Geim AK, Novoselov KS (2010) From one electron to one hole: quasiparticle counting in graphene quantum dots determined by electrochemical and plasma etching. Small 6(14):1469–1473

Nurunnabi M, Khatun Z, Huh KM, Park SY, Lee DY, Cho KJ, Lee Y-k (2013) In vivo biodistribution and toxicology of carboxylated graphene quantum dots. ACS Nano 7(8): 6858–6867

Ou L, Song B, Liang H, Liu J, Feng X, Deng B, Sun T, Shao L (2016) Toxicity of graphene-family nanoparticles: a general review of the origins and mechanisms. Part Fibre Toxicol 13(1):57

Pan D, Zhang J, Li Z, Wu M (2010) Hydrothermal route for cutting graphene sheets into blue-luminescent graphene quantum dots. Adv Mater 22(6):734–738

Park E-J, Lee G-H, Han BS, Lee B-S, Lee S, Cho M-H, Kim J-H, Kim D-W (2015) Toxic response of graphene nanoplatelets in vivo and in vitro. Arch Toxicol 89(9):1557–1568

Ponomarenko LA, Schedin F, Katsnelson MI, Yang R, Hill EW, Novoselov KS, Geim AK (2008) Chaotic dirac billiard in graphene quantum dots. Science 320(5874):356–358

Raja IS, Lee JH, Hong SW, Shin DM, Lee JH, Han DW (2021) A critical review on genotoxicity potential of low dimensional nanomaterials. J Hazard Mater 409:124915

Rauti R, Lozano N, León V, Scaini D, Musto M, Rago I, Ulloa Severino FP, Fabbro A, Casalis L, Vázquez E, Kostarelos K, Prato M, Ballerini L (2016) Graphene oxide nanosheets reshape synaptic function in cultured brain networks. ACS Nano 10(4):4459–4471

Ritter KA, Lyding JW (2009) The influence of edge structure on the electronic properties of graphene quantum dots and nanoribbons. Nat Mater 8(3):235–242

Russier J, Leon V, Orecchioni M, Hirata E, Virdis P, Fozza C, Sgarrella F, Cuniberti G, Prato M, Vazquez E, Bianco A, Delogu LG (2017) Few-layer graphene kills selectively tumor cells from myelomonocytic leukemia patients. Angew Chem Int Ed Engl 56(11):3014–3019

Saeed RM, Schlegel JP, Castano C, Sawafta R (2018) Preparation and enhanced thermal performance of novel (solid to gel) form-stable eutectic PCM modified by nano-graphene platelets. J Energy Storage 15:91–102

Sasidharan A, Swaroop S, Chandran P, Nair S, Koyakutty M (2016) Cellular and molecular mechanistic insight into the DNA-damaging potential of few-layer graphene in human primary endothelial cells. Nanomedicine 12(5):1347–1355

Schinwald A, Murphy FA, Jones A, MacNee W, Donaldson K (2012) Graphene-based nanoplatelets: a new risk to the respiratory system as a consequence of their unusual aerodynamic properties. ACS Nano 6(1):736–746

Seabra AB, Paula AJ, de Lima R, Alves OL, Duran N (2014) Nanotoxicity of graphene and graphene oxide. Chem Res Toxicol 27(2):159–168

Shao Y, Wang J, Wu H, Liu J, Aksay IA, Lin Y (2010) Graphene based electrochemical sensors and biosensors: a review. Electroanalysis 22(10):1027–1036

Shin YC, Lee JH, Lee IS, Han DW (2018) Chapter 12 Biosafety of carbon-based nanoparticles and nanocomposites. Part V Biosafety and clinical translation of nanobiomaterials. In: Wang X, Ramanlingan M, Kong X, Zhao L (eds) Nanobiomaterials: classification, fabrication and biomedical applications. Wiley-VCH Verlag GmbH & Co. KGaA, Weinheim, pp 431–458. https://doi.org/10.1002/9783527698646.ch18

Shinde DB, Pillai VK (2012) Electrochemical preparation of luminescent graphene quantum dots from multiwalled carbon nanotubes. Chemistry 18(39):12522–12528

Tabish TA, Pranjol MZI, Hayat H, Rahat AAM, Abdullah TM, Whatmore JL, Zhang S (2017) In vitro toxic effects of reduced graphene oxide nanosheets on lung cancer cells. Nanotechnology 28(50):504001

Terrones M, Botello-Méndez AR, Campos-Delgado J, López-Urías F, Vega-Cantú YI, Rodríguez-Macías FJ, Elías AL, Muñoz-Sandoval E, Cano-Márquez AG, Charlier J-C, Terrones H (2010) Graphene and graphite nanoribbons: morphology, properties, synthesis, defects and applications. Nano Today 5(4):351–372

Tian X, Xiao B-B, Wu A, Yu L, Zhou J, Wang Y, Wang N, Guan H, Shang Z-F (2016) Hydroxylated-graphene quantum dots induce cells senescence in both p53-dependent and -independent manner. Toxicol Res 5(6):1639–1648

Wang G, Yang J, Park J, Gou X, Wang B, Liu H, Yao J (2008) Facile synthesis and characterization of graphene nanosheets. J Phys Chem C 112(22):8192–8195

Wang T, Zhu S, Jiang X (2015) Toxicity mechanism of graphene oxide and nitrogen-doped graphene quantum dots in RBCs revealed by surface-enhanced infrared absorption spectroscopy. Toxicol Res 4(4):885–894

Wu C, Wang C, Han T, Zhou X, Guo S, Zhang J (2013) Insight into the cellular internalization and cytotoxicity of graphene quantum dots. Adv Healthc Mater 2(12):1613–1619

Wu W, Yan L, Wu Q, Li Y, Li Q, Chen S, Yang Y, Gu Z, Xu H, Yin ZQ (2016) Evaluation of the toxicity of graphene oxide exposure to the eye. Nanotoxicology 10(9):1329–1340

Xu S, Zhang Z, Chu M (2015) Long-term toxicity of reduced graphene oxide nanosheets: effects on female mouse reproductive ability and offspring development. Biomaterials 54:188–200

Xue Q, Huang H, Wang L, Chen Z, Wu M, Li Z, Pan D (2013) Nearly monodisperse graphene quantum dots fabricated by amine-assisted cutting and ultrafiltration. Nanoscale 5(24): 12098–12103

Yadav MD, Dasgupta K, Kushwaha A, Srivastava AP, Patwardhan AW, Srivastava D, Joshi JB (2017) Few layered graphene by floating catalyst chemical vapour deposition and its extraordinary H2O2 sensing property. Mater Lett 199:180–183

Yan X, Cui X, Li B, L-s L (2010) Large, solution-processable graphene quantum dots as light absorbers for photovoltaics. Nano Lett 10(5):1869–1873

Zhang M, Bai L, Shang W, Xie W, Ma H, Fu Y, Fang D, Sun H, Fan L, Han M, Liu C, Yang S (2012) Facile synthesis of water-soluble, highly fluorescent graphene quantum dots as a robust biological label for stem cells. J Mater Chem 22(15):7461–7467

Zhang D, Zhang Z, Liu Y, Chu M, Yang C, Li W, Shao Y, Yue Y, Xu R (2015) The short- and long-term effects of orally administered high-dose reduced graphene oxide nanosheets on mouse behaviors. Biomaterials 68:100–113

Part II

Graphene-Based Nanomaterials as Scaffolds for Stem Cell Fate Modulation and Tissue Regeneration

Graphene-Based Materials for Efficient Neurogenesis

Yeon-Woo Cho, Kwang-Ho Lee, and Tae-Hyung Kim

Abstract

Graphene, a two-dimensional plane-structured carbon allotrope, has outstanding properties. Owing to their unique features, graphene-based materials including graphene derivatives have recently emerged as an ideal material and been used in various fields. Especially, in terms of specific advantages of graphene including great electrical conductivity, high potential to conjugate with biomolecules, and applicability to three-dimensional structures, neurogenesis-based stem cell therapies using graphene-based materials have been reported to be a candidate of treatment for neurodegenerative disease (e.g., Parkinson's disease, Alzheimer's disease, and Huntington's disease). To date, extensive studies on neurogenesis-based stem cell therapies including enhanced neural differentiation and monitoring stem cells behavior have been conducted using graphene-based materials. Herein, we have summarized recent various studies of neurogenesis using graphene-based materials in depth and focused on effect of graphene on functional improvement of neural stem cells and monitoring of differentiation into neural linages.

Keywords

Graphene · Graphene-based materials · Neurogenesis

Y.-W. Cho · K.-H. Lee · T.-H. Kim (✉)
School of Integrative Engineering, Chung-Ang University, Seoul, South Korea
e-mail: thkim0512@cau.ac.kr

D.-W. Han, S. W. Hong (eds.), *Multifaceted Biomedical Applications of Graphene*, Advances in Experimental Medicine and Biology 1351,
https://doi.org/10.1007/978-981-16-4923-3_3

1 Introduction

All species of animal have their own nervous systems which have vital role in communicating and controlling among organs. Nervous system cells of most are produced and reproduced by neurogenesis in which neural stem cells differentiate into various types of neural cells including neurons, astrocytes, and oligodendrocytes (Gage 2000). Neurogenesis is generally categorized to into two types: development neurogenesis and adult neurogenesis. Not only neurogenesis occurs during embryo development (development neurogenesis), but new nervous system cells are also generated throughout life (adult neurogenesis) (Götz and Huttner 2005; Alvarez-Buylla and Garcıa-Verdugo 2002). Provided that the specific nervous system cells decrease aberrantly, various neurological diseases such as Alzheimer's disease (AD), Huntington disease, and Parkinson's disease can occur following loss of specific function of nervous system (Good et al. 1996; Reiner et al. 1988; Janezic et al. 2013). For this reason, numerous studies for treatments of neurodegenerative diseases via neurogenesis have been carried out (Kim et al. 2013, 2015a; Solanki et al. 2013; Li et al. 2013). Many previous studies have concluded that the cellular loss of nervous system in models of neurological diseases can be replaced theoretically by induction of neurogenesis; however, it is still not clear about exact mechanism of human adult neurogenesis and behavior of endogenous neural stem cells (Kempermann et al. 2018). There are many reasons for limitations of human adult neurogenesis studies; for instance, the common cell labeling analysis method is not able to apply to human; moreover, studying adult stem cells in humans is subject to ethical limitations (Kempermann et al. 2018; Kim et al. 2020b). Therefore, the needs for further studies on the mechanisms of human adult neurogenesis haven. Additionally, control of neural stem cell behavior is necessary since the regulation of differentiation of endogenous neural stem cells is essential for therapeutic application of neural stem cells toward the treatment of neurological diseases (Grawish et al. 2020; Jeong et al. 2020).

Graphene, a two-dimensional plane-structured carbon allotrope, has emerged as an ideal material and been used in various fields, since it has diverse outstanding properties, including good thermal conductivity, high electron mobility, and strong tensile strength (Balandin et al. 2008; Bolotin et al. 2008; Papageorgiou et al. 2017). Owing to these advantages, graphene has also been applied in widespread biomedical fields such as biomedical sensing, drug delivery system, tissue engineering, and bioimaging. To date, graphene-based materials have been also applied to neurogenesis study, in which it is important that functions of nervous system cells in vitro should be regulated and maintained similarly to native cells in real tissue and monitoring non-invasively cellular function such as differentiation into specific cell types is crucial in therapeutic approaches (Geckil et al. 2010; Suhito et al. 2018). Specifically, the graphene family (e.g., two-dimensional graphene, graphene oxide, reduced graphene oxide) has been modified with various materials including nanoparticles and conductive materials to induce neural differentiation of cells and detect their functions (Fang et al. 2020; Polo et al. 2021; Guo et al. 2021; Rodriguez-Losada et al. 2020; Ma et al. 2019).

Herein, we have summarized recent various studies of neurogenesis using graphene-based materials in depth (Table 1) and focused on effect of graphene on functional improvement of neural stem cells and monitoring of differentiation into neural lineages, as summarily described in Fig. 1.

2 Graphene Derivatives for Neurogenesis

To date, various graphene derivatives have been reported (e.g., graphene oxide (GO), reduced graphene oxide (rGO), and fluorographene (FG)) (Dasari et al. 2017; Polo et al. 2021; Zaoralová et al. 2020; Sumisha and Haribabu 2020). The graphene derivatives are generated via reduction reactions or chemical functionalization, and individuated unique properties as compared to original graphene. For instance, GO which is formed by oxidation of graphene contains various oxygen-functional groups such as epoxy, hydroxyl, and carboxyl, which means that GO is more stable than graphene in aqueous environment and can be easily combined with other biomolecules (Jiang et al. 2017; Foroutan et al. 2020). Additionally, rGO can be formed by reduction of GO and has an excellent electrical conductivity, whereas GO has relatively low electrical conductivity (Liu et al. 2013; Jaafar and Kashif 2018). Since neurogenesis through neural differentiation of neural stem cells (NSCs) is known to be related to its electrical activity, highly conductive rGO has especially been an effective candidate material to stimulate neurogenesis via electrical stimulation (Zhu et al. 2019). Fluorographene, a form of fluorinated graphene, lacks π–π interactions in its structure (Sharma et al. 2020). Owing to this structure, fluorographene has unique properties such as a wide bandgap, enhanced electrical properties, and broad dispersibility (Zhu et al. 2013). Through previous studies, various graphene derivatives having unique their own properties have been used to improve neurogenesis (Tu et al. 2013; Lv et al. 2012; Sánchez-González et al. 2018; Kim et al. 2015a; Wang et al. 2012; Yang et al. 2016). Tu et al. developed functionalized graphene oxide (GO) with 2-methacryloyloxyethyl phosphorylcholine (MPC) and dimethylaminoethyl methacrylate (DMAEMA) to mimic environment in nervous system; phosphorylcholine and acetylcholine, respectively (Tu et al. 2013). They confirmed that neurite extension and outgrowth of primary rat hippocampal neurons were promoted on functionalized GO substrates. In addition, expression level of regenerating neuronal marker was higher on functionalized GO substrates than on non-functionalized GO, which indicates functionalization of GO has an effect on improvement of neural differentiation. Lv et al. reported the potential of GO as biomolecule-loading system for neurogenesis (Lv et al. 2012). They cultured neuroblastoma cells on GO nanosheet and induced neural differentiation with retinoic acid (RA). This study showed that neurite extension facilitated on GO nanosheet, as well as neuronal differentiation induced by RA was enhanced on GO nanosheet. A recent report by Sánchez et al. have shown that GO and rGO with poly(ε-caprolactone) (PCL) which is a biocompatible polymer are effective to differentiate and mature human iPSC (Induced pluripotent stem cells)-derived neural progenitor cells (NPCs) (Sánchez-González et al. 2018).

Table 1 Neurogenesis studies using graphene-based materials

Materials	Species/cell line	Enhanced neurogenesis type	References
GO–DMAEMA, GO-MPC	Rat/hippocampal neuron	Neuronal differentiation	Tu et al. (2013)
GO	Human/SH-SY5Y	Neuronal differentiation	Lv et al. (2012)
PCL-GO, PCL-rGO	Human/iPSC-derived NPCs	Neuronal differentiation	Sánchez-González et al. (2018)
NGO	Human/ADMSCs	Neuronal differentiation	Kim et al. (2015a)
GO-pattern	Human/NSCs	Neuronal differentiation	Yang et al. (2016)
FG	Human/MSCs	Neuronal/glial differentiation	Wang et al. (2012)
Graphene- PCL hybrid	Rat/NSCs	Glial differentiation	Shah et al. (2014)
SiNP-GO hybrid	Human/NSCs	Neuronal differentiation	Solanki et al. (2013)
GO-microfibers	Rat/embryonic neural progenitor cells (ENPCs)	Neuronal/glial differentiation	González-Mayorga et al. (2017)
rGO-microfiber patterns	Rat/PC-12	Neuronal differentiation	Wang et al. (2020)
GO-encapsulated magnetic nanoparticles	Human/SH-SY5Y	Neural growth/ synaptogenesis	Min et al. (2017)
GO- PLGA electrospun nanofibrous mats	Mouse/NSCs	Neuronal differentiation/long-term survival	Qi et al. (2019)
3D-GF	Mouse/NSCs	Neuronal/glial differentiation	Li et al. (2013)
3D-GF	Human/hESC-derived cortical neurons	Long-term survival	D'Abaco et al. (2018)
3D-GF	Mouse/MSCs	Dopaminergic neuronal differentiation	Tasnim et al. (2018)
rGO-PEDOT hybrid microfiber	Mouse/MSCs	Neuronal/glial differentiation	Guo et al. (2016)
SF-rGO	Rat/PC-12	Neurite formation	Aznar-Cervantes et al. (2017)
Gr-AP	Rat/PC-12	Neural proliferation	Golafshan et al. (2018)
Graphene monolayer intracortical probe	Mouse/primary hippocampal neurons	Neurorehabilitation	Bourrier et al. (2019)
3D pGO-GNP-pGO	Mouse/NSCs	Neural differentiation	Kim et al. (2013)
LHONA	Human/NSCs	Dopaminergic neural differentiation	Kim et al. (2015a)

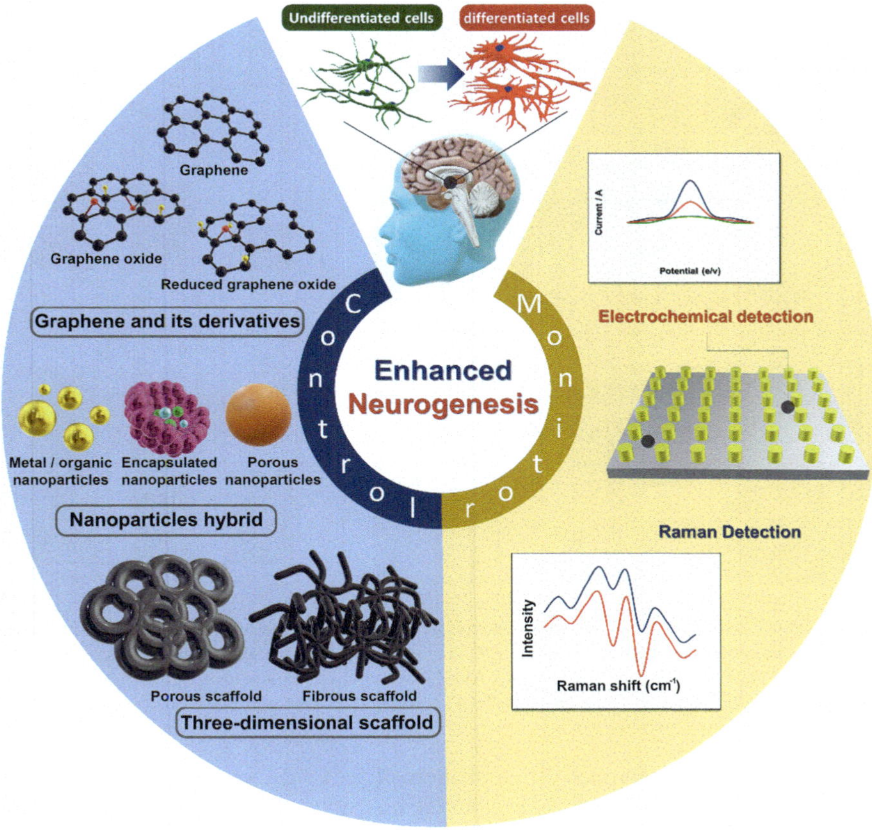

Fig. 1 Schematic illustration of studies to enhance neurogenesis using graphene-based materials

In this study, the effects of GO and rGO to cell adhesion and proliferation as well as differentiation were assessed. The cell adhesion on PCL-GO coated substrates was slightly increased as compared to both PCL and PCL-GO coated substrates, and PCL-GO and PCL-rGO coated substrates presented higher cell proliferation. It was also confirmed that the Tuj1 expression on PCL-rGO coated substrates was higher than TCP substrates, which is rGO enhanced neuronal differentiation of NPCs. All up, this study has also proved that graphene derivatives-based materials are useful for neurogenesis, including improvement of neural differentiation, cell adhesion, and proliferation. The other strategy to apply graphene derivatives to neurogenesis is an alternation of physical array of graphene derivatives. Kim, et al. have demonstrated that nano-sized GO (NGO) hybrid patterns can be used to direct specific differentiation of human adipose-derived mesenchymal stem cells (hADMSCs) (Kim et al. 2015a).

In this study, GO line pattern and grid pattern were fabricated for osteogenic differentiation and neuronal differentiation, respectively. To evaluate the effect of

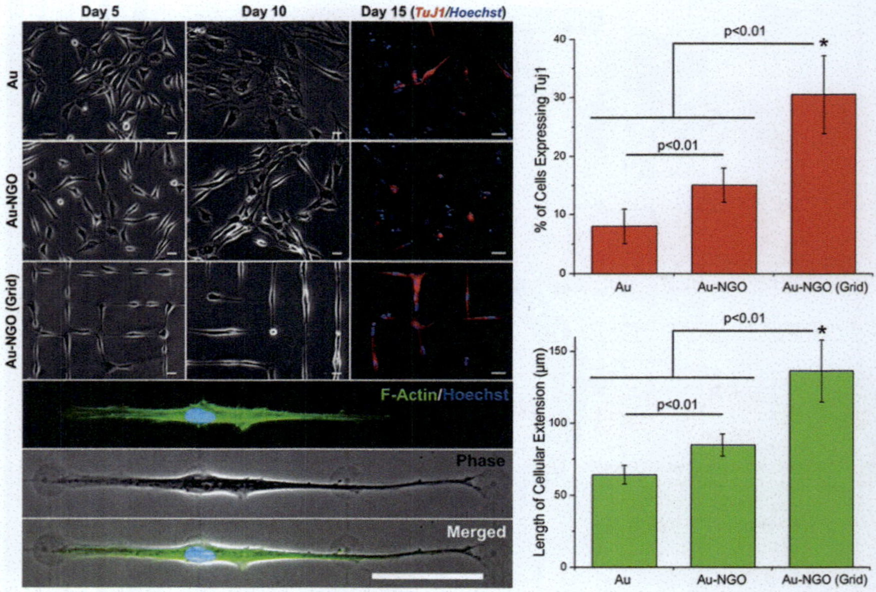

Fig. 2 Nano-sized GO hybrid patterns enhancing neuronal differentiation and neurite extension of hADMSCs (Reproduced from Kim et al. 2015a, Copyright (2015) American Chemical Society Nano). Abbreviations: GO graphene oxide, hADMSCs human adipose-derived mesenchymal stem cells, Au-NGO NGO-coated substrates, Au-NGO (Grid) NGO grid-patterned substrates

GO grid pattern on neuronal differentiation of hADMSCs, expression levels of neuronal marker in the stem cells cultured on Au-coated, Au/NGO-coated, and Au/NGO grid pattern-coated substrates, respectively, were quantified (Fig. 2). The gene expression was highest in Au/NGO grid pattern-coated substrates. Similarly, the length of hADMSCs cultured on Au/NGO grid pattern-coated substrates clearly increased, as compared with the experimental groups. Likewise, Yang et al. developed a culture platform composed of graphene oxide microgroove-structured patterns which can induce to generate functional neuron-like cells from human neural stem cells (hNSCs) (Yang et al. 2016).

The GO-based patterned substrates (GPS) were fabricated via sequential photolithography, graphene coating, and lift-off process. Thereby, micro-sized grooves and nano-scaled ridges were generated and used to culture the cells within the pattern area. Herein, several markers including ß1 integrin, focal adhesion kinase (FAK), vinculin or paxillin which indicate the level of F-actin alignment, integrin clustering, and focal adhesion, respectively (Fig. 3). Especially, it was observed that not only does expression of neuronal marker on the GPS increased, but also neurite formation was facilitated. Moreover, a whole-cell patch clamp analysis was performed to investigate the differentiation of hNSCs into functional neurons. According to the analysis data, voltage-dependent ionic currents and action potentials which are features of matured neurons were confirmed on the GPS. All the data taken together,

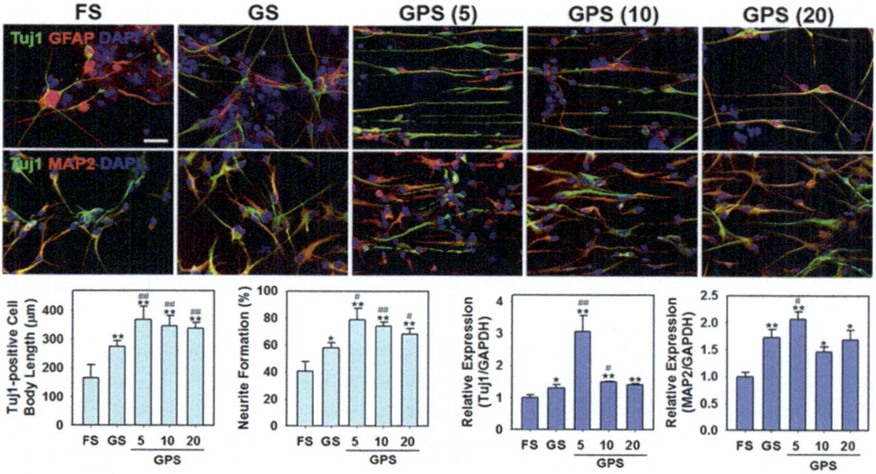

Fig. 3 Micro-scaled GO groove patterns inducing neuronal differentiation of hNSCs (Reproduced from Yang et al. 2016, Copyright (2016) American Chemical Society Applied Materials and Interfaces). Abbreviations: GO graphene oxide, hNSCs human neural stem cells, FS flat substrates, GS GO-coated substrates, GPS (5, 10, 20) GO-based patterned substrates with groove with 5 μm, 10 μm, and 20 μm, respectively

the topographical cues and physical structure provided by GO can enhance neuronal differentiation as well as cell alignment and focal adhesion. Above all, GO-based culture platform for neural stem cells could induce the neuron maturation. Likewise, Wang et al. have reported that fluorographene (FG) and its micro-patterned structure is useful for neuronal differentiation of mesenchymal stem cells (Wang et al. 2012). The neuronal differentiation was drastically increased in fluorographene substrate, whereas glial differentiation was slighted enhanced as compared to both normal tissue culture plate and graphene substrates. Moreover, FG-micro-patterned substrates which were fabricated via ink-jet printing of polydimethylsiloxane (PDMS) on the FG substrates induced the neuronal differentiation and neuronal maturation.

To sum up, it has been demonstrated that graphene and its derivatives are biocompatible materials and have high potential to be utilized for cell cultivation. In addition, each graphene derivative has its own specific features, which can have a beneficial effect on various cell functions (e.g., cell adhesion, proliferation, differentiation) and eventually neurogenesis.

3 Graphene-Nanoparticle Hybrid for Neurogenesis

Notably, nanoparticles have been widely used in nano-bio field for their unique characteristics and tremendous surface area. As the size scale dwindled to nano range, compared to bulk material, the ratio of surface effect surged. And this led to show unexpected physical and chemical alterations, explained as a quantum effect in that they were able to confine their electrons. To date, nanoparticles have been developed for various purposes (e.g., labeling, drug delivery, activation of signaling pathway) and functions (Wang et al. 2018; Ahn et al. 2020; Shin et al. 2020; Ghasemi Goorbandi et al. 2020). By manipulating its physical cues, nanoparticles are utilized in labeling in which they are implemented as a fixation agent and also promote detection of single cell level (Hsiao et al. 2007). Also, encapsulated type helps release chemical substances in a controlled manner and play a critical role in activating stem cell differentiation. For example, encapsulating retinoic acid on silica nanoparticle stimulates neural growth (Park et al. 2017). Furthermore, iron oxide nanoparticles incorporated with nerve growth factor (NGF) showed potential of neural regeneration (Marcus et al. 2015). Indeed, graphene-nanoparticle hybrids have the connection between graphene and nanoparticles (electrostatic force, hydrophobic interaction and covalent bonds); it prevents π-π accumulation between graphene sheets as well as aggregation between nanoparticles which improves the surface area of its hybrid structures. To repeat, nanoparticles have their own unexpected characteristics which are distinct from macro and visible area. For instance, their small size allows easy access into cells by crossing the plasma membrane. Moreover, the surface chemistry environment can be adjusted such that the nanoparticles selectively respond to other biomolecules found either on the cell membrane or within the cytoplasm, normal and diseased tissues, or bodily fluids (Khan et al. 2017). As such, these nanoparticle-hybrids have various functionalities originated from their graphene properties and specific nanoparticle features, which might be suitable environment for neural stem cells (NSCs) differentiation. Therefore, these materials could be utilized for various purposes (e.g., biomedical, electronic, catalytic, energy). Regarding its synergistic effects, unavailable for bulk materials like graphene, it can dramatically augment the chance of biomedical application (González-Mayorga et al. 2017). Generally, graphene-nano particle hybrids could be divided into two classes. One is graphene-nanoparticle composites (GNC), where the nanoparticles decorate or are placed on graphene nano sheets, and the other is graphene-encapsulated nanoparticles wherein the nanoparticle surface is wrapped or coated with graphene. The first type mentioned above functions as a catalyst whereas the latter type is known for its cargo delivery (e.g., small drugs, nucleic acids). However, inducing neural stem cell differentiation toward a desired cell type requires elaborate control over the biochemical and physical cues. For its biomedical application, surface modification of graphene is important for improving bioactivity and promoting endogenous repair mechanisms for nerve regeneration. Particularly, Shah et al. deposited 3 disparate concentrations of GO on PCL nanofibers. Comparing GO-coated glass and PCL nanofiber, the value of oligodendrocyte marker increased about a 2-fold. Whereas, neuronal marker showed 1.3-fold increase and astrocyte marker exhibited 0.5-fold decrease. Taken together, these data support the improved oligodendrocyte differentiation of NSCs in hybrid scaffolds (Fig. 4) (Shah et al. 2014).

In controlling stem cell fate, the ability to use physical signals like nano topographic features, substrate stiffness, and dimensions of ECM protein pattern has a

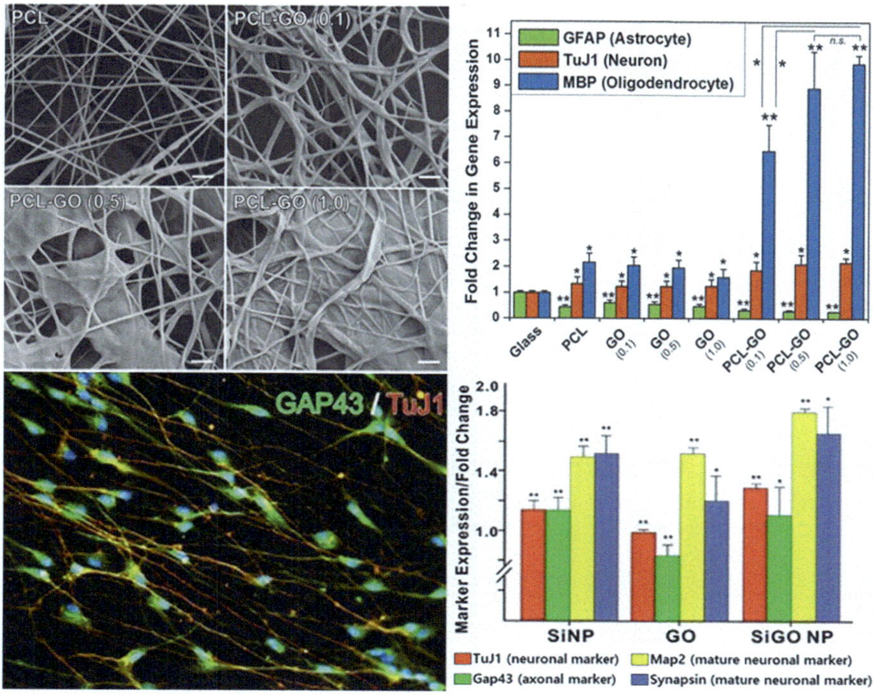

Fig. 4 The comparison of neuronal marker expression regarding individual graphene nanoparticle substrates (Reproduced from Shah et al. 2014, Copyright (2014) Advanced materials, Solanki et al. 2013, copyright (2013) Advanced materials). Abbreviations: PCL electrospun polycaprolactone, GO graphene oxide, SINP silica nanoparticles

great potential in regenerative medicine. Therefore, through the development and application of new GNPs, the challenge is to provide an engineered microenvironment that specifically controls the axon alignment and growth of NSC-derived neurons. Specifically, Solanki et al. showed that NSCs cultured on different substrates clearly reveal that differentiated hNSCs on SiNP-GO (silica nanoparticles-graphene oxide) substrates exhibit over 30% improved expression of neuronal marker. Altogether, SiNP-GO hybrid structures could be ideal for regulating axon alignment that could possibly enhance communication by providing a suitable microenvironment and this result might give a clue to repair the damaged spinal cord (Fig. 4) (Solanki et al. 2013).

Furthermore, the treatment and modeling of brain disease and spinal cord injury requires precisely controlled axon direction and growth to simulate synaptic network between each neuron. To generate GO hybrid pattern with different morphology, Min et al. firstly synthesize GO-capsulated magnetic nanoparticles and facilitate migration of GO films to desirable substrates with external magnetic forces. At first, GO is functionalized with Fe_3O_4 nanoparticles through electrostatic interaction to

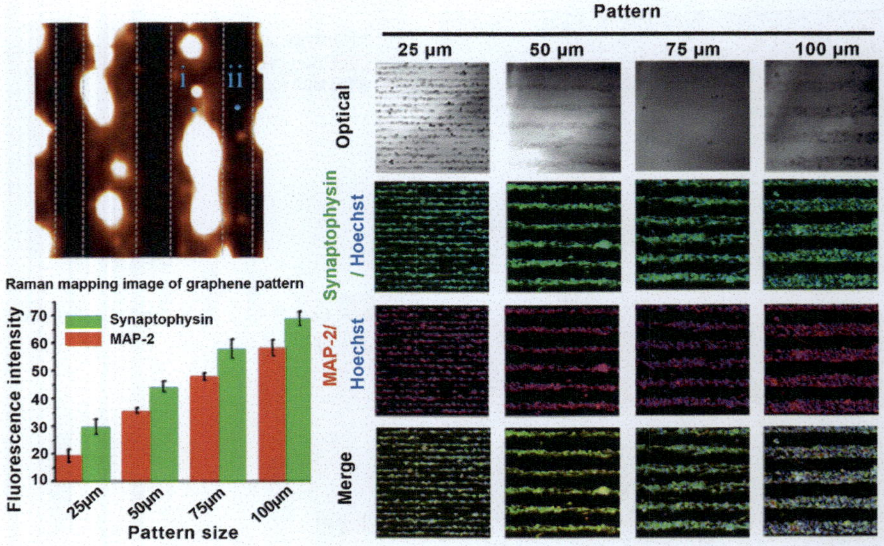

Fig. 5 Effect of GO hybrid line pattern dimension on synaptogenesis (Reproduced from Min et al. 2017)

produce hybrid magnetic nanoparticles, which can help transfer the GO layer to the target substrate in the presence of external magnetic forces. As shown in Fig. 5, neuron expression levels were found to increase with increasing pattern. In 25 μm line pattern, the value of synapse and mature neuron markers was, respectively, 29.8 and 19.3 intensity. In contrast, 100 μm showed 68.7 and 60.2, over 2-fold higher. Therefore, using a 100 μm pattern, the expression was very stable and evenly distributed, which is important for directed signal transduction of neurons. Accordingly, the MF-driven GO hybrid pattern was effective in controlling the behavior and function of neuronal cells, especially both cell morphology and the functioning of the synaptic junctions (Min et al. 2017). In addition, Mayorga et al. focused on the possibility of reduced graphene oxide (rGO) microfibers as a substrate for nerve growth in wounded. Moreover, the mature neuronal markers, rGO-PLL (poly-L-lysine) on 21 days showed over 3-fold expression compared to Glass-PLL. This result revealed the possible transplantation of these rGO microfibers as a guide platform for the damaged rat spinal cord (González-Mayorga et al. 2017). Besides, the successful growth factor delivery by immobilization on graphene oxide (GO)-linked poly (lactic-co-glycolic acid) (PLGA) biodegradable electrospun nanofibers was done by Qi et al. Based on the cell viability assay, the cell proliferation was observed in all nanofibers as 7 days culture. The OD value of the PLGA-GO group was about a 20% higher than the OD value of the PLGA group on day 7. To sum up, PLGA-GO nanofibers improved NSC survival in microenvironments to some extent. Thus, the immobilization of IGF-1 on PLGA-GO nanofibers has indicated that it has an enormous potential to enhance neurogenesis of nerve implants (Qi et al. 2019).

Thus, graphene nanoparticles could be a promising alternative for nano-material platform for neurogenesis in that it has shown to accelerate and direct the differentiation of neural stem cells into osteoblasts and neurons.

4 Three-Dimensional Graphene Scaffolds for Neurogenesis

Conventionally, two-dimensional cultures, adherent culture systems providing monolayer for adhesion of cells, have been used as in vitro tool, owing to several advantages: i) simple and low-priced maintenance of cultivation and ii) allowing constant supply of growth factors or nutrients on culture surfaces (Kapałczyńska et al. 2018). However, two-dimensional cultures have many critical limitations, in terms of native structures of tissues. In fact, native tissues consist of three-dimensional structures, dynamic environment, which are referred to extracellular matrix (ECM) that provides not only structural support but also topographical cues to regulate most of cell functions (e.g., proliferation, cell adhesion, and differentiation) with cell-cell interactions and cell-ECM interactions (Borza and Pozzi 2012; Miller et al. 2020). For this reason, culture results from two-dimensional culture which provide only adherent surfaces to cells are different from native environment, including morphological features, cell functions, and metabolism (Antoni et al. 2015; Krishnamurthy and Nör 2013; Duval et al. 2017). Therefore, the needs for three-dimensional cultures (3D-cultures) mimicking natural environment of tissues have been highlighted (Hsiao et al. 2007; Kawai et al. 2020; Wang et al. 2020).

To date, various three-dimensional scaffolds for tissue engineering which reproduce native ECM have been reported to address limitations of two-dimensional cultures (Loh and Choong 2013; Ajiteru et al. 2020). To mimic structure of native ECM, three-dimensional scaffolds are required to have great biocompatibility and biodegradability, as well as provide specific mechanical properties and architectures which are similar with that of native ECM (Lv et al. 2012; Loh and Choong 2013; Grossemy et al. 2020). Above all, it is crucial that growth factors or nutrients are supplied to cells which locate in three-dimensional cell clusters via connection with its scaffold (Loh and Choong 2013; Freyman et al. 2001). In this regard, various porous scaffolds have been developed, since porosity forms interconnection of cell-cell and cell-scaffold, as well as high surface area. Additionally, since various native tissues have their own unique three-dimensional structures of ECM and topographical cues, different structures of scaffolds have to design and fabricate depending on types of cells cultured (Sampath and Reddi 1981; Celio and Blumcke 1994).

Specifically, ECM in nervous tissues mostly comprises proteins such as proteoglycans and laminin which involve in essential function of nervous tissues (Barros et al. 2011). In nervous tissues, neurons have known to have electrical activity and communicate electrically to other cells using electrical synapse, a gap junction between two adjacent neurons (Zhang and Poo 2001). In addition, it has been proven that axons of mature neurons grow and extend directionally during neurogenesis; thus, fibrous scaffolds which can guide axonal growth and extension have widely developed for neurogenesis in vitro/in vivo (Gage 2002). In the sense,

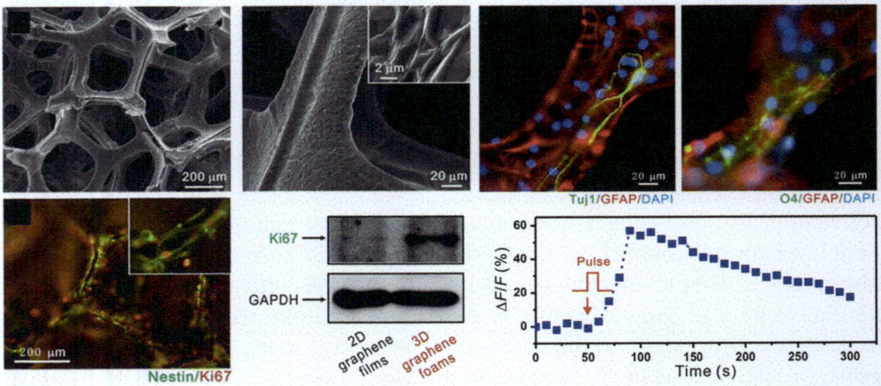

Fig. 6 3D graphene foam scaffold having porous structure and potential for electrical stimulation of NSCs (Reproduced from Li et al. 2013). Abbreviations: NSCs neural stem cells

graphene and its derivatives (e.g., doped graphene, graphene oxides and porous graphene) are suitable materials for three-dimensional scaffolds which enhance neurogenesis, owing to their properties: (i) excellent electrical conductivity, (ii) high biocompatibility and biodegradability, and (iii) great machinability to other forms (e.g., porous, fibrous, and nanostructured structures) (Choo et al. 2017; Kim et al. 2015a; Zhou et al. 2013). The graphene-based three-dimensional scaffolds are mostly classified depending on structural types of scaffolds: (i) graphene foam-based porous scaffold and (ii) electrospun graphene nano/micro-fiber-based fibrous scaffold. In this section, we focused on neurogenesis studies using these types of graphene-based three-dimensional scaffolds.

Graphene foam is referred to three-dimensional (3D) porous structures of graphene fabricated using chemical vapor deposition (CVD). In previous studies, it has been proven that 3D-graphene foam can guide differentiation of NSCs and enhance cellular function through electrical stimulation (Li et al. 2013; D'Abaco et al. 2018; Tasnim et al. 2018). In 2013, Li et al. developed three-dimensional graphene foams (3D-GFs) platform for efficient neural differentiation of NSCs (Li et al. 2013). The structure of 3D-CFs was confirmed as 2D-graphene films. It was confirmed that 3D-GFs had porous structure (pore size: 100–300 μm), sophisticated 3D structure (width of graphene frame: 100–200 μm), and many wrinkles on its surface. They demonstrated that 3D-GFs can enhance formation of neural network, cell adhesion, and cell proliferation as compared to two-dimensional (2D) graphene films (Fig. 6).

Moreover, the expression of neuronal marker and glial marker of NSCs cultured on 3D-GFs increases by about 2.5 and 1.5 times, respectively, compared to 2D-graphene films. In addition to enhanced cell functions of NSCs, according to results from cyclic voltammetry, 3D-GFs showed higher current than 2D-graphene films, indicating that 3D-GFs have high potential to be an electrode for electric stimulation. The effects of 3D-graphene foam scaffold on neurogenesis were proven in not only NSCs but also other types of cells. D'Abaco et al. reported high potential of 3D-graphene foam scaffold as a culture platform that support human embryonic stem cells (hESCs)-derived cortical neurons which mature toward GABAergic or glutamatergic neuronal phenotypes (D'Abaco et al. 2018). In the study, they first cultured hESCs-derived cortical neurons on normal culture plate for 14 days. Subsequently, the neurons were transferred and cultured on the 3D-graphene foams scaffold (pore size: 580 μm; thickness: 200 μm) for further 3 weeks in media without any growth factors. They confirmed the maturation of hESCs-derived cortical neurons maintained on 3D-graphene foam for long period, indicating that the scaffold has potential to be a platform for neurogenesis of hESCs. In 2018, it was demonstrated that 3D graphene foam scaffold supports the differentiation of mesenchymal stem cells (MSCs) into dopaminergic neurons (Tasnim et al. 2018). In the study, neuronal differentiation of MSCs cultured on collagen-coated 3D graphene foam scaffold was compared with one cultured on collagen gels. According to differentiation outcomes, the expression levels of both neuronal marker and dopaminergic neuronal marker of MSCs cultured on collagen-coated 3D graphene foam scaffold were confirmed to increase. Hereby, 3D graphene foam scaffold can improve the neuronal differentiation of MSCs into dopaminergic neurons.

Electrospun fibrous types of scaffolds have been used in vivo transplantation, since fibrous scaffolds can provide high flexibility, biodegradability, and 3D structures mimicking native tissues, as well as fiber scale of scaffolds is controllable (micro/nano-scale) (Guo et al. 2016; Aznar-Cervantes et al. 2017; Al-Dhahebi et al. 2020). In 2016, reduced graphene oxide (rGO)-poly(3,4-ethylenedioxythiophene) (PEDOT) hybrid microfibers for enhanced neural differentiation of MSCs were developed (Guo et al. 2016). The MSCs were stimulated electrically with triboelectric nanogenerator. With synergic effect of electrical stimulation and 3D-graphene-based fibrous scaffolds, expression levels of both neuronal marker and glial marker increased in MSCs cultured on the scaffolds, as well as aligned morphological features and enhanced cell growth were observed on the scaffolds. Taken together, it was verified that structural effects of 3D-graphene-based fibrous scaffolds and electrical stimulation lead to improve cellular functions for neurogenesis, including proliferation and differentiation. In 2017, Aznar-Cervantes et al. reported effect of silk fibroin fibrous scaffolds coated with reduced graphene (SF-rGO) on neurite formation of PC-12 cells (Aznar-Cervantes et al. 2017). According to their results, neurite formation of PC-12cells was enhanced on the fibrous scaffold with electrical stimulation. Golafshan et al. developed aligned fibers composed of graphene embedded sodium alginate polyvinyl alcohol (Gr-AP scaffolds). (Golafshan et al. 2018). They confirmed that cell growth on aligned Gr-AP scaffolds was higher than on random Gr-AP scaffolds. As well as, the cell proliferation of aligned Gr-AP scaffolds was enhanced after electrical stimulation, which proved that electrical stimulation could increase cell proliferation.

Overall, it has been reported that graphene can be applied to manufacture various types of 3D scaffold. Specifically, porous scaffold and fiber scaffold have been used in enhancement of neurogenesis.

5 Graphene-Based Sensor for Detecting Neurogenesis

Conventionally, sensors are based on numerous approaches, plasmonics, optics, electro-analysis, magnetics and there are two main measurement sensors that play crucial role in medical area (Intille et al. 2012; Liu et al. 2021; Ratnam et al. 2020; Sethi et al. 2020; Bae et al. 2020; Kim et al. 2020a). For example, sensors that are used for direct insertion into human, in other words, to be in direct contact with blood, represent invasive sensor. This type of sensor could be potentially hazardous due to its interaction between the sensor itself and the outer environment, including blood or interstitial fluid. Furthermore, the continuous sample should be needed to monitor real time especially for the drug efficiency or the level of cell incubation and the degree of infection. It might be impractical in that mentioned above and showed clear limitations. On the other hand, non-invasive sensor constitutes external position to the body surface, which allows safer analyses and no interference on target measurements or the process being held. Comparably, the latter one exhibits lasting monitoring for its unnecessary need to cater sample media. These obvious advantages for biomedical sensing devices further augment the trend over development of non-invasive measurement devices (Downes et al. 2010). Combined with non-invasive measurement devices and owing to graphene's physical and chemical features, it has been emerged as an ideal material for biosensing. In sensing applications, efficient biosensor requires excellent electron transfer velocity between electrodes and biomolecules. In this context, graphene has an electrical conductivity of order of 1000 siemens per meter and thermal conductivities between 1500 and 2500 $Wm^{-1} K^{-1}$ which is incomparably higher than silicon we used these days. As a consequence, its upper value of mechanical strength and electrical conductivity facilitates electron transfer between target and sensor and that is the main interest in nano-biosensor (Wang et al. 2011).

Particularly, there are several configurations of graphene biosensors, including electronic sensor, electrochemical sensors, and optical sensor (Yin et al. 2015). The example of electronic sensor, field-effect transistor (FET) based biosensor is in direct contact with target sample, which helps elevate sensitivity that proceeds at the channel surface and regulating the channel conductance (Allen et al. 2007; Park et al. 2020). Thus, it can preserve the electrical characteristic of graphene by linkage between probe and nanoparticle. And it is expected to detect protein particles which is important indicator of pathological conditions (Mao et al. 2010). Next, electrochemical sensors are divided into two main types: affinity-based sensors and catalytic sensors (Chikkaveeraiah et al. 2012). Affinity sensors are based on the selective binding effect that occurs between biomolecules and its target analyte, which leads to the production of an electric signal. Conversely, catalytic sensors mostly contain nanoparticles or catalytic enzymes that recognize the target analyte and emit electroactive species (Ronkainen et al. 2010). In addition, in optical sensors, graphene-nanoparticle hybrid structures can be widely utilized for optical device applications. It can fall into two specific categories including fluorescence-based and SERS-based biosensors. In case of the fluorescence-based biosensors, the unique ability of wide range of fluorescent wavelength and quenching fluorescent molecules enable the superior optical detection in a way of nondestructive method (Roy et al. 2008; Smitha et al. 2020). On the other hand, SERS is a surface-sensitive technology that enables the improvement of Raman scattering with sensitivities that can detect as low as single molecules (Ahijado-Guzmán et al. 2012; Choi et al. 2020).

Nowadays, graphene-nanoparticle hybrid sensors are fabricated in the biomedical filed. Several merits exist including large surface area for detection and sufficient sites for bioconjugation (Myung et al. 2012). Altogether, graphene biosensors provide a chance to detect biomolecules in the arena of tissue engineering or regenerative medicine. Initially, Bourrier et al. pioneered the innovate way to roll intracortical probe in a sequentially single layer graphene sheet. They showed the effect of graphene on the detection efficiency of probes in the cortex. To be specific, graphene-coated probe proved healthier neuron network, and the lower density of astrocyte enables improved signaling. Through this experiment, it allowed for the detection of single neuron spikes with highly excellent efficiency even after 5 weeks of transplantation. Conclusively, it supported the reliable monitoring of mouse NSCs (Bourrier et al. 2019).

Similarly, Kim et al. used graphene-coated nanoparticles to detect neural stem cell differentiation. Particularly, monitoring stem cell differentiation and pluripotency is an essential process for stem cell use in regenerative medicine. Therefore, there is an urgent need for a new nondestructive detection method that can monitor stem cell differentiation without further modification. In this study, encapsulated 3D graphene oxide or gold nanoparticles were fabricated to detect the potential for differentiation of neural stem cells according to SERS signal. The reason for using gold nanoparticles was specifically developed to induce a double synergistic effect of graphene oxide and gold nanoparticles on SERS signals which is effective for undifferentiated NSC. This study attempted to detect differentiation potential using spectro-electrochemical methods, as shown in Fig. 7, which clearly distinguished differentiation and undifferentiated states. In 154 mv, there was a weak oxidation peak in undifferentiated single mNSC, but no redox peak in GO-GNP substrate. Furthermore, differentiated single mNSC indicated no oxidation peak and higher current intensities which clearly revealed an effective method for monitoring neural activity. Therefore, the proposed technique can be used as an excellent nondestructive tool for determining the differentiation potential of various stem cells such as neural stem cells, various mesenchymal and hematopoietic stem cells (Kim et al. 2013). Subsequently, quantitative techniques are required to realize the potential for stem cell therapy. Following this concept, Kim et al. reported a cell-based sensing platform (large-scale homogeneous nanocup-electrode arrays (LHONA)) that can detect neurotransmitters from dopaminergic cells in real time and detect very sophisticated electrochemical signals. LHONA was formed by repeating cup-shaped nanostructures on indium tin oxide (ITO) by laser interference lithography (LIL) and electrochemical vapor deposition (ECD) methods. In Fig. 7, only matured dopaminergic neurons exhibited distinct redox potential compared to other progenitor cells (e.g., neurospheres, hNSCs and premature neurons). Also, I_{pc} value of DA neurons was 0.13 uA and it was found to be the only signal derived.

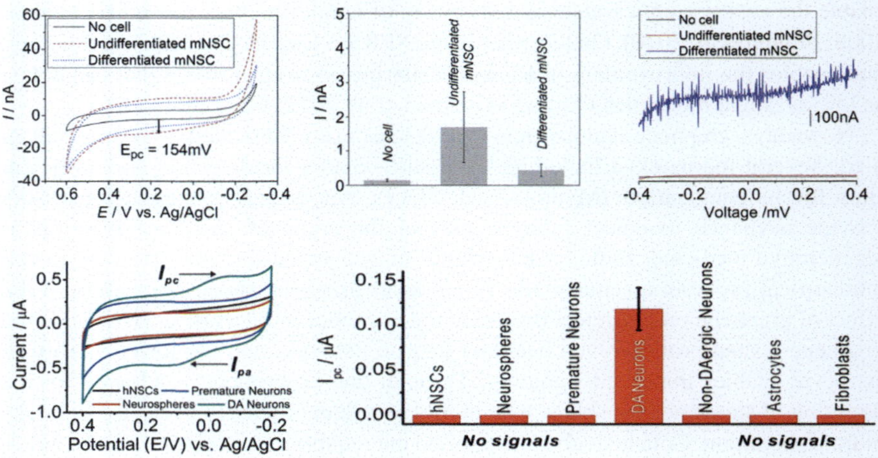

Fig. 7 Detection of electrical/electrochemical signal generated from undifferentiated/differentiated single mNSC and DA neuronal cells (Reproduced from Kim et al. 2013, Copyright (2013), Kim et al. 2015, Copyright (2015)). Abbreviations: mNSC mouse neural stem cells, DA dopamine, LHONA large-scale homogeneous nanocup-electrode arrays

In conclusion, it can be expected that it will be a superior biomaterial in terms of dopamine detection than other transparent electrodes as well as a successful monitoring of hNSCs into Dopaminergic neurons (Kim et al. 2015b). Overall, graphene-based materials are fabricated to biosensors for the detection of neurogenesis in the form of nanoparticle hybrid and distinct nano-scaled structure. Since this study is quite up-to-date field, lots of approaches could be manipulated to reach immediate and real-time analysis of NSCs.

6 Conclusions and Perspectives

In this study, we summarized and described various applications of graphene-based materials including graphene derivatives, graphene-nanoparticle hybrid, and three-dimensional graphene scaffold for enhanced neurogenesis. According to the previous studies, it has been proved that the graphene-based materials are biocompatible and greatly useful to enhanced neurogenesis without regard for the forms/types of graphene. To be specific, graphene-based materials can enhance cellular functions such as cell adhesion, proliferation, and eventually neural differentiation of stem cells. In addition to enhancement of cellular functions, utilizing graphene-based materials is a good approach to achieve non-invasive monitoring of stem cell behaviors. In terms of that, graphene-based materials have been reported to be able to solve limitations of potential stem cell therapies, including poor efficiency of neural differentiation, toxicity, and involvement of invasive monitoring of differentiation. Moreover, graphene-based materials can be readily applied to foam three-dimensional scaffold. This advantage of graphene increases the possibility that graphene can be used in highly efficient stem cell therapy aimed at treating neurodegenerative diseases through stem cell transplantation. Hereafter, further studies on neurogenesis using graphene-based materials would be conducted, as well as more various forms of graphene would be developed.

Acknowledgments This work was supported by National Research Foundation of Korea (NRF) grants funded by the Korean government (Grant Nos. NRF-2019R1C1C1007633, NRF-2019M3A9H2031820, NRF-2019R1A4A1028700).

References

Ahijado-Guzmán R, Gõmez-Puertas P, Alvarez-Puebla RA, Rivas G, Liz-Marzãn LM (2012) Surface-enhanced Raman scattering-based detection of the interactions between the essential cell division FtsZ protein and bacterial membrane elements. ACS Nano 6:7514–7520

Ahn JH, Kim I-R, Kim Y, Kim D-H, Park S-B, Park B-S, Bae M-K, Kim Y-I (2020) The effect of mesoporous bioactive glass nanoparticles/graphene oxide composites on the differentiation and mineralization of human dental pulp stem cells. Nanomaterials 10:620

Ajiteru O, Sultan MT, Lee YJ, Seo YB, Hong H, Lee JS, Lee H, Suh YJ, Ju HW, Lee OJ (2020) A 3D printable electroconductive biocomposite bioink based on silk fibroin-conjugated graphene oxide. Nano Lett 20:6873–6883

Al-Dhahebi AM, Gopinath SCB, Saheed MSM (2020) Graphene impregnated electrospun nanofiber sensing materials: a comprehensive overview on bridging laboratory set-up to industry. Nano Converg 7:27

Allen BL, Kichambare PD, Star A (2007) Carbon nanotube field-effect-transistor-based biosensors. Adv Mater 19:1439–1451

Alvarez-Buylla A, Garcıa-Verdugo JM (2002) Neurogenesis in adult subventricular zone. J Neurosci 22:629–634

Antoni D, Burckel H, Josset E, Noel G (2015) Three-dimensional cell culture: a breakthrough in vivo. Int J Mol Sci 16:5517–5527

Aznar-Cervantes S, Pagãn A, Martïnez JG, Bernabeu-Esclapez A, Otero TF, Meseguer-Olmo L, Paredes JI, Cenis JL (2017) Electrospun silk fibroin scaffolds coated with reduced graphene promote neurite outgrowth of PC-12 cells under electrical stimulation. Mater Sci Eng C 79:315–325

Bae J, Shin D-S, Ha J-H, Hwang Y, Lee C-S, Kim HJ, Jang A, Park S-H (2020) Flexible chemical sensors using signal generation from cyclodextrin-analyte interactions on polymer composites. Biochip J 14:251–257

Balandin AA, Ghosh S, BAO W, Calizo I, Teweldebrhan D, Miao F, Lau CN (2008) Superior thermal conductivity of single-layer graphene. Nano Lett 8:902–907

Barros CS, Franco SJ, Mþller U (2011) Extracellular matrix: functions in the nervous system. Cold Spring Harb Perspect Biol 3:a005108–a005108

Bolotin KI, Sikes KJ, Jiang Z, Klima M, Fudenberg G, Hone J, Kim P, Stormer HL (2008) Ultrahigh electron mobility in suspended graphene. Solid State Commun 146:351–355

Borza CM, Pozzi A (2012) The role of cell-extracellular matrix interactions in glomerular injury. Exp Cell Res 318:1001–1010

Bourrier A, Shkorbatova P, Bonizzato M, Rey E, Barraud Q, Courtine G, Othmen R, REITA V, Bouchiat V, Delacour C (2019) Monolayer graphene coating of intracortical probes for long-lasting neural activity monitoring. Adv Healthc Mater 8:1801331

Celio MR, Blumcke I (1994) Perineuronal nets—a specialized form of extracellular matrix in the adult nervous system. Brain Res Rev 19:128–145

Chikkaveeraiah BV, Bhirde AA, Morgan NY, Eden HS, Chen X (2012) Electrochemical immunosensors for detection of cancer protein biomarkers. ACS Nano 6:6546–6561
Choi J-H, Kim T-H, El-Said WA, Lee J-H, Yang L, Conley B, Choi J-W, Lee K-B (2020) In situ detection of neurotransmitters from stem cell-derived neural Interface at the single-cell level via graphene-hybrid SERS nanobiosensing. Nano Lett 20:7670–7679
Choo S-S, Kang E-S, Song I, Lee D, Choi J-W, Kim T-H (2017) Electrochemical detection of dopamine using 3D porous graphene oxide/gold nanoparticle composites. Sensors (Basel, Switzerland) 17:861
D'Abaco GM, Mattei C, Nasr B, Hudson EJ, Alshawaf AJ, Chana G, Everall IP, Nayagam B, Dottori M, Skafidas E (2018) Graphene foam as a biocompatible scaffold for culturing human neurons. R Soc Open Sci 5:171364
Dasari BL, Nouri JM, Brabazon D, Naher S (2017) Graphene and derivatives – synthesis techniques, properties and their energy applications. Energy 140:766–778
Downes A, Mouras R, Elfick A (2010) Optical spectroscopy for noninvasive monitoring of stem cell differentiation. J Biomed Biotechnol 2010:10
Duval K, Grover H, Han L-H, Mou Y, Pegoraro AF, Fredberg J, Chen Z (2017) Modeling physiological events in 2D vs. 3D cell culture. Physiology (Bethesda) 32:266–277
Fang Q, Zhang Y, Chen X, Li H, Cheng L, Zhu W, Zhang Z, Tang M, Liu W, Wang H, Wang T, Shen T, Chai R (2020) Three-dimensional graphene enhances neural stem cell proliferation through metabolic regulation. Front Bioeng Biotechnol 7:436
Foroutan T, Ahmadi F, Moayer F, Khalvati S (2020) Effects of intraperitoneal injection of magnetic graphene oxide on the improvement of acute liver injury induced by CCl4. Biomater Res 24:14
Freyman TM, Yannas IV, Gibson LJ (2001) Cellular materials as porous scaffolds for tissue engineering. Prog Mater Sci 46:273–282
Gage FH (2000) Mammalian neural stem cells. Science 287:1433–1438
Gage FH (2002) Neurogenesis in the adult brain. J Neurosci 22:612–613
Geckil H, Xu F, Zhang X, Moon S, Demirci U (2010) Engineering hydrogels as extracellular matrix mimics. Nanomedicine (Lond) 5:469–484
Ghasemi Goorbandi R, Mohammadi MR, Malekzadeh K (2020) Synthesizing efficacious genistein in conjugation with superparamagnetic Fe3O4 decorated with bio-compatible carboxymethylated chitosan against acute leukemia lymphoma. Biomater Res 24:9
Golafshan N, Kharaziha M, Fathi M, Larson BL, Giatsidis G, Masoumi N (2018) Anisotropic architecture and electrical stimulation enhance neuron cell behaviour on a tough graphene embedded PVA: alginate fibrous scaffold. RSC Adv 8:6381–6389
González-Mayorga A, Lõpez-Dolado E, Gutiërrez MC, Collazos-Castro JE, Ferrer ML, Del Monte F, Serrano MC (2017) Favorable biological responses of neural cells and tissue interacting with graphene oxide microfibers. ACS Omega 2:8253–8263
Good PF, Werner P, Hsu A, Olanow CW, Perl DP (1996) Evidence of neuronal oxidative damage in Alzheimer's disease. Am J Pathol 149:21
Götz M, Huttner WB (2005) Developmental cell biology: the cell biology of neurogenesis. Nat Rev Mol Cell Biol 6:777
Grawish ME, Grawish LM, Grawish HM, Grawish MM, El-Negoly SA (2020) Challenges of engineering biomimetic dental and paradental tissues. Tissue Eng Regen Med 17:403–421
Grossemy S, Chan PP, Doran PM (2020) Stimulation of cell growth and neurogenesis using protein-functionalized microfibrous scaffolds and fluid flow in bioreactors. Biochem Eng J 159:107602
Guo W, Zhang X, Yu X, Wang S, Qiu J, Tang W, Li L, Liu H, Wang ZL (2016) Self-powered electrical stimulation for enhancing neural differentiation of mesenchymal stem cells on graphene–poly(3,4-ethylenedioxythiophene) hybrid microfibers. ACS Nano 10:5086–5095
Guo R, Li J, Chen C, Xiao M, Liao M, Hu Y, Liu Y, Li D, Zou J, Sun D, Torre V, Zhang Q, Chai R, Tang M (2021) Biomimetic 3D bacterial cellulose-graphene foam hybrid scaffold regulates neural stem cell proliferation and differentiation. Colloids Surf B: Biointerfaces 200:111590

Hsiao J-K, Tai M-F, Chu H-H, Chen S-T, Li H, Lai D-M, Hsieh S-T, Wang J-L, Liu H-M (2007) Magnetic nanoparticle labeling of mesenchymal stem cells without transfection agent: cellular behavior and capability of detection with clinical 1.5 T magnetic resonance at the single cell level. Magn Reson Med 58:717–724

Intille SS, Lester J, Sallis JF, Duncan G (2012) New horizons in sensor development. Med Sci Sports Exerc 44:S24–S31

Jaafar E, Kashif M (2018) Study on morphological, optical and electrical properties of graphene oxide (GO) and reduced graphene oxide (rGO). Mater Sci Forum 917:112–116

Janezic S, Threlfell S, Dodson PD, Dowie MJ, Taylor TN, Potgieter D, Parkkinen L, Senior SL, Anwar S, Ryan B (2013) Deficits in dopaminergic transmission precede neuron loss and dysfunction in a new Parkinson model. Proc Natl Acad Sci U S A 110:E4016–E4025

Jeong H-J, Jimenez Z, Mukhambetiyar K, Seo M, Choi J-W, Park T-E (2020) Engineering human brain organoids: from basic research to tissue regeneration. Tissue Eng Regen Med 17:747–757

Jiang Y, Raliya R, Liao P, Biswas P, Fortner JD (2017) Graphene oxides in water: assessing stability as a function of material and natural organic matter properties. Environ Sci Nano 4: 1484–1493

Kapałczyńska M, Kolenda T, Przybyła W, Zajączkowska M, Teresiak A, Filas V, Ibbs M, Bliźniak R, Łuczewski Ł, Lamperska K (2018) 2D and 3D cell cultures - a comparison of different types of cancer cell cultures. Arch Med Sci 14:910–919

Kawai S, Yamazaki M, Shibuya K, Yamazaki M, Fujii E, Nakano K, Suzuki M (2020) Three-dimensional culture models mimic colon cancer heterogeneity induced by different microenvironments. Sci Rep 10:1–11

Kempermann G, Gage FH, Aigner L, Song H, Curtis MA, Thuret S, Kuhn HG, Jessberger S, Frankland PW, Cameron HA, Gould E, HEN R, Abrous DN, Toni N, Schinder AF, Zhao X, Lucassen PJ, Frisën J (2018) Human adult neurogenesis: evidence and remaining questions. Cell Stem Cell 23:25–30

Khan I, Saeed K, Khan I (2017) Nanoparticles: properties, applications and toxicities. Arab J Chem 12:908–931

Kim T-H, Lee K-B, Choi J-W (2013) 3D graphene oxide-encapsulated gold nanoparticles to detect neural stem cell differentiation. Biomaterials 34:8660–8670

Kim T-H, Shah S, Yang L, Yin PT, Hossain MK, Conley B, Choi J-W, Lee K-B (2015a) Controlling differentiation of adipose-derived stem cells using combinatorial graphene hybrid-pattern arrays. ACS Nano 9:3780–3790

Kim T-H, Yea C-H, Chueng S-TD, Yin PT-T, Conley B, Dardir K, Pak Y, Jung GY, Choi J-W, Lee K-B (2015b) Large-scale nanoelectrode arrays to monitor the dopaminergic differentiation of human neural stem cells. Adv Mater 27:6356–6362

Kim SM, Kim J, Noh S, Sohn H, Lee T (2020a) Recent development of aptasensor for influenza virus detection. Biochip J 14:327–339

Kim T-Y, Lim JW, Lim M-C, Song N-E, Woo M-A (2020b) Aptamer-based fluorescent assay for simple and sensitive detection of fipronil in liquid eggs. Biotechnol Bioprocess Eng 25:246–254

Krishnamurthy S, Nör JE (2013) Orosphere assay: a method for propagation of head and neck cancer stem cells. Head Neck 35:1015–1021

Li N, Zhang Q, Gao S, Song Q, Huang R, Wang L, Liu L, Dai J, Tang M, Cheng G (2013) Three-dimensional graphene foam as a biocompatible and conductive scaffold for neural stem cells. Sci Rep 3:1604

Liu H, Zhang L, Guo Y, Cheng C, Yang L, Jiang L, Yu G, Hu W, Liu Y, Zhu D (2013) Reduction of graphene oxide to highly conductive graphene by Lawesson's reagent and its electrical applications. J Mater Chem C 1:3104–3109

Liu N, Xiang X, Fu L, Cao Q, Huang R, Wu L (2021) Regenerative field effect transistor biosensor for in vivo monitoring of dopamine in fish brains. Biosens Bioelectron 188:113340

Loh QL, Choong C (2013) Three-dimensional scaffolds for tissue engineering applications: role of porosity and pore size. Tissue Eng Part B Rev 19:485–502

Lv M, Zhang Y, Liang L, Wei M, Hu W, Li X, Huang Q (2012) Effect of graphene oxide on undifferentiated and retinoic acid-differentiated SH-SY5Y cells line. Nanoscale 4:3861–3866
Ma X, Xiao M, Hao Y, Cheng G (2019) Precisely controllable hybrid graphene scaffold reveals size effects on differentiation of neural progenitor cells in mimicking neural network. Carbon 145: 90–99
Mao S, Lu G, Yu K, Bo Z, Chen J (2010) Specific protein detection using thermally reduced graphene oxide sheet decorated with gold nanoparticle-antibody conjugates. Adv Mater 22: 3521–3526
Marcus M, Skaat H, Alon N, Margel S, Shefi O (2015) NGF-conjugated iron oxide nanoparticles promote differentiation and outgrowth of PC12 cells. Nanoscale 7:1058–1066
Miller AE, Hu P, Barker TH (2020) Feeling things out: bidirectional signaling of the cell–ECM Interface, implications in the mechanobiology of cell spreading, migration, proliferation, and differentiation. Adv Healthc Mater 9:1901445
Min K-J, Kim T-H, Choi J-W (2017) Magnetic force-driven graphene patterns to direct synaptogenesis of human neuronal cells. Materials 10:1151
Myung S, Yin PT, Kim C, Park J, Solanki A, Reyes PI, Lu Y, Kim KS, Lee K-B (2012) Label-free polypeptide-based enzyme detection using a graphene-nanoparticle hybrid sensor. Adv Mater (Deerfield Beach, Fla) 24:6081–6087
Papageorgiou DG, Kinloch IA, Young RJ (2017) Mechanical properties of graphene and graphene-based nanocomposites. Prog Mater Sci 90:75–127
Park S-J, Kim S, Kim S-Y, Jeon NL, Song JM, Won C, Min D-H (2017) Highly efficient and rapid neural differentiation of mouse embryonic stem cells based on retinoic acid encapsulated porous nanoparticle. ACS Appl Mater Interfaces 9:34634–34640
Park D, Kim JH, Kim HJ, Lee D, Lee DS, Yoon DS, Hwang KS (2020) Multiplexed femtomolar detection of Alzheimer's disease biomarkers in biofluids using a reduced graphene oxide field-effect transistor. Biosens Bioelectron 167:112505
Polo Y, Luzuriaga J, Iturri J, Irastorza I, Toca-Herrera JL, Ibarretxe G, Unda F, Sarasua J-R, Pineda JR, Larraóaga A (2021) Nanostructured scaffolds based on bioresorbable polymers and graphene oxide induce the aligned migration and accelerate the neuronal differentiation of neural stem cells. Nanomedicine 31:102314
Qi Z, Guo W, Zheng S, Fu C, Ma Y, Pan S, Liu Q, Yang X (2019) Enhancement of neural stem cell survival, proliferation and differentiation by IGF-1 delivery in graphene oxide-incorporated PLGA electrospun nanofibrous mats. RSC Adv 9:8315–8325
Ratnam KV, Manjunatha H, Janardan S, Naidu KCB, Ramesh S (2020) Nonenzymatic electrochemical sensor based on metal oxide, MO (M= cu, Ni, Zn, and Fe) nanomaterials for neurotransmitters: an abridged review. Sensors Int 1:100047
Reiner A, Albin RL, Anderson KD, D'Amato CJ, Penney JB, Young AB (1988) Differential loss of striatal projection neurons in Huntington disease. Proc Natl Acad Sci U S A 85:5733–5737
Rodriguez-Losada N, Wendelbob R, Ocaóa MC, Casares AD, Guzman De Villoría R, Aguirre Gomez JA, Arraez MA, Gonzalez-Alegre P, Medina MA, Arenas E, Narvaez JA (2020) Graphene oxide and reduced derivatives, as powder or film scaffolds, differentially promote dopaminergic neuron differentiation and survival. Front Neurosci 14:570409
Ronkainen NJ, Halsall HB, Heineman WR (2010) Electrochemical biosensors. Chem Soc Rev 39: 1747–1763
Roy R, Hohng S, Ha T (2008) A practical guide to single-molecule FRET. Nat Methods 5:507–516
Sampath T, Reddi AH (1981) Dissociative extraction and reconstitution of extracellular matrix components involved in local bone differentiation. Proc Natl Acad Sci 78:7599–7603
Sánchez-González S, Diban N, Bianchi F, Ye H, Urtiaga A (2018) Evidences of the effect of GO and rGO in PCL membranes on the differentiation and maturation of human neural progenitor cells. Macromol Biosci 18:1800195
Sethi J, Van Bulck M, Suhail A, Safarzadeh M, Perez-Castillo A, Pan G (2020) A label-free biosensor based on graphene and reduced graphene oxide dual-layer for electrochemical determination of beta-amyloid biomarkers. Microchim Acta 187:1–10

Shah S, Yin PT, Uehara TM, Chueng S-TD, Yang L, Lee K-B (2014) Guiding stem cell differentiation into oligodendrocytes using graphene-nanofiber hybrid scaffolds. Adv Mater (Deerfield Beach, Fla) 26:3673–3680

Sharma R, Asmara TC, Sahoo KR, Grage SL, Zhang R, Sun J, Das S, Ulrich AS, Rusydi A, Aryasomayajula S (2020) Structural and electronic transport properties of fluorographene directly grown on silicates for possible biosensor applications. ACS Appl Nano Mater 3: 5399–5409

Shin J-H, Lee M-J, Choi J-H, Song J-A, Kim T-H, Oh B-K (2020) Electrochemical H2O2 biosensor based on horseradish peroxidase encapsulated protein nanoparticles with reduced graphene oxide-modified gold electrode. Nano Converg 7:39

Smitha PK, Bathula C, Kn C, Das M (2020) Usage of graphene oxide in fluorescence quenching-linked immunosorbent assay for the detection of Cry2Ab protein present in transgenic plants. J Agric Food Chem 68:3656–3662

Solanki A, Chueng S-TD, Yin PT, Kappera R, Chhowalla M, Lee K-B (2013) Axonal alignment and enhanced neuronal differentiation of neural stem cells on graphene-nanoparticle hybrid structures. Adv Mater (Deerfield Beach, Fla) 25:5477–5482

Suhito IR, Angeline N, Choo S-S, Woo HY, Paik T, Lee T, Kim T-H (2018) Nanobiosensing platforms for real-time and non-invasive monitoring of stem cell pluripotency and differentiation. Sensors (Basel, Switzerland) 18:2755

Sumisha A, Haribabu K (2020) Nanostructured polypyrrole as cathode catalyst for Fe (III) removal in single chamber microbial fuel cell. Biotechnol Bioprocess Eng 25:78–85

Tasnim N, Thakur V, Chattopadhyay M, Joddar B (2018) The efficacy of graphene foams for culturing mesenchymal stem cells and their differentiation into dopaminergic neurons. J Stem Cells Int 2018:12

Tu Q, Pang L, Wang L, Zhang Y, Zhang R, Wang J (2013) Biomimetic choline-like graphene oxide composites for neurite sprouting and outgrowth. ACS Appl Mater Interfaces 5:13188–13197

Wang Y, Li Z, Wang J, Li J, Lin Y (2011) Graphene and graphene oxide: biofunctionalization and applications in biotechnology. Trends Biotechnol 29:205–212

Wang Y, Lee WC, Manga KK, Ang PK, Lu J, Liu YP, Lim CT, Loh KP (2012) Fluorinated graphene for promoting neuro-induction of stem cells. Adv Mater 24:4285–4290

Wang C, Zhang Z, Chen B, Gu L, Li Y, Yu S (2018) Design and evaluation of galactosylated chitosan/graphene oxide nanoparticles as a drug delivery system. J Colloid Interface Sci 516: 332–341

Wang J, Wang H, Mo X, Wang H (2020) Reduced graphene oxide-encapsulated microfiber patterns enable controllable formation of neuronal-like networks. Adv Mater (Deerfield Beach, Fla) 32: 2004555

Yang K, Lee J, Lee JS, Kim D, Chang G-E, Seo J, Cheong E, Lee T, Cho S-W (2016) Graphene oxide hierarchical patterns for the derivation of electrophysiologically functional neuron-like cells from human neural stem cells. ACS Appl Mater Interfaces 8:17763–17774

Yin PT, Shah S, Chhowalla M, Lee K-B (2015) Design, synthesis, and characterization of graphene–nanoparticle hybrid materials for bioapplications. Chem Rev 115:2483–2531

Zaoralová D, Hrubý VTZ, Šedajová V, Mach R, Kupka VC, Ugolotti J, Bakandritsos A, Medved' M, Otyepka M (2020) Tunable synthesis of nitrogen doped graphene from fluorographene under mild conditions. ACS Sustain Chem Eng 8:4764–4772

Zhang LI, Poo M-M (2001) Electrical activity and development of neural circuits. Nat Neurosci 4: 1207–1214

Zhou G, Yin L-C, Wang D-W, Li L, Pei S, Gentle IR, Li F, Cheng H-M (2013) Fibrous hybrid of graphene and sulfur nanocrystals for high-performance lithium–sulfur batteries. ACS Nano 7: 5367–5375

Zhu M, Xie X, Guo Y, Chen P, Ou X, Yu G, Liu M (2013) Fluorographene nanosheets with broad solvent dispersibility and their applications as a modified layer in organic field-effect transistors. Phys Chem Chem Phys 15:20992–21000

Zhu R, Sun Z, Li C, Ramakrishna S, Chiu K, He L (2019) Electrical stimulation affects neural stem cell fate and function in vitro. Exp Neurol 319:112963

Functional Graphene Nanomaterials-Based Hybrid Scaffolds for Osteogenesis and Chondrogenesis

Moon Sung Kang, Hee Jeong Jang, Seok Hyun Lee, Yong Cheol Shin, Suck Won Hong, Jong Hun Lee, Bongju Kim, and Dong-Wook Han

Abstract

With the emerging trends and recent advances in nanotechnology, it has become increasingly possible to overcome current hurdles for bone and cartilage regeneration. Among the wide type of nanomaterials, graphene (G) and its derivatives (graphene-based materials, GBMs) have been highlighted due to the specific physicochemical and biological properties. In this review, we present the recent development of GBM-based scaffolds for bone and cartilage engineering, focusing on the formulation/shape/size-dependent characteristics, types of scaffold and modification, biocompatibility, bioactivity and underlying mechanism, drawback and prospect of each study. From the findings described herein, mechanical property, biocompatibility, osteogenic and chondrogenic property of GBM-based scaffolds could be significantly enhanced through various scaffold fabrication methods and conjugation with polymers/nanomaterials/drugs. In conclusion, the results presented in this review support the promising prospect of using

Moon Sung Kang and Hee Jeong Jang equally contributed to this work.

M. S. Kang · H. J. Jang · S. H. Lee · S. W. Hong · D.-W. Han (✉)
Department of Cogno-Mechatronics Engineering, College of Nanoscience and Nanotechnology, Pusan National University, Busan, South Korea
e-mail: swhong@pusan.ac.kr; nanohan@pusan.ac.kr

Y. C. Shin
Department of Biomedical Engineering, The University of Texas at Austin, Austin, TX, USA

J. H. Lee
Department of Food Science and Biotechnology, Gachon University, Seongnam, South Korea

B. Kim
Dental Life Science Research Institute/Innovation Research and Support Center for Dental Science, Seoul National University Dental Hospital, Seoul, South Korea

D.-W. Han, S. W. Hong (eds.), *Multifaceted Biomedical Applications of Graphene*,
Advances in Experimental Medicine and Biology 1351,
https://doi.org/10.1007/978-981-16-4923-3_4

GBM-based scaffolds for improved bone and cartilage tissue engineering. Although GBM-based scaffolds have some limitations to be overcome by future research, we expect further developments to provide innovative results and improve their clinical potential for bone and cartilage regeneration.

Keywords

Graphene · Biocompatibility · Osteogenesis · Chondrogenesis · bone and cartilage tissue engineering

1 Introduction

Recently, much attention has been paid to the bone and cartilage regeneration because of the soaring middle and old aged population. However, the efficacy of conventional bone and cartilage grafts is hindered because of that: (a) grafted artificial bone and cartilage are not mechanically synchronized with natural tissues, (b) do not support functional growth of tissue, and (c) do not provide appropriate cell-matrix interaction (Petite et al. 2000; Dimitriou et al. 2011; Mistry and Mikos 2005). These have led to several side effects such as weakening of original tissues by the shear stress and friction, or necrosis and inflammatory reaction of surrounding tissues. Bone and cartilage tissue engineering, which aims to regenerate functionalized tissues, is a promising solution for these problems. Tissue engineering is a therapeutic approach that combines cells, biomaterials, and biological cues to induce appropriate cellular growth and differentiation. The engineered tissues can be surgically transplanted and promote tissue repair and functional regeneration. In particular, the fabrication of an artificial scaffold is the most important part, which controls the cell growth, proliferation, and maintenance of cell functionality. Multiple cells sources have been explored for bone and cartilage tissue engineering, which are mesenchymal stem cells (MSCs) (Shin et al. 2004; Li et al. 2005a), induced pluripotent stem cells (iPSCs) (Diekman et al. 2012; Liu et al. 2013), embryonic stem cells (ESCs) (Tang et al. 2012; Toh et al. 2011), and the kinds of progenitor cells (Guilak et al. 2004; Qiao et al. 2013), osteocytes (Tetteh et al. 2014), and chondrocytes (Wang et al. 2006). Proper mechanical and biochemical interactions within scaffold and cells are the most challenging part of tissue engineering. Therefore, researches on bone and cartilage tissue engineering have employed a wide range of materials to promote structural support with desired osteogenic and chondrogenic potentials (Burg et al. 2000; Kuo et al. 2006) (Fig. 1).

With the emerging trends and recent advances in nanotechnology, it has become increasingly possible to overcome current hurdles that tissue engineering researches encounter. Among the wide type of nanomaterials, graphene (G) and its derivatives (graphene-based materials, GBMs) have been highlighted due to the specific characteristics such as exceptional electric conductivity, large surface area, good elasticity, mechanical strength, and extraordinary biological properties (Bitounis

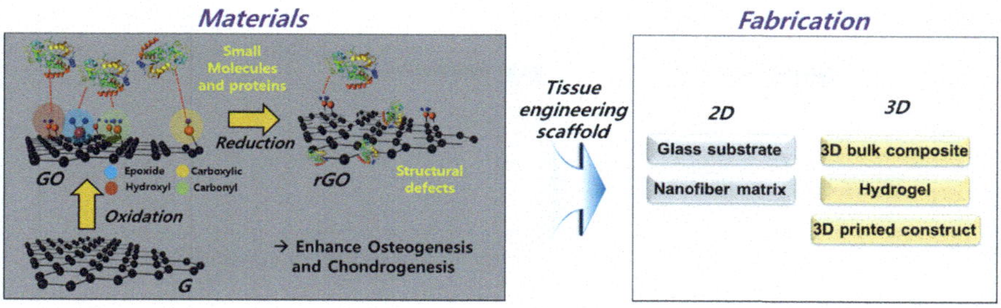

Fig. 1 Schematic diagram of GBM and types of fabricated scaffolds

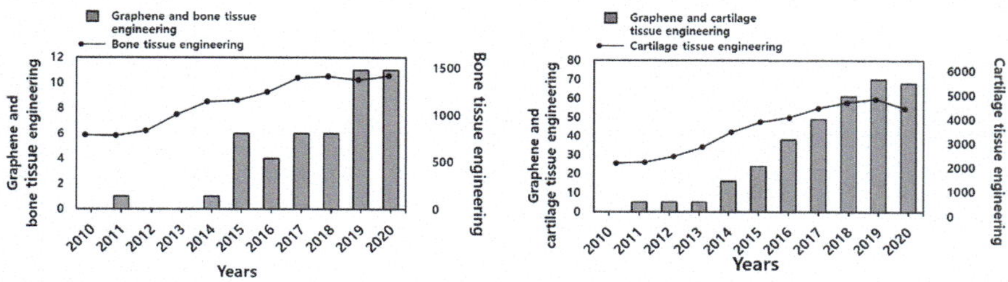

Fig. 2 Numbers of recent publications featuring the terms graphene, bone tissue engineering, and cartilage tissue engineering. Y-axis of graphs represents the numbers of published papers containing each of the indicated words. All values were obtained from PubMed

et al. 2013). G is a two-dimensional (2D) nanomaterial composed of single-layered sp^2-bonded carbon atoms arranged in a honeycomb structure. G is obtained by physicochemical exfoliation of graphite, whereas graphene oxide (GO), a highly oxidative form of graphene, and reduced graphene (rGO), prepared by chemical or thermal reduction of GO, exist as graphene derivatives family (Sanchez et al. 2012). GBMs feature strong carbon-carbon bonding in the basal plane, an aromatic structure, presence of π electrons, and the reactive sites for surface reaction. Meanwhile, GO has oxygen-containing moieties such as hydroxyl groups, carbonyl groups, carboxylic groups, and epoxide, while rGO has structural defects and retains a few residual oxygen moieties. The biological activity of GBMs can be associated with their unique chemical structure. It is known that the GBMs induce strong hydrophobic interactions with cells and tissues (Liao et al. 2018). The biomolecules such as DNA, proteins, and small molecules are easily absorbed by the structural defects and functional moieties. Moreover, the exceptional biological characteristics of GBMs significantly enhanced cellular behaviors such as adhesion, migration, and proliferation (Shin et al. 2015a, 2016), and even induced the spontaneous differentiation of progenitor cells toward chondrogenesis and osteogenesis (Lee et al. 2015a; Olate-Moya et al. 2020; Shin et al. 2015b, 2018; Jin et al. 2015). Herein, we intend to provide an overview of the current state of GBMs-based tissue engineering scaffold for cartilage and bone tissue engineering (Fig. 2). Recent studies will be discussed focusing on the formulation/shape/size-dependent characteristics of GBMs, types of scaffold and modification, bioactivity, and underlying mechanism, and the prospect of each study (Table 1).

2 Biocompatibility

GBMs have the extensive potential for biomedical engineering and biotechnology. However, existing studies regarding the biological effects of GBMs often show contradictory or inconclusive results. That is because the biocompatibility of GBMs depends on the intrinsic physicochemical properties of themselves, which are highly variable by the raw materials and production methods (Syama and Mohanan 2016). Moreover, GBMs exhibits time, dose, and size-dependent toxicity (Ou et al. 2016). For this reason, it is sometimes difficult to extract the exact conclusion from the different kinds of literature. Therefore, we summarized the toxicity of GBMs on different conditions and hope to provide a guideline for later in-depth studies.

Most of the in vitro studies suggest GBMs exhibit cytotoxicity at the concentration range of over 10–100 μg/ml. Liao et al. investigated the hemolytic and cytotoxic effect of GO (GO sheet, hydrazine-free hydrothermal route) while varying the physicochemical properties such as particle size, particulate state, surface charge, and oxygen content (Liao et al. 2011). The hemolysis of red blood cells (RBCs) suggested the strong effect of preparation methods of GO, suggesting the concentration of GO leading to 50% lysis of RBCs are varying between 20.2 and 142 μg/ml. Meanwhile, the cell viability of human fibroblasts was significantly different per preparation methods. 24 h exposed to GO indicated that 50–200 μg/ml GO induced cell viability under 80%, by generating reactive oxygen species (ROS) which is a common reason for the toxicity of the carbon-based and nanoscale materials. On the other hand, Chung and Pumera investigated that the cytotoxicity of GO depends on the oxidative methods, which are Staudenmaier, Hoffman, Hummers, and Tour methods (Chng and Pumera 2013). GO prepared by Hoffman, Hummers, and Tour methods exhibited significant cytotoxicity at concentration $\geq$8 μg/ml, while GO prepared by Staudenmaier methods showed no cytotoxicity at the concentration range of 0–125 μg/ml. The authors suggested that cytotoxicity of GO is differed by the ratio of C/O contents and the number of carbonyl groups. Akhavan et al. presented size-dependent cytotoxicity of rGO nanoplatelets (rGONPs) in hMSCs (Akhavan et al. 2012). The authors suggested that oxidative stress and cell membrane damage are the main reasons for GO cytotoxicity. The cell viability test showed 1.0 μg/mL rGONPs with an average lateral diameter of 11 $\pm$ 4 nm damaged cell membrane integrity, while the rGO sheets with a diameter of 3.8 $\pm$ 0.4 μm exhibited significant cytotoxicity at a higher concentration of 100 μg/mL after 1 h exposure time. However, only 0.1 μg/ml of rGONPs has induced genotoxicity through the DNA fragmentation and chromosomal aberrations.

The in vivo toxicity of GBMs also has been intensively studied. To understand the potential in vivo toxicity of GBMs, the behaviors and biodistribution of GBMs in animal models should be elucidated. The accumulation of GO in targeted organs and

Table 1 Classification of recent research on the use of GBM in bone and cartilage tissue engineering and their scaffold type, use of GBM, key materials, and cell sources

Tissue type	Scaffold type	GBMs used	Key materials	Cell sources	Ref.
Cartilage	Glass/substrate	rGO	Chitosan	Human chondrocyte	Cao et al. (2017a)
	Nanofiber matrix	G	Polycarprolactone (PCL), poly(L-lysine)	Human MSC (hMSC)	Holmes et al. (2016)
		GO	Chitosan, poly(vinyl alcohol) (PVA)	ATDC5	Cao et al. (2017b)
	3D bulk/porous structure/ composite	G	PCL	Mouse bone-marrow derived MSC (bMSC)	Deliormanlı (2019)
		GO	Chitosan	Human chondrocyte	Shamekhi et al. (2019)
			Methacrylated chondroitin sulfate (CSMA), oly (ethylene glycol) methyl ether-ε-caprolactone-acryloyl chloride (MPEG-PCL-AC, PECA)	3T3 cell	Liao et al. (2015)
		GO	PCL, 13-93 bioactive glass	MC3T3-E1	Deliormanlı and Atmaca (2018)
			Acellular ECM	Rabbit bMSC	Gong et al. (2021)
		G, GO	Cell-assembled composite	hMSC	Lee et al. (2015b)
	Hydrogel	GO	Gelatin	Primary rat chondrocyte	Satapathy et al. (2019)
			Photopolymerizable poly-D,L-lactic acid/ polyethylene glycol (PDLLA)	bMSC	Shen et al. (2018)
			TGF-β, PDLLA	bMSC	Shen et al. (2020)
			TGF-β, collagen	hMSC	Zhou et al. (2019b)
			FEFKFEFK (β-sheet forming self-assembling peptide) hydrogels	Nucleus pulposus cell	Ligorio et al. (2019)
	3D printing	G	Chitosan, gelatin, hydroxyapatite (HA)	bMSC	Hu et al. (2019)
		GO	Alginate, gelatin, chondroitin sulfate	Human adipose-derived stem cell (hADSC)	Olate-Moya et al. (2020)
			Gelatin methacrylate (GelMA), polyethylene (glycol) diacrylate (PEGDA)	hMSC	Zhou et al. (2017b)

(continued)

Table 1 (continued)

Tissue type	Scaffold type	GBMs used	Key materials	Cell sources	Ref.
Bone	Glass/substrate	G	Borate-based bioactive glass	MC3T3-E1	Turk and Deliormanlı (2017)
		GO	Mesoporous bioactive glass (MBG)	Rat bMSC	Wang et al. (2019b)
			Polyetheretherketone (PEEK), HA	MG-63	Peng et al. (2017)
			Chitosan, HA, Au, polyvinyl alcohol (PVA)	Mouse mesenchymal cells (C3H10T1/2)	Prakash et al. (2020)
		rGO	Chitosan	hMSC	Kim et al. (2013)
			PCL, Cu	MC3T3-E1	Jaidev et al. (2017)
	Nanofiber matrix	GO	Silica, HA	bMSC	Dalgic et al. (2018)
			Bone morphogenetic protein 2 (BMP-2), silk fibroin (SF)	bMSC	Wu et al. (2019)
			PCL, chitosan, collagen	MG-63	Aidun et al. (2019)
			PCL, Simvastatin polyurethane	MG-63	Rezaei et al. (2021)
			Poly(3-hydroxybutyrate-co-4-hydroxybutyrate) (p_{34}HB)	bMSC	Zhou et al. (2017a)
		rGO	SF	bMSC	Nalvuran et al. (2018)
	3D bulk/porous structure/ composite	GO	Carboxymethyl chitosan (CMC).	hADSC	Ruan et al. (2016)
			Chitosan	hADSC	Dinescu et al. (2019)
			Alginate, chitosan, collagen	MC3T3-E1	Kolanthai et al. (2018)
			Chitosan, gelatin	Rat calvarial osteoprogenitor cell, C3H10T1/2	Saravanan et al. (2017)
			Collagen, aerogel	Rat bMSC	Liu et al. (2019)
			HA, SF	Mouse MSC	Wang et al. (2017)
			BMP-9, Citrate	ASC	Zhao et al. (2018)
			Bone-like apatite (AP), collagen	Rat BMSC	Zhou et al. (2018)

			Salvianolic acid B, SF	Rat BMSC	Wang et al. (2020)
			CaP	bMSC	Arnold et al. (2019)
			Poly(propylene fumarate) (PPF)	MC3T3-E1	Farshid et al. (2019)
			CuO/Cu_2O, CaP	bMSC	Zhang et al. (2016)
		rGO	HA	bMSC	Zhou et al. (2019a)
			PDMS	hADSC	Li et al. (2017)
			Bovine serum albumin (BSA), bredigite	G-292	Askari et al. (2021)
		rGO, GO	PCL	Rat MSC	Evlashin et al. (2019)
	Hydrogel	G	PCL, polyaniline	MG-63	Khorshidi and Karkhaneh (2018)
		GO	HA, Chitosan genipin (GNP), sodium ascorbate (NaVC)	Rabbit bMSC	Yu et al. (2017)
			HA	MSC	Xie et al. (2015)
	3D printed	G	PCL	hADSC	Wang et al. (2019a)
		GO	BMP-2 liked peptide, PLGA, β-TCP	Rat bMSC	Zhang et al. (2019)
			TCP, alginic acid (AA)	hOB	Boga et al. (2018)

oxidative stress by the generated ROS may play important roles in inducing GO toxicity. In mice, GO could be accumulated in the targeted organs of the lung, liver, spleen, and kidney. Duch et al. demonstrated that GO induced severe and persistent lung injury of mice and increased rate of mitochondrial respiration and the generation of ROS, activating the inflammatory and apoptotic pathways (Duch et al. 2011). Generally, in vivo behaviors and toxicology of GBMs are closely associated with their surface coatings, size, and administration routes. Yang et al. investigated the in vivo behaviors and toxicity of GO and PEGylated GO after oral and intraperitoneal injection. The results indicated that PEGylated GO was not absorbed in the organs and was rapidly excreted, while the pristine GO was significantly accumulated in the liver. The formation of granulomas, which could be induced by intraperitoneal injected long multi-walled carbon nanotubes (MWNTs), was not observed in mice injected with a high dose of GO or PEGylated GO (Yang et al. 2013a). Yang et al. demonstrated that there were no toxic effects to the vital organs of rats after 4 weeks of intravenous injection of graphene quantum dots (GQDs) although there was a slight reduction of platelet number, monocyte reduction, and eosinophil fraction occurred, which were normalized after (Yang et al. 2013b).

3 Graphene for Chondrogenesis

3.1 Glass/Substrate/Nanofiber Scaffold for Chondrogenesis

Conventionally, kinds of 2D substrates have been utilized for cartilage tissue engineering scaffolds because of the ease of fabrication, culture, and observation of seeded cells. Chitosan is a partially deacetylated derivative of chitin. Due to the biodegradability, biocompatibility, non-antigenicity, non-toxicity, and novel biofunctionality, chitosan has been widely adopted as a bone tissue engineering scaffold (Levengood and Zhang 2014). However, chitosan is mechanically weak and easily deformed by the swelling nature (Li et al. 2005b). The incorporation of GBMs can enhance the mechanical property, as well as the hydrophilic nature of chitosan can be strengthened by GBMs to promote cell adhesion and proliferation. Cao et al. fabricated rGO/chitosan film for cartilage tissue engineering (Cao et al. 2017a). The synthesized rGO was uniformly coated onto the chitosan film. Further, the in vitro cytotoxicity results showed the biocompatible nature of prepared rGO/chitosan film toward the human articular chondrocytes indicating the advantage of a combination of chitosan and rGO for cartilage tissue engineering. On the other hand, electrospun nanofiber has several advantages as a tissue engineering scaffold because the highly porous structure can mimic the structural characteristics of natural ECM, and therefore can mimic the cell-matrix interaction at in vitro. The natural cartilage ECM is also a nano-scaled structure; therefore Holmes et al. investigated the chondrogenic differentiation-inducing ability of GBM-containing PCL nanofiber matrices on human bMSCs (Holmes et al. 2016). All of the fabricated nanofibers in the different compositions (G nanoplatelets, CNT, and poly-l-lysine coated/

uncoated) showed enhanced chondrogenesis of bMSCs by upregulating the expression level of GAG and collagen type II, which are two of the major components of cartilage ECM. The study conducted by Lei et al. also proved the efficacy of GBM-containing nanofiber matrices for cartilage tissue engineering by combining PVA, chitosan, and GO into the nanofiber matrix (Cao et al. 2017b). Incorporation of GO into electrospun nanofiber also can enhance the physicochemical properties of the fabricated nanofiber. The results showed that GO-containing nanofiber has exhibited smaller size owing to the electrical conductivity of GO; moreover, the tensile strength of nanofiber has been increased by the incorporation of GO. Consequently, the in vitro tests revealed that the chitosan/PVA/GO nanofiber matrices supported a suitable environment for the growth of chondrogenic cell line ATDC5.

3.2 3D Bulk/Porous Structure/Composite for Chondrogenesis

Originally, cells grow while interacting with the surrounding ECM in the 3D environment. Therefore, by mimicking the natural 3D environment, the scaffold can give the cells appropriate mechanical and biochemical signals. Meanwhile, the synthetic polymer PCL is often used in biomedical engineering due to its superior properties such as biodegradability, non-immunogenicity, biocompatibility, and suitable mechanical properties. However, the lack of the cell adhesion site and the non-biofunctionality limits its application for tissue engineering scaffold. Using the chondrogenesis-inducing properties of GBMs, PCL-GBMs 3D hybrid scaffolds are widely studied. Deliormanlı fabricated lattice-structured 3D PCL scaffolds in the presence of graphene nanoplatelets for cartilage tissue engineering applications (Deliormanlı 2019). Mouse bMSCs showed enhanced cell viability after 7–21 days of incubation. Moreover, after the bMSCs are cultured on the G-containing PCL lattice with the presence of chondrogenesis-inducing media, the expression level of glycosaminoglycan (GAG) significantly enhanced, which is an important ECM component of cartilage tissue and provides biological signals to stem cells and chondrocytes for development and functional regeneration of cartilage (Deliormanlı 2019; Yaylaci et al. 2016).

As mentioned above, chitosan has been used for a 3D cartilage engineering scaffold because it is structurally similar to the N-acetyl glucosamine group and GAGs found in cartilage ECM. Therefore, the utilization of chitosan for cartilage regeneration would be advantageous because of the similar structure of chitosan to GAGs (Suh and Matthew 2000). Shamekhi et al. studied the effectiveness of GO-containing chitosan 3D scaffolds for chondrocyte culture (Shamekhi et al. 2019). The results indicated that human articular chondrocytes on the nanocomposite scaffolds showed an increased proliferation with augmentation of the GO percentage, particularly in prolonged cultivation periods. The authors suggest that the enhanced surface roughness and stiffness by the incorporation of GO caused stimulation of human articular chondrocyte proliferation and improved morphology. Another study conducted by Lee et al. used GO–cell biocomposites to pre-concentrate growth factors for chondrogenic differentiation of MSCs through the direct contact of GO and growth factors with the laden cells (Lee et al. 2015b). GO in this study was modified to porous structure because the defective structure and the oxygen-containing moieties of GO are known to enhance the absorbance of serum proteins and growth factors which are essential for cell survival and proliferation. The direct contact of GO and cell has served as "growth factor factories," and hence enhanced the formation and maturation of chondrogenic tissues.

To reappear the biological functions of natural cartilage, various biomolecules have been incorporated into the cartilage scaffolds. Chondroitin sulfate is one of the essential structural components of cartilage and provides much of its resistance to compression (Ko et al. 2009; Li et al. 2004). Liao et al. introduced the hybrid scaffold composed of methacrylated chondroitin sulfate (CSMA), poly(ethylene glycol) methyl ether-ε-caprolactone-acryloyl chloride (MPEG-PCL-AC, PECA was used as an abbreviation for MPEG-PCL-AC) and graphene oxide (GO) and evaluated its potential application in cartilage tissue engineering (Liao et al. 2015). The prepared scaffolds could reappear the structural characteristics of natural cartilage ECM as pore size, porosity, swelling ability, compression modulus, and conductivity. In the in vivo implantation in the rabbit, Micro-CT and histology observation showed the CSMA/PECA/GO scaffold with cellular supplementation showed continuous subchondral bone and much thicker newly formed cartilage compared with the control groups, indicating the great potential for the implantation as articular cartilage.

Deliormanlı and Atmaca further studied the osteochondral dual-inducing activities of G-containing PCL/bioactive glass bilayered (G/PCL) scaffolds for the co-culture of MC3T3-E1 mouse fibroblasts and ATDC5 chondrogenic cells (Deliormanlı and Atmaca 2018). The advantage of the bilayer scaffold and co-culture of osteoblasts and chondrocytes is that the scaffolds can mimic the physical structure required for repairing the tissue by endowing cells with the environment of articular cartilage and subchondral bone regions. The results showed the dual inductivity of G/PLC scaffold for osteochondral maturation by enhancing the cell adhesion, proliferation, mineralization of osteoblast, and GAG expression of chondrocytes. On the other hand, PCL has been widely applied for the fabrication of nanofibers.

3.3 Hydrogel/3D Printing for Chondrogenesis

Hydrogels are three-dimensional polymeric networks filled with water and have a similar structure with tissue environments. Therefore, they are considered as an optimal scaffold to deliver cells and tissue engineering scaffolds. Hydrogel scaffold can be served as a carrier for GBMs for chondrogenesis of laden cells (Satapathy et al. 2019; Shen et al. 2018, 2020; Zhou et al. 2019b; Ligorio et al. 2019). Satapathy et al. introduced GO-reinforced gelatin hydrogel prepared by the microplasma-assisted crosslinking method for cartilage tissue engineering application (Satapathy et al. 2019). There are kinds of crosslinking methods to endow the hydrogel suitable mechanical property. However, non-invasive and biocompatible crosslinking is needed for tissue engineering scaffold because the laden cells can be damaged. The study utilized reactive radicals from the nonthermal plasma for the crosslinking of gelatin hydrogel. The process resulted in high cytocompatibility, moderate surface characteristics, pore size, viscoelasticity, and degradability. In vitro studies demonstrated excellent cell-matrix interaction between rat chondrocyte and gelatin hydrogel. Moreover in vivo studies showed the formation of healthy hyaline cartilage after the microfracture was enhanced by the fabricated hydrogel implant.

So far, we discussed the chondrogenesis-enhancing properties of GBMS; however, the direct effect of GBMs itself on stem cells without the presence of chondrogenesis-inducing factors has not been discussed. Shen et al. introduced GO nanosheets and human bMSCs into a photopolymerizable poly-D, L-lactic acid/polyethylene glycol (PDLLA) hydrogel, a robust chondrosupportive scaffold, and assessed bMSCs differentiation to the chondrogenic lineage without supplemental chondroinductive factors (Shen et al. 2018). The higher expression of cartilage matrix genes such as aggrecan and collagen type II demonstrated the spontaneous chondrogenic differentiation of bMSCs by the GO. Moreover, the immunohistochemical results suggested that GO-induced chondrogenesis was correlated with enriched sequestration of insulin, which is a necessary supplement to have pro-chondrogenesis effects on MSCs.

On the other hand, one of the concerns of hydrogel scaffolds is that the matured chondrogenesis takes more than weeks, while the incorporated chondrogenesis-inducing biomolecules such as transforming growth factor β3 (TGF- β3) are cleared fast. However, the incorporation of GBM into hydrogel can increase the mechanical strength and slow down the degradation rates. A further study conducted by Shen et al. suggested the incorporation of GO into hydrogel has shown to increase the mechanical strength and compressive modulus, and therefore supported long-term sustained release of TGF-β3 for up to 4 weeks (Shen et al. 2020). These findings support the potential application of GO and TGF-β3 bMSC hydrogel-based cartilage engineering scaffolds for chondrogenesis of bMSCs. Not only the increased mechanical strength of hydrogel but the chemical interactions between GO and TGF- β3 also can support the sustained release of TGF- β3. Zhou et al. utilized graphene oxide (GO) flakes to adsorb TGF-β3, which were then incorporated into a collagen hydrogel (Zhou et al. 2019b) (Fig. 3). The study showed that GO flakes were adsorbed more than 99% with TGF-β3 with less than 1.7% released. Adsorbed TGF-β3 retained a similar conformation to its dissolved free protein state but exhibited greater conformational stability. The GO/TGF-β3 containing hydrogel induced higher chondrogenic gene expression and cartilage-specific ECM matrix deposition within cells, especially in long-termed culture, compared to exogenously delivered TGF- β3 groups. These two studies suggest that the sustained release of chondrogenesis-inducing factor such as TGF- β3 by the GO/hydrogel significantly increases the chondrogenesis of MSCs.

Meanwhile, with the superior characteristics of GBMs-containing hydrogels, they have been utilized for bioink for 3D bioprinting. As an emerging tissue engineering scaffold fabrication technology, 3D bioprinting has been highlighted due to the cell-

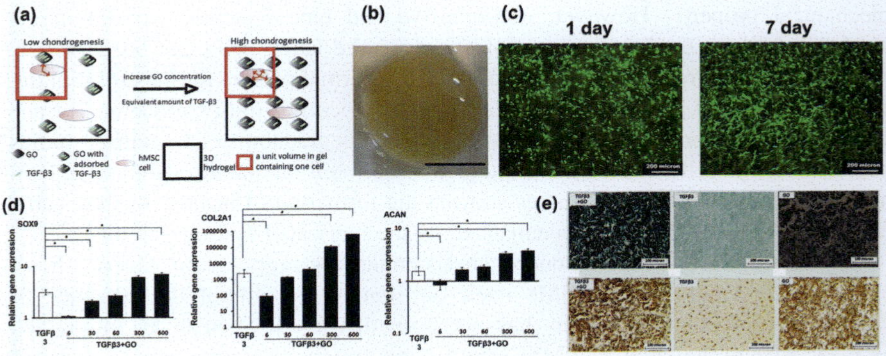

Fig. 3 (**a**) Schematic diagram of GO/TGF-ß3 conjugate in the hydrogel to promote chondrogenesis. (**b**) Digital images of fabricated scaffold and (**c**) live dead assay of laden hMSCs at 1 and 7 days cultured. (**d**) Western blots indicating chondrogenic gene expression (SOX9, COL2A1, and ACAN). (**e**) Immunohistochemical staining of GAG (Upper) and collagen type II (Lower), respectively. All these results indicated the enhanced chondrogenesis of hMSCs in GO/TGF-ß3-incorporated hydrogel (Reproduced from Zhou et al. 2019b, copyright open accessed)

friendly methods and feasibility of customizable 3D hierarchical structures (Olate-Moya et al. 2020; Hu et al. 2019; Zhou et al. 2017b). Especially hydrogel-based 3D bioprinting can further increase the bioactivity of cells. However, most of the hydrogels present problems related to their processability that limit their use to produce tailor-made scaffolds. Recently developed bioconjugated hydrogel bioinks for 3D printed scaffold fabrication focus on both biocompatibility and processability. Olate-Moya et al. developed alginate/gelatin/chondroitin sulfate/GO composite bioink to mimic the cartilage ECM (Olate-Moya et al. 2020). The results show that the incorporation of GO into the bioinks considerably improved the shape fidelity and resolution of 3D printing because of a faster viscosity recovery after the bioink extruded. In vitro proliferation and live/dead assay proved the high cytocompatibility of 3D bioprinting process and prepared bioink to hADSC. Moreover, the incorporated GO led to spontaneous chondrogenesis of hADSC without any exogenous pro-chondrogenesis factors that were observed from immunostaining after 28 days of culture. These results indicate the conjugation of biochemical cues and GBMs can enhance the 3D printability and biofunctionality of the 3D bioprinted cartilage. In a similar study conducted by Zhou et al. fabricated GO-GelMA–PEGDA hydrogel as a biocompatible photopolymerizable bioink (Zhou et al. 2017b). The photo-medicated crosslinking showed high cytocompatibility due to the absence of toxic exogenous crosslinker and non-invasiveness. The results indicated GO/GelMA/PEGDA hydrogel promoted the GAG, and collagen levels after GO induced chondrogenic differentiation of MSCs. Moreover, the collagen type II, SOX 9, and aggrecan gene expressions, which are associated with chondrogenesis of MSCs, were greatly promoted on the scaffolds.

4 Graphene for Osteogenesis

4.1 Glass/Substrate/Nanofiber for Osteogenesis

Up to now, bone grafts and titanium implants have been the general therapeutic approaches used for bone repair and regeneration. However, these types of treatment have several shortcomings, like limited availability, risk of donor-to-recipient infection, and tissue morbidity (Boga et al. 2018). Therefore, many novel strategies are focusing on the biofunctional materials and scaffolds have been introduced to overcome the hurdles. Bioactive glasses are a group of surface reactive materials that are often used for bone and soft tissue repair (Bi et al. 2012; Chen et al. 2006). Within the kinds of bioactive glasses, borated-based 13-93B3 bioactive glass (mol%: 54.6% B2O3, 22.1% CaO, 7.9% K2O, 7.7%, MgO, 6.0% Na2O, and 1.7% P2O5) (Turk and Deliormanlı 2017) and mesoporous bioactive glasses (MBG) which are similar to the porous structure of subchondral bone, because of their highly interconnected large pores (300–500 μm) and ordered structure nanopores (2–50 nm) (Wang et al. 2019b), have been developed as the bone tissue engineering scaffolds. The 13-93B3 bioactive glass scaffold containing G nanoplatelets introduced by Turk and Deliormanlı showed enhanced the electrical conductivity that promotes kinds of cell-matrix interactions (Turk and Deliormanlı 2017). The results indicated MC3T3-E1 cells grown on the prepared glass showed enhanced cell viability and HA forming ability. On the other hand, the GO-containing MBG bioactive glasses (MBG-GO) introduced by Wang et al. demonstrate that the MBG-GO scaffolds have better cytocompatibility and higher osteogenesis differentiation ability with rat BMSCs than the purely MBG scaffold, suggesting the promoting effects of GO (Wang et al. 2019b). Further, MBG-GO enhanced the in vivo bone repair at the defect site in a rat cranial defect model. The new bone was fully integrated both in the periphery and center of the scaffold. These two studies suggest the potential use of GBM-containing bioactive glasses for bone tissue scaffolds (Fig. 4).

CaP ceramics are applied as a bone graft because of their similar chemical composition with natural bone minerals. Kinds of CaP subfamily exist such as tri-calcium phosphate (TCP), bi-calcium phosphate (BCP), and hydroxyapatite (HA). Within the CaP subfamily, HA has a hexagonal structure, and a stoichiometric Ca/P ratio of 1.67, which is identical to bone apatite (Kalita et al. 2007; Teixeira et al. 2009; Guo et al. 2003). Moreover, HA is the thermodynamically stable CaP subfamily under the physiological temperature, pH, and composition of body fluid. Therefore, many studies have utilized HA and GBMs in bone tissue engineering substrate scaffolds by their synergistic osteogenesis-inducing effects. Peng et al. fabricated PEEK-HA/GO scaffolds for bone tissue engineering application (Peng et al. 2017). The PEEK-HAP/GO scaffolds induced the formation of bone-like apatite by supporting MG-63 human osteoblast-like cellular adhesion, proliferation, as well as osteogenic differentiation. More importantly, in vivo bone defect repair experiments showed that new bone formed throughout the scaffolds at 60 days after implantation. Besides, Prakash et al. fabricated a chitosan film containing GO/HA/

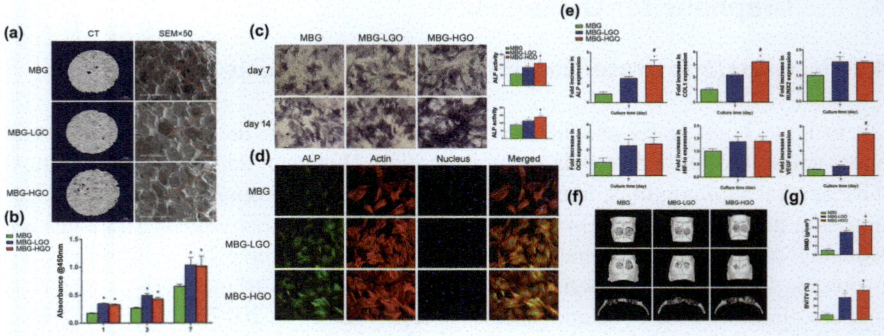

Fig. 4 (**a**) MBG/GO scaffold for bone regeneration. Computed tomography (CT) and scanning electron microscopy (SEM) of pristine MBG, MBG containing a low density of GO (MBG-LGO), and MBG containing a high density of GO (MBG-HGO). (**b**) Cell proliferation, (**c**) ALP staining, (**d**) immunocytochemical analysis of bMSCs cultured on each sample. (**e**) Osteogenic gene expression in vitro. (**f**) Indicates the results of the in vivo bone-repairing model (front, back, and micro-CT). (**g**) Quantitative analysis of bone mineral density (BMD) and bone volume relative to tissue volume (BV/TV) (Reproduced from Wang et al. 2019b, copyright open accessed)

Au (Prakash et al. 2020). Au was incorporated due to the exceptional antiseptic and antibacterial activity which helps to prevent bacterial growth in the surgical wound. The GO/HA/Au nanocomposite incorporated Chitosan-PVA film showed antibacterial effects against pathogenic microorganisms and led to improved cell proliferation and osteogenic differentiation of mouse mesenchymal cells C3H10T1/2. Thus, it can be suggested biofunctional substrates incorporating GBMs and HA are highly biocompatible and potentially used for bone regeneration applications. To evaluate the synergistic effect of HA and GBMs, Lee et al. examined whether nanocomposites of rGO and HA (rGO/HAp NCs) could enhance the osteogenesis of MC3T3-E1 preosteoblasts and promote new bone formation (Lee et al. 2015a). The results indicated that rGO/HA NCs promoted spontaneous osteodifferentiation of MC3T3-E1 osteoblasts which was confirmed by a significant increase of ALP activity, calcium deposition, and expression level of osteopontin and osteocalcin. Furthermore, in vivo implanted rGO/HA NCs enhanced new bone formation in full-thickness calvarial defects in mice without inflammatory responses. These results could be explained that the cells were exposed to rGO/HA NCs and subsequent effectiveness of interaction between cells and those NCs was increased, then, leading to promoted intracellular signaling.

On the other hand, electrospinning is a versatile technique that produces nanofiber matrices easily and continuously. The advantage of electrospun nanofiber matrices for tissue engineering scaffolds is that several biomolecules and GBMs can be easily combined to the fiber and can endow seeded cells nano-topographical signals that lead to enhanced cellular morphology and behaviors. For osteogenesis, many studies have utilized nanofiber matrices containing GBMs. Aidun et al. introduced the GO-containing PCL/chitosan/collagen nanofiber matrices for enhanced cellular behaviors of MG-63 osteoblast-like cells (Aidun et al. 2019).

The incorporation of GO increased the electrical conductivity of electrospinning solution; therefore, fiber diameter was reduced, and porosity and mechanical properties of fabricated nanofibers were enhanced. The resulting high porous fiber makes the scaffold have a more favorable microenvironment for cells to attach and grow in. Moreover, the hydrophilicity and bioactivity of the surface were improved by the incorporation of GO; hence the cell attachment, proliferation, and osteogenesis activity has been significantly increased. Similarly, another report is suggesting the GO-containing nanofiber matrices can enhance the biomineralization and osteogenic differentiation of MSCs. Liu et al. fabricated GO-incorporated cellulose acetate (CA) nanofiber matrices for enhancement of biomineralization and osteogenic differentiation of human mesenchymal stem cells (hMSCs) (Liu et al. 2017). Higher formation of calcium phosphate nanocrystals due to biomineralization of GO was observed, indicating the potential to provide a more biomimetic environment to induce the higher expression level of osteogenic factor such as ALP, and thereby resulting in enhanced osteogenic differentiation of hMSCs. Furthermore, the osteogenesis-promoting effects of GO-containing nanofiber matrices were evaluated for in vivo bone repair in rats. Zhou et al. fabricated poly(3-hydroxybutyrate-co-4-hydroxybutyrate)/graphene oxide (P34HB/GO) nanofiber matrices and tested the efficacy for bone regeneration in rats (Zhou et al. 2017a). New bone formation in the calvarial defects of rats was significantly higher (19.11% more) in the P34HB/GO group than in the P34HB group, indicating the superior osteogenic capability of GO in vivo.

4.2 3D Bulk/Porous Structure/Composite for Osteogenesis

Chitosan, as mentioned above, has several biological advantages as a tissue engineering scaffold and is also often used for bone tissue scaffold. The studies conducted by Ruan et al. developed biocompatible bone scaffold through the covalent crosslinking of GO and CMC (Ruan et al. 2016). The steric hindrance effect derived from GO enhanced the water retention of the scaffold, and modulus and hardness were significantly increased by incorporation of GO. In vitro studies indicated GO-CMC upregulated osteogenesis-related genes, including osteopontin, bone sialoprotein, osterix, osteocalcin, and alkaline phosphatase. On the other hand, Dinescu et al. fabricated a 3D GO-chitosan scaffold with high porous and interconnected microstructure. The addition of GO determined the formation of more ordered morphologies with higher total porosity and greater surface available for cell attachment. The prepared scaffold also proved to support in vitro hASC osteogenic differentiation and in vivo upregulated expression levels of osteogenic markers (Dinescu et al. 2019).

The connection between angiogenesis and osteogenesis is critical during the healing process of bone fractures, as a timely and coordinated angiogenic response is of vital importance for successful bone repair (Dickson et al. 1995). Blood vessels in the bone play critical roles in the biomechanical stability of implanted bone analogs because the new blood vessels bring the oxygen and nutrients to the actively generating bones and act as a route for inflammatory cells and bone precursor cells to reach the regeneration site (Hankenson et al. 2011). However, conventional bone scaffolds are not able to support sufficient angiogenesis; therefore, recent studies have explored angiogenesis-promoting scaffolds. Zhang et al. fabricated GO-Copper (CU) NCs-coated porous CaP scaffold to enhance the angiogenesis-inducing ability of scaffold (Zhang et al. 2016). The GO-Cu BCs activated Hif-1α (hypoxia-inducible factor-1α, a critical mediator in accelerating bone regeneration) in the rat bMSCs, and sequentially enhanced the expression level of vascular endothelial growth factor (VEGF) and bone morphogenetic protein 2 (BMP-2), two main regulation factors of angiogenesis and osteogenesis, via the Erk1/2 signaling pathway (Tang et al. 2010; Jiang et al. 2016). Through the promoting effects, the prepared scaffold enhanced the adhesive and osteogenic capacities of the rat bMSCs when seeded on the GO-Cu coating, and further facilitated vascularized bone regeneration in a critical-sized rat cranial bone defect model. On the other hand, the extraordinary surface-volume ratio and a large number of surface moieties make the GO as a suitable angiogenic factor carrier into the new bone. Salvianolic acid B (SB), a traditional Chinese herbal medicine, has been shown to promote spinal fusion by promoting osteogenesis and angiogenesis (Lin et al. 2019). Wang et al. introduced the silk fibroin (SF)/GO/SB scaffolds for stable and sustained SB release at the bone regeneration site (Wang et al. 2020). The results indicated that the released SB significantly promoted cell migration, osteogenesis, and angiogenesis of rat bMSCs, and moreover promoted bone regeneration through increasing the efficacy of in vivo osteogenesis and angiogenesis.

4.3 Hydrogel and 3D Printing for Osteogenesis

Amphiphilic peptides such as HA can be self-assembled by generating physical crosslinking via hydrogen bonds and electrostatic interactions with divalent ions. The derived hydrogels have promising properties due to their biocompatibility, reversibility, trigger capability, and tunability (Rivas et al. 2019). Moreover, due to the osteogenic property of HA and GBM, they are often combined as self-assembled hydrogel is for bone tissue engineering scaffolds. Yu et al. introduced GO, HA, and chitosan self-assembled 3D hydrogel with the assistance of crosslinking GNP for chitosan and reducing agent NaVC for GO simultaneously (Yu et al. 2017). Such a simultaneous crosslinking and reduction are promising strategies to produce many 3D G-polymer-based nanocomposites for tissue engineering scaffold, because the dense and oriented microstructure of the resulted hydrogel endows it with high mechanical strength, high fixing capacity of HA, and high porosity. Consequently, rat bMSC revealed enhanced cell viability and proliferation on the sample indicating the biocompatibility of GO/HA/chitosan hydrogel. On the other hand, Xie et al. fabricated G and HA self-assembled nanocomposite hydrogel that entraps colloidal HA NPs into the 3D graphene network by a self-assembled graphite-like shell formed around it (Xie et al. 2015). The resulting G-HA hydrogels exhibited suitable properties such as highly porous, strong, electrically conductive, and biocompatible, making them promising scaffolds for bone tissue engineering.

The use of 3D printing technologies in bone tissue engineering has been clinically utilized in recent years. 3D printed bone scaffolds feature the ability to directly print bone analogs with designed shape, controlled chemistry, and interconnected porosity. The biodegradability can be modulated, and printed bone scaffold has site-specific growth factor/drug delivery abilities. Zhang et al. introduced PLGA/β-TCP-based cryogenic 3D printing technology with in situ loading of GO and BMP-2-like osteogenic peptides (Zhang et al. 2019). The conceptual idea of the cryogenic method is that it allows inks in a solution state to transform into a solid-state, thus allowing stable structures without the need for a support bath (Tan et al. 2017). The incorporation of GO further improved the scaffold wettability and mechanical strength, and 3D printed scaffolds were mechanically comparable to the human cancellous bone and hierarchically porous, hence could support favorable microenvironment in vitro and in vivo. In vitro study showed that the 3D printed scaffold promoted laden rat bMSCs growth and enhanced osteogenic differentiation. The in vivo study indicated that the 3D printed constructs exhibited sustained delivery of the peptide, which consequently promote bone regeneration in a critical bone defect. To provide laden cells more favorable 3D microenvironment, Choe et al. introduced GO/alginate bioink enabling the creation of 3D constructs containing hydrogel and cells in the desired shape or pattern (Choe et al. 2019). MSCs printed with alginate/GO showed good proliferation and higher survival in vitro. The 3D printed MSCs and GO/alginate analogs demonstrated significantly enhanced osteogenic differentiation compared with those printed with MSCs and alginate, indicating the osteogenic efficacy of GO into 3D printed constructs.

5 Conclusion

GBMs are the most attractive material in the field of developing bone and cartilage regeneration owing to their extraordinary physicochemical properties that provide several clinical benefits. This review has summarized the recent progress of GBM-based functional scaffolds for enhanced osteogenesis and chondrogenesis of cells. Biocompatibility and physicochemical characteristics for consideration in the development of the GBM-based scaffolds have been discussed. From the findings described herein, mechanical property, biocompatibility, the osteogenic and chondrogenic property could be significantly enhanced through various scaffold fabrication methods and conjugation with polymers/nanomaterials/drugs. In conclusion, the results presented in this review support the promising prospect of using GBM-based scaffolds for improved bone and cartilage tissue engineering. Although GBM-based scaffolds have some limitations to be overcome by future research, we expect further developments to provide innovative results and improve its clinical potential for bone and cartilage regeneration.

Acknowledgments This research was supported by National Research Foundation of Korea (NRF) funded by the Ministry of Science (NRF-2021R1A2C2006013) and by Korea Evaluation Institute of Industrial Technology (KEIT) grant funded by the Ministry of Trade, Industry and Energy (MOTIE, Korea) (No. 20014399).

References

Aidun A, Safaei Firoozabady A, Moharrami M, Ahmadi A, Haghighipour N, Bonakdar S, Faghihi S (2019) Graphene oxide incorporated polycaprolactone/chitosan/collagen electrospun scaffold: enhanced osteogenic properties for bone tissue engineering. Artif Organs 43(10):E264–E281

Akhavan O, Ghaderi E, Akhavan A (2012) Size-dependent genotoxicity of graphene nanoplatelets in human stem cells. Biomaterials 33(32):8017–8025

Arnold AM, Holt BD, Daneshmandi L, Laurencin CT, Sydlik SA (2019) Phosphate graphene as an intrinsically osteoinductive scaffold for stem cell-driven bone regeneration. PNAS 116(11): 4855–4860

Askari E, Rasouli M, Darghiasi SF, Naghib SM, Zare Y, Rhee KY (2021) Reduced graphene oxide-grafted bovine serum albumin/bredigite nanocomposites with high mechanical properties and excellent osteogenic bioactivity for bone tissue engineering. Bio-Des Manuf 4(2):243–257

Bi L, Jung S, Day D, Neidig K, Dusevich V, Eick D, Bonewald L (2012) Evaluation of bone regeneration, angiogenesis, and hydroxyapatite conversion in critical-sized rat calvarial defects implanted with bioactive glass scaffolds. J Biomed Mater Res A 100(12):3267–3275

Bitounis D, Ali-Boucetta H, Hong BH, Min DH, Kostarelos K (2013) Prospects and challenges of graphene in biomedical applications. Adv Mater 25(16):2258–2268

Boga JC, Miguel SP, de Melo-Diogo D, Mendonça AG, Louro RO, Correia IJ (2018) In vitro characterization of 3D printed scaffolds aimed at bone tissue regeneration. Colloid Surf B 165: 207–218

Burg KJ, Porter S, Kellam JF (2000) Biomaterial developments for bone tissue engineering. Biomaterials 21(23):2347–2359

Cao L, Lu C, Wang Q, Li F (2017a) Biocompatibility and fabrication of RGO/chitosan film for cartilage tissue recovery. Environ Toxicol Pharmacol 54:199–203

Cao L, Zhang F, Wang Q, Wu X (2017b) Fabrication of chitosan/graphene oxide polymer nanofiber and its biocompatibility for cartilage tissue engineering. Mater Sci Eng C 79:697–701

Chen QZ, Thompson ID, Boccaccini AR (2006) 45S5 Bioglass®-derived glass–ceramic scaffolds for bone tissue engineering. Biomaterials 27(11):2414–2425

Chng ELK, Pumera M (2013) The toxicity of graphene oxides: dependence on the oxidative methods used. Chem Eur J 19(25):8227–8235

Choe G, Oh S, Seok JM, Park SA, Lee JY (2019) Graphene oxide/alginate composites as novel bioinks for three-dimensional mesenchymal stem cell printing and bone regeneration applications. Nanoscale 11(48):23275–23285

Dalgic AD, Alshemary AZ, Tezcaner A, Keskin D, Evis Z (2018) Silicate-doped nano-hydroxyapatite/graphene oxide composite reinforced fibrous scaffolds for bone tissue engineering. J Biomater Appl 32(10):1392–1405

Deliormanlı AM (2019) Direct write assembly of graphene/poly (ε-caprolactone) composite scaffolds and evaluation of their biological performance using mouse bone marrow mesenchymal stem cells. Appl Biochem Biotechnol 188(4):1117–1133

Deliormanlı AM, Atmaca H (2018) Biological response of osteoblastic and chondrogenic cells to graphene-containing PCL/bioactive glass bilayered scaffolds for osteochondral tissue engineering applications. Appl Biochem Biotechnol 186(4):972–989

Dickson K, Katzman S, Paiement G (1995) The importance of the blood supply in the healing of tibial fractures. Contemp Orthop 30(6):489–493

Diekman BO, Christoforou N, Willard VP, Sun H, Sanchez-Adams J, Leong KW, Guilak F (2012) Cartilage tissue engineering using differentiated and purified induced pluripotent stem cells. PNAS 109(47):19172–19177

Dimitriou R, Jones E, McGonagle D, Giannoudis PV (2011) Bone regeneration: current concepts and future directions. BMC Med 9(1):1–10

Dinescu S, Ionita M, Ignat S-R, Costache M, Hermenean A (2019) Graphene oxide enhances chitosan-based 3D scaffold properties for bone tissue engineering. Int J Mol Sci 20(20): 5077–5097

Duch MC, Budinger GS, Liang YT, Soberanes S, Urich D, Chiarella SE, Campochiaro LA, Gonzalez A, Chandel NS, Hersam MC (2011) Minimizing oxidation and stable nanoscale dispersion improves the biocompatibility of graphene in the lung. Nano Lett 11(12):5201–5207

Evlashin S, Dyakonov P, Tarkhov M, Dagesyan S, Rodionov S, Shpichka A, Kostenko M, Konev S, Sergeichev I, Timashev P (2019) Flexible polycaprolactone and polycaprolactone/graphene scaffolds for tissue engineering. Materials 12(18):2991

Farshid B, Lalwani G, Mohammadi MS, Sankaran JS, Patel S, Judex S, Simonsen J, Sitharaman B (2019) Two-dimensional graphene oxide-reinforced porous biodegradable polymeric nanocomposites for bone tissue engineering. J Biomed Mater Res A 107(6):1143–1153

Gong M, Sun J, Liu G, Wu S, Xiang Z (2021) Graphene oxide–modified 3D acellular cartilage extracellular matrix scaffold for cartilage regeneration. Mat Sci Eng C 119:111603–111617

Guilak F, Awad HA, Fermor B, Leddy HA, Gimble JM (2004) Adipose-derived adult stem cells for cartilage tissue engineering. Biorheology 41(3-4):389–399

Guo L, Huang M, Zhang X (2003) Effects of sintering temperature on structure of hydroxyapatite studied with Rietveld method. J Mater Sci Mater Med 14(9):817–822

Hankenson KD, Dishowitz M, Gray C, Schenker M (2011) Angiogenesis in bone regeneration. Injury 42(6):556–561

Holmes B, Fang X, Zarate A, Keidar M, Zhang LG (2016) Enhanced human bone marrow mesenchymal stem cell chondrogenic differentiation in electrospun constructs with carbon nanomaterials. Carbon 97:1–13

Hu X, Man Y, Li W, Li L, Xu J, Parungao R, Wang Y, Zheng S, Nie Y, Liu T (2019) 3D Bio-printing of CS/Gel/HA/Gr hybrid osteochondral scaffolds. Polymers 11(10):1601

Jaidev L, Kumar S, Chatterjee K (2017) Multi-biofunctional polymer graphene composite for bone tissue regeneration that elutes copper ions to impart angiogenic, osteogenic and bactericidal properties. Colloid Surf B 159:293–302

Jiang X, Zhang Y, Fan X, Deng X, Zhu Y, Li F (2016) The effects of hypoxia-inducible factor (HIF)-1α protein on bone regeneration during distraction osteogenesis: an animal study. Int J Oral Maxillofac Surg 45(2):267–272

Jin L, Lee JH, Jin OS, Shin YC, Kim MJ, Hong SW, Lee MH, Park J-C, Han D-W (2015) Stimulated osteogenic differentiation of human mesenchymal stem cells by reduced graphene oxide. J Nanosci Nanotechnol 15(10):7966–7970

Kalita SJ, Bhardwaj A, Bhatt HA (2007) Nanocrystalline calcium phosphate ceramics in biomedical engineering. Mater Sci Eng C 27(3):441–449

Khorshidi S, Karkhaneh A (2018) Hydrogel/fiber conductive scaffold for bone tissue engineering. J Biomed Mater Res A 106(3):718–724

Kim J, Kim Y-R, Kim Y, Lim KT, Seonwoo H, Park S, Cho S-P, Hong BH, Choung P-H, Chung TD (2013) Graphene-incorporated chitosan substrata for adhesion and differentiation of human mesenchymal stem cells. J Mater Chem B 1(7):933–938

Ko C-S, Huang J-P, Huang C-W, Chu I-M (2009) Type II collagen-chondroitin sulfate-hyaluronan scaffold cross-linked by genipin for cartilage tissue engineering. J Biosci Bioeng 107(2): 177–182

Kolanthai E, Sindu PA, Khajuria DK, Veerla SC, Kuppuswamy D, Catalani LH, Mahapatra DR (2018) Graphene oxide—a tool for the preparation of chemically crosslinking free alginate–chitosan–collagen scaffolds for bone tissue engineering. ACS Appl Mater Interfaces 10(15): 12441–12452

Kuo CK, Li W-J, Mauck RL, Tuan RS (2006) Cartilage tissue engineering: its potential and uses. Curr Opin Rheumatol 18(1):64–73

Lee JH, Shin YC, Lee S-M, Jin OS, Kang SH, Hong SW, Jeong C-M, Huh JB, Han D-W (2015a) Enhanced osteogenesis by reduced graphene oxide/hydroxyapatite nanocomposites. Sci Rep 5(1):1–13

Lee WC, Lim CH, Su C, Loh KP, Lim CT (2015b) Cell-assembled graphene biocomposite for enhanced chondrogenic differentiation. Small 11(8):963–969

Levengood SKL, Zhang M (2014) Chitosan-based scaffolds for bone tissue engineering. J Mater Chem B 2(21):3161–3184

Li Q, Williams CG, Sun DD, Wang J, Leong K, Elisseeff JH (2004) Photocrosslinkable polysaccharides based on chondroitin sulfate. J Biomed Mater Res A 68(1):28–33

Li W-J, Tuli R, Okafor C, Derfoul A, Danielson KG, Hall DJ, Tuan RS (2005a) A three-dimensional nanofibrous scaffold for cartilage tissue engineering using human mesenchymal stem cells. Biomaterials 26(6):599–609

Li Z, Ramay HR, Hauch KD, Xiao D, Zhang M (2005b) Chitosan–alginate hybrid scaffolds for bone tissue engineering. Biomaterials 26(18):3919–3928

Li J, Liu X, Crook JM, Wallace GG (2017) Development of a porous 3D graphene-PDMS scaffold for improved osseointegration. Colloids Surf B Biointerfaces 159:386–393

Liao K-H, Lin Y-S, Macosko CW, Haynes CL (2011) Cytotoxicity of graphene oxide and graphene in human erythrocytes and skin fibroblasts. ACS Appl Mater Interfaces 3(7):2607–2615

Liao J, Qu Y, Chu B, Zhang X, Qian Z (2015) Biodegradable CSMA/PECA/graphene porous hybrid scaffold for cartilage tissue engineering. Sci Rep 5:9879–9894

Liao C, Li Y, Tjong SC (2018) Graphene nanomaterials: synthesis, biocompatibility, and cytotoxicity. Int J Mol Sci 19(11):3564–3599

Ligorio C, Zhou M, Wychowaniec JK, Zhu X, Bartlam C, Miller AF, Vijayaraghavan A, Hoyland JA, Saiani A (2019) Graphene oxide containing self-assembling peptide hybrid hydrogels as a potential 3D injectable cell delivery platform for intervertebral disc repair applications. Acta Biomater 92:92–103

Lin S, Cui L, Chen G, Huang J, Yang Y, Zou K, Lai Y, Wang X, Zou L, Wu T (2019) PLGA/β-TCP composite scaffold incorporating salvianolic acid B promotes bone fusion by angiogenesis and osteogenesis in a rat spinal fusion model. Biomaterials 196:109–121

Liu J, Chen W, Zhao Z, Xu HH (2013) Reprogramming of mesenchymal stem cells derived from iPSCs seeded on biofunctionalized calcium phosphate scaffold for bone engineering. Biomaterials 34(32):7862–7872

Liu X, Shen H, Song S, Chen W, Zhang Z (2017) Accelerated biomineralization of graphene oxide–incorporated cellulose acetate nanofibrous scaffolds for mesenchymal stem cell osteogenesis. Colloids Surf B Biointerfaces 159:251–258

Liu S, Zhou C, Mou S, Li J, Zhou M, Zeng Y, Luo C, Sun J, Wang Z, Xu W (2019) Biocompatible graphene oxide–collagen composite aerogel for enhanced stiffness and in situ bone regeneration. Mater Sci Eng C 105:110137–110147

Mistry AS, Mikos AG (2005) Tissue engineering strategies for bone regeneration. In: Regenerative medicine II. Springer, pp 1–22

Nalvuran H, Elçin AE, Elçin YM (2018) Nanofibrous silk fibroin/reduced graphene oxide scaffolds for tissue engineering and cell culture applications. Int J Biol Macromol 114:77–84

Olate-Moya F, Arens L, Wilhelm M, Mateos-Timoneda MA, Engel E, Palza H (2020) Chondroinductive alginate-based hydrogels having graphene oxide for 3D printed scaffold fabrication. ACS Appl Mater Interfaces 12(4):4343–4357

Ou L, Song B, Liang H, Liu J, Feng X, Deng B, Sun T, Shao L (2016) Toxicity of graphene-family nanoparticles: a general review of the origins and mechanisms. Part Fibre Toxicol 13(1):57–80

Peng S, Feng P, Wu P, Huang W, Yang Y, Guo W, Gao C, Shuai C (2017) Graphene oxide as an interface phase between polyetheretherketone and hydroxyapatite for tissue engineering scaffolds. Sci Rep 7:46604–46617

Petite H, Viateau V, Bensaid W, Meunier A, de Pollak C, Bourguignon M, Oudina K, Sedel L, Guillemin G (2000) Tissue-engineered bone regeneration. Nat Biotechnol 18(9):959–963
Prakash J, Prema D, Venkataprasanna K, Balagangadharan K, Selvamurugan N, Venkatasubbu GD (2020) Nanocomposite chitosan film containing graphene oxide/hydroxyapatite/gold for bone tissue engineering. Int J Biol Macromol 154(1):62–71
Qiao P, Wang J, Xie Q, Li F, Dong L, Xu T (2013) Injectable calcium phosphate–alginate–chitosan microencapsulated MC3T3-E1 cell paste for bone tissue engineering in vivo. Mater Sci Eng C 33(8):4633–4639
Rezaei H, Shahrezaee M, Monfared MJ, Karkan SF, Ghafelehbash R (2021) Simvastatin-loaded graphene oxide embedded in polycaprolactone-polyurethane nanofibers for bone tissue engineering applications. J Polym Eng 41(5):375–386
Rivas M, Del Valle LJ, Alemán C, Puiggalí J (2019) Peptide self-assembly into hydrogels for biomedical applications related to hydroxyapatite. Gels 5(1):14–42
Ruan J, Wang X, Yu Z, Wang Z, Xie Q, Zhang D, Huang Y, Zhou H, Bi X, Xiao C (2016) Enhanced physiochemical and mechanical performance of chitosan-grafted graphene oxide for superior osteoinductivity. Adv Funct Mater 26(7):1085–1097
Sanchez VC, Jachak A, Hurt RH, Kane AB (2012) Biological interactions of graphene-family nanomaterials: an interdisciplinary review. Chem Res Toxicol 25(1):15–34
Saravanan S, Chawla A, Vairamani M, Sastry T, Subramanian K, Selvamurugan N (2017) Scaffolds containing chitosan, gelatin and graphene oxide for bone tissue regeneration in vitro and in vivo. Int J Biol Macromol 104:1975–1985
Satapathy MK, Manga YB, Ostrikov KK, Chiang W-H, Pandey A, Nyambat B, Chuang E-Y, Chen C-H (2019) Microplasma cross-linked graphene oxide-gelatin hydrogel for cartilage reconstructive surgery. ACS Appl Mater Interfaces 12(1):86–95
Shamekhi MA, Mirzadeh H, Mahdavi H, Rabiee A, Mohebbi-Kalhori D, Eslaminejad MB (2019) Graphene oxide containing chitosan scaffolds for cartilage tissue engineering. Int J Biol Macromol 127:396–405
Shen H, Lin H, Sun AX, Song S, Zhang Z, Dai J, Tuan RS (2018) Chondroinductive factor-free chondrogenic differentiation of human mesenchymal stem cells in graphene oxide-incorporated hydrogels. J Mater Chem B 6(6):908–917
Shen H, Lin H, Sun AX, Song S, Wang B, Yang Y, Dai J, Tuan RS (2020) Acceleration of chondrogenic differentiation of human mesenchymal stem cells by sustained growth factor release in 3D graphene oxide incorporated hydrogels. Acta Biomater 105:44–55
Shin M, Yoshimoto H, Vacanti JP (2004) In vivo bone tissue engineering using mesenchymal stem cells on a novel electrospun nanofibrous scaffold. Tissue Eng 10(1-2):33–41
Shin YC, Lee JH, Jin L, Kim MJ, Kim Y-J, Hyun JK, Jung T-G, Hong SW, Han D-W (2015a) Stimulated myoblast differentiation on graphene oxide-impregnated PLGA-collagen hybrid fibre matrices. J Nanobiotechnol 13(1):21–31
Shin YC, Lee JH, Jin OS, Kang SH, Hong SW, Kim B, Park J-C, Han D-W (2015b) Synergistic effects of reduced graphene oxide and hydroxyapatite on osteogenic differentiation of MC3T3-E1 preosteoblasts. Carbon 95:1051–1060
Shin SR, Zihlmann C, Akbari M, Assawes P, Cheung L, Zhang K, Manoharan V, Zhang YS, Yüksekkaya M, Wan K (2016) Reduced graphene oxide-gelMA hybrid hydrogels as scaffolds for cardiac tissue engineering. Small 12(27):3677–3689
Shin YC, Song S-J, Jeong SJ, Kim B, Kwon IK, Hong SW, Oh J-W, Han D-W (2018) Graphene-based nanocomposites as promising options for hard tissue regeneration. In: Cutting-edge enabling technologies for regenerative medicine. Springer, New York, pp 103–117
Suh J-KF, Matthew HW (2000) Application of chitosan-based polysaccharide biomaterials in cartilage tissue engineering: a review. Biomaterials 21(24):2589–2598
Syama S, Mohanan P (2016) Safety and biocompatibility of graphene: a new generation nanomaterial for biomedical application. Int J Biol Macromol 86:546–555
Tan Z, Parisi C, Di Silvio L, Dini D, Forte AE (2017) Cryogenic 3D printing of super soft hydrogels. Sci Rep 7(1):1–11

Tang X, Zhang Q, Shi S, Yen Y, Li X, Zhang Y, Zhou K, Le AD (2010) Bisphosphonates suppress insulin-like growth factor 1-induced angiogenesis via the HIF-1α/VEGF signaling pathways in human breast cancer cells. Int J Cancer 126(1):90–103

Tang M, Chen W, Weir MD, Thein-Han W, Xu HH (2012) Human embryonic stem cell encapsulation in alginate microbeads in macroporous calcium phosphate cement for bone tissue engineering. Acta Biomater 8(9):3436–3445

Teixeira S, Rodriguez M, Pena P, De Aza A, De Aza S, Ferraz M, Monteiro F (2009) Physical characterization of hydroxyapatite porous scaffolds for tissue engineering. Mater Sci Eng C 29(5):1510–1514

Tetteh G, Khan A, Delaine-Smith R, Reilly G, Rehman I (2014) Electrospun polyurethane/hydroxyapatite bioactive scaffolds for bone tissue engineering: the role of solvent and hydroxyapatite particles. J Mech Behav Biomed Mater 39:95–110

Toh WS, Lee EH, Cao T (2011) Potential of human embryonic stem cells in cartilage tissue engineering and regenerative medicine. Stem Cell Rev Rep 7(3):544–559

Turk M, Deliormanlı AM (2017) Electrically conductive borate-based bioactive glass scaffolds for bone tissue engineering applications. J Biomater Appl 32(1):28–39

Wang Y, Blasioli DJ, Kim H-J, Kim HS, Kaplan DL (2006) Cartilage tissue engineering with silk scaffolds and human articular chondrocytes. Biomaterials 27(25):4434–4442

Wang Q, Chu Y, He J, Shao W, Zhou Y, Qi K, Wang L, Cui S (2017) A graded graphene oxide-hydroxyapatite/silk fibroin biomimetic scaffold for bone tissue engineering. Mater Sci Eng C 80:232–242

Wang W, Junior JRP, Nalesso PRL, Musson D, Cornish J, Mendonça F, Caetano GF, Bártolo P (2019a) Engineered 3D printed poly (ε-caprolactone)/graphene scaffolds for bone tissue engineering. Mater Sci Eng C 100:759–770

Wang W, Liu Y, Yang C, Qi X, Li S, Liu C, Li X (2019b) Mesoporous bioactive glass combined with graphene oxide scaffolds for bone repair. Int J Biol Sci 15(10):2156

Wang W, Liu Y, Yang C, Jia W, Qi X, Liu C, Li X (2020) Delivery of salvianolic acid B for efficient osteogenesis and angiogenesis from silk fibroin combined with graphene oxide. ACS Biomater Sci Eng 6(6):3539–3549

Wu J, Zheng A, Liu Y, Jiao D, Zeng D, Wang X, Cao L, Jiang X (2019) Enhanced bone regeneration of the silk fibroin electrospun scaffolds through the modification of the graphene oxide functionalized by BMP-2 peptide. Int J Nanomed 14:733–751

Xie X, Hu K, Fang D, Shang L, Tran SD, Cerruti M (2015) Graphene and hydroxyapatite self-assemble into homogeneous, free standing nanocomposite hydrogels for bone tissue engineering. Nanoscale 7(17):7992–8002

Yang K, Gong H, Shi X, Wan J, Zhang Y, Liu Z (2013a) In vivo biodistribution and toxicology of functionalized nano-graphene oxide in mice after oral and intraperitoneal administration. Biomaterials 34(11):2787–2795

Yang K, Li Y, Tan X, Peng R, Liu Z (2013b) Behavior and toxicity of graphene and its functionalized derivatives in biological systems. Small 9(9-10):1492–1503

Yaylaci SU, Sen M, Bulut O, Arslan E, Guler MO, Tekinay AB (2016) Chondrogenic differentiation of mesenchymal stem cells on glycosaminoglycan-mimetic peptide nanofibers. ACS Biomater Sci Eng 2(5):871–878

Yu P, Bao R-Y, Shi X-J, Yang W, Yang M-B (2017) Self-assembled high-strength hydroxyapatite/graphene oxide/chitosan composite hydrogel for bone tissue engineering. Carbohydr Polym 155:507–515

Zhang W, Chang Q, Xu L, Li G, Yang G, Ding X, Wang X, Cui D, Jiang X (2016) Graphene oxide-copper nanocomposite-coated porous CaP scaffold for vascularized bone regeneration via activation of Hif-1α. Adv Healthc Mater 5(11):1299–1309

Zhang Y, Wang C, Fu L, Ye S, Wang M, Zhou Y (2019) Fabrication and application of novel porous scaffold in situ-loaded graphene oxide and osteogenic peptide by cryogenic 3D printing for repairing critical-sized bone defect. Molecules 24(9):1669–1688

Zhao C, Zeng Z, Qazvini NT, Yu X, Zhang R, Yan S, Shu Y, Zhu Y, Duan C, Bishop E (2018) Thermoresponsive citrate-based graphene oxide scaffold enhances bone regeneration from BMP9-stimulated adipose-derived mesenchymal stem cells. ACS Biomater Sci Eng 4(8): 2943–2955

Zhou T, Li G, Lin S, Tian T, Ma Q, Zhang Q, Shi S, Xue C, Ma W, Cai X (2017a) Electrospun poly (3-hydroxybutyrate-co-4-hydroxybutyrate)/graphene oxide scaffold: enhanced properties and promoted in vivo bone repair in rats. ACS Appl Mater Interfaces 9(49):42589–42600

Zhou X, Nowicki M, Cui H, Zhu W, Fang X, Miao S, Lee S-J, Keidar M, Zhang LG (2017b) 3D bioprinted graphene oxide-incorporated matrix for promoting chondrogenic differentiation of human bone marrow mesenchymal stem cells. Carbon 116:615–624

Zhou C, Liu S, Li J, Guo K, Yuan Q, Zhong A, Yang J, Wang J, Sun J, Wang Z (2018) Collagen functionalized with graphene oxide enhanced biomimetic mineralization and in situ bone defect repair. ACS Appl Mater Interfaces 10(50):44080–44091

Zhou K, Yu P, Shi X, Ling T, Zeng W, Chen A, Yang W, Zhou Z (2019a) Hierarchically porous hydroxyapatite hybrid scaffold incorporated with reduced graphene oxide for rapid bone ingrowth and repair. ACS Nano 13(8):9595–9606

Zhou M, Lozano N, Wychowaniec JK, Hodgkinson T, Richardson SM, Kostarelos K, Hoyland JA (2019b) Graphene oxide: a growth factor delivery carrier to enhance chondrogenic differentiation of human mesenchymal stem cells in 3D hydrogels. Acta Biomater 96:271–280

Role of Graphene Family Nanomaterials in Skin Wound Healing and Regeneration

Iruthayapandi Selestin Raja, Hee Jeong Jang, Moon Sung Kang, Ki Su Kim, Yu Suk Choi, Jong-Rok Jeon, Jong Hun Lee, and Dong-Wook Han

Abstract

Owing to astonishing properties such as the large surface area to volume ratio, mechanical stability, antimicrobial property, and collagen crosslinking, graphene family nanomaterials (GFNs) have been widely used in various biomedical applications including tissue regeneration. Many review literatures are available

Iruthayapandi Selestin Raja and Hee Jeong Jang equally contributed to this work.

I. S. Raja
BIO-IT Fusion Technology Research Institute, Pusan National University, Busan, South Korea

H. J. Jang · M. S. Kang · D.-W. Han (✉)
Department of Cogno-Mechatronics Engineering, College of Nanoscience and Nanotechnology, Pusan National University, Busan, South Korea
e-mail: nanohan@pusan.ac.kr

K. S. Kim
School of Chemical Engineering, College of Engineering, Pusan National University, Busan, South Korea

Institute of Advanced Organic Materials, Pusan National University, Busan, South Korea
e-mail: kisukim@pusan.ac.kr

Y. S. Choi
School of Human Sciences, The University of Western Australia, Crawley, WA, Australia
e-mail: yusuk.choi@uwa.edu.au

J.-R. Jeon
Department of Agricultural Chemistry and Food Science & Technology, Institute of Agriculture and Life Science, Gyeongsang National University, Jinju, South Korea
e-mail: jrjeon@gnu.ac.kr

J. H. Lee
Department of Food Science and Biotechnology, Gachon University, Seongnam, South Korea
e-mail: foodguy@gachon.ac.kr

D.-W. Han, S. W. Hong (eds.), *Multifaceted Biomedical Applications of Graphene*, Advances in Experimental Medicine and Biology 1351,
https://doi.org/10.1007/978-981-16-4923-3_5

to compile the role of GFNs in cardiac, bone, and neuronal tissue regeneration. However, the contribution of GFNs in skin wound healing and tissue regeneration was not yet discussed. In the present review, we have highlighted the properties of GFNs and their application in skin wound healing. In addition, we have included challenges and future directions of GFNs in skin tissue regeneration in the portion of conclusion and perspectives.

Keywords

Graphene family nanomaterials · Skin tissue regeneration · Normal skin wound · Diabetic wound · Infection wound

1 Introduction

Graphene (GF) is an allotrope of carbon, which is the primary starting material for producing carbon nanotubes and fullerenes (Lakshmanan and Maulik 2018). The sp^2-hybridized 2D structural arrangement of carbon atoms in graphene forms a honey-comb lattice with a high chemical reactivity at the edges than the center (Geim and Novoselov 2007; Geim 2009; Lee et al. 2018). The graphene layers intercalate in either a rhombohedral or hexagonal arrangement with weak van der Waals forces. The average distance between two carbon atoms and the interlayer distance between the graphene sheets are approximately 0.142 nm and 0.34 nm, respectively (Lakshmanan and Maulik 2018). The physical and chemical modifications transform graphene sheets into single and multilayered graphene, graphene oxide (GO), and reduced GO (rGO), which are collectively called as graphene family nanomaterials (GFNs) (Shin et al. 2013, 2016). There were reports that an increasing number of layers in graphene significantly changed the properties of the material. The graphene stacked by more than 10 layers behaves like graphite (Geim and Novoselov 2007). The methods, including chemical vapor deposition, electrochemical exfoliation, and mechanical cleavage of graphite, are conventionally used to produce graphene (Novoselov et al. 2004; Liu et al. 2008; Shang et al. 2008). Generally, multilayer graphenes or graphene oxides are widely used for biomedical applications, as the synthesis of single-layer and defect-free graphene in bulk is difficult (Goenka et al. 2014). Graphene oxide (GO) contains numerous carboxylic, epoxide, and hydroxyl groups in its plane (Kim et al. 2010a). The peripheral carboxylate groups provide the pH-dependent negative surface charge and colloidal stability, whereas hydrogen bonding and other surface reactions are possible by the epoxide and hydroxyl groups on the basal plane (Guo et al. 2011; Goenka et al. 2014). GO is considered as an amphiphilic sheet-like molecule to stabilize hydrophobic molecules with π-π interactions in a solution (Kim et al. 2010b). Reduced graphene oxide is produced when GO is subjected to chemical, thermal, or UV treatment in the presence of reducing agents like hydrazine (Park et al. 2009). It was demonstrated that the planar configuration of sp^2 hybrid orbitals in graphene provides a high specific surface area of about 2630 m^2/g, and C-C bond strength endows strong mechanical properties with Young's modulus of 1100 GPa (Shang et al. 2019).

Due to unique tunable properties, GFNs have been extensively used in electronics (Eda et al. 2008), electrochemical sensors (Shao et al. 2010), energy storage devices (Pumera 2011), drug carriers (Kakran et al. 2011), tissue regeneration scaffolds (Hong et al. 2014), and cancer treatment materials (Raja et al. 2020). The graphene nanomaterials can act as reinforcing materials to enhance their composites' mechanical strength (Sreenivasulu et al. 2018). They are useful in neutralizing free radicals and hence possess substantial antioxidant properties (Qiu et al. 2014). The planar structure of graphene helps for loading a large amount of various pharmaceutical drugs and biomolecules (Tran et al. 2015). GFNs can interact with biomolecules such as DNA, enzymes, proteins, or peptides, and they show remarkable properties in tissue regeneration (Shin et al. 2013; Lee et al. 2014). The excellent optical properties and large surface area with free π electrons make them used for biomedical imaging and drug delivery, respectively (Yang et al. 2011; Cha et al. 2014). Generally, the cellular intake of GO nanosheets (NSs) occurs through macropinocytosis; however, the route can be modulated depending on the alterations in surface chemistry and the nanocarrier's overall size (Linares et al. 2014). The internalization of graphene into the cells causes depolarization of the mitochondrial membrane and, therefore, increases intracellular reactive oxygen species (ROS) (Li et al. 2012). It was reported that GFNs generate ROS, which causes lipid peroxidation, DNA damage, caspase activation, and chromatin condensation to induce cell death via apoptosis (Pinto et al. 2013). The internalized hydrophilic and small GFNs can accumulate into cytoplasmatic, perinuclear, and nuclear portion. Meanwhile, hydrophobic GFNs were hindered from entering the cell membrane due to agglomeration in physiological medium and repulsive interactions (Sasidharan et al. 2011). When intravenously administered, GO sheets in mice cause platelet aggregation and pulmonary thromboembolism, whereas rGO had suffered less aggregation effect (Sasidharan et al. 2012). The properties of GFNs in tissue engineering and regenerative medicine are multifaceted. Some literature works revealed that GFNs could act as stem cell inducers, leading to mesenchymal stem cells (MSCs) differentiation (Lee et al. 2015a, b; Shin et al. 2015; Weaver and Cui 2015). GFNs are good candidates in neuronal (Bramini et al. 2016; Qu et al. 2017), bone (Jakus and Shah 2017; Saravanan et al. 2017), and cardiac tissue (Park et al. 2015; Bao et al. 2017) regeneration due to their outstanding mechanical strength and electrical conductivity. There are scientific reports to prove the efficacy of GFNs in skin wound healing and tissue regeneration (Frontiñán-Rubio et al. 2018; Daisy et al. 2020). In the present review, we compile the literature works which demonstrate the contribution of GFNs in skin wound healing. As shown in Fig. 1, GFNs (2D and 3D forms) exhibit excellent physicochemical properties such as a large surface area to volume, mechanical stability, and hydrophilicity. The in vitro biological studies prove that GFNs express cell attachment and proliferation, antimicrobial property, angiogenesis, and vascularization. Collagen deposition, re-epithelialization, and wound contraction are the functions of GFNs during in vivo wound healing.

The skin is a complex structure, which is mainly composed of the epidermis and dermis, that protects the underlying organs for the survival of the organism (Balañá

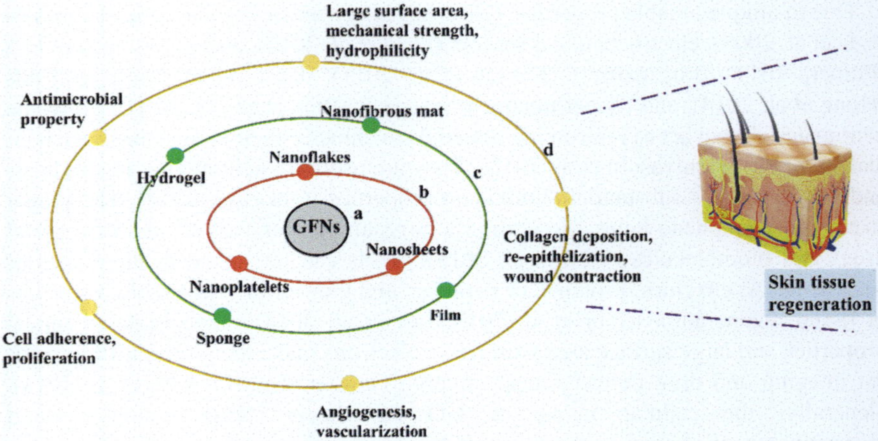

Fig. 1 (a) Graphene family nanomaterials (GFNs), including graphene, graphene oxide, and reduced graphene oxide, are prepared in different 2D forms, such as nanosheets, nanoflakes, and nanoplatelets (b) following various synthetic approaches. The 2D GFNs are embedded into 3D structures (c) viz. hydrogel, electrospun nanofibrous mat, film, and sponge to apply for skin tissue regeneration. These nanomaterials exhibit many physicochemical and biological properties, which include a large surface area, mechanical strength, antimicrobial property, cell attachment, angiogenesis, collagen deposition, and wound contraction during the in vitro and in vivo studies, which are prerequisites of skin wound healing

et al. 2015; Takeo et al. 2015). Wounds can be classified as acute and chronic ones depending on their healing potential. Acute injuries can heal without any issue unless there are any significant concerns like age, obesity, diabetes, deprived blood circulation, and stressed environmental situations. Meanwhile, chronic wounds do not heal in an organized set of normal wound healing stages and consume more than 12 weeks for healing (Anisha et al. 2013; Mohandas et al. 2015). Leg or foot ulcers are the primary concerns in diabetic patients. Diabetes declines the ability to metabolize glucose. As a result, the wound healing process is delayed with a negative long-term impact in each phase of wound healing (Arya et al. 2014; Patel et al. 2019). Diabetic wounds exhibit a persistent inflammatory phase and impede the formation of mature granulation tissue. Ischemia possibly occurred due to vascular damage and reduced wound tensile strength (Galkowska et al. 2006; Alavi et al. 2014). It was reported that diabetic wound healing was connected with the overexpression of proinflammatory cytokines, including IL-1β, IL-6, and TNF-α. Diabetic patients suffer declined production of collagen and growth factors during wound healing, due to high blood glucose (Qiu et al. 2007; Costa and Soares 2013; Tan et al. 2019). Wounds can be classified into two, i.e., external and internal, based on its origin. External origin wounds include injuries, cuts, burns, and bruises, whereas internal wounds are skin ulcers and calluses (Patel et al. 2019).

Wound healing is a dynamic process that involves the restoration of anatomic integrity, and the prime requirement for wound management is rapid and complete healing without infection. Wound healing follows a cascade of complicated cellular and biochemical processes to restore damaged tissue into its original state (Thu et al. 2012). Generally, healing a normal wound is initiated by hemostasis, which involves preventing blood loss and microbial attack. The second wound healing stage is an inflammatory phase, in which neutrophils and macrophages remove debris and pathogens along with cytokines and growth factors. Formation of new blood vessels (angiogenesis), new tissue and matrix construction occurs in the subsequent proliferative phase. At the final stage of wound healing, the remodeling phase reduces the blood supply to the wound area and increases the tensile strength of the extracellular matrix (ECM) (Komarcević 2000; Sharp and Clark 2011). An ideal scaffold should provide a platform for interacting with living cells, carrying active biomolecules, and generating proper physiological signals. As GFNs can perform these activities, they are considered as excellent candidates for tissue regeneration (Zhao et al. 2017; Shang et al. 2019).

2 Applications

2.1 Excisional or Incisional Normal Skin Wound Healing

Excisional wounds are the most common in literature reports. The size of the wound varies from 2 to 20 mm in diameter. Biopsy punches lacerate, surgical scissors crush, and lasers cauterize are generally used to generate the wound. In contrast to excisional wounds, the incisional wound is created from 10 to 15 mm in length, and the scalpel is used to create the injury (Ansell et al. 2014). The role of GFNs in skin wound healing are listed out in Table 1.

Electrospun nanofibrous scaffolds exhibit high surface area to volume ratio, tunable mechanical properties, and high porosity. Also, they possess essential topographical features of ECM to promote cell adhesion, migration, and proliferation. Hence, electrospun nanofibers bear advantages over other types of tissue scaffolds (Tamayol et al. 2013). Lu et al. prepared electrospun chitosan-PVA-graphene nanofibers with antimicrobial property for enhanced wound healing (Lu et al. 2012). It was noted that CS-PVA-graphene nanofibers had no effect on the multiplication of eukaryotic cells (yeast) but prevented the propagation of prokaryotic cells (*Escherichia coli* cells and Agrobacterium). The reason was attributed to the fact that the electron movement from graphene to living cells due to the potential of the cell membrane. The graphene electrons entering a eukaryotic cell hardly permeate the nuclear membrane and hence do not damage genetic materials, including DNA. As prokaryotic cells do not have a nuclear membrane, the genetic elements are disrupted by the influence of graphene. Reflecting on the in vitro studies, the CS-PVA + graphene group showed an improved wound healing rate compared to CS-PVA and control groups.

A three-dimensional graphene foam loaded with bone marrow-derived mesenchymal stem cells (MSCs) was developed to improve skin wound healing with reduced scarring (Li et al. 2015). The scaffold proved excellent biocompatibility

Table 1 In vivo skin tissue regeneration application of GFNs and their composites. The number of wounds in each group is denoted by n

GFNs and its dimension	3-D composite of GFNs	Wound model	Animal model	Features of GFNs	Main outcomes	References
GF, 3 layers	Electrospun nanofiber (GF + chitosan (CS) + polyvinyl alcohol (PVA))	Excision (normal)	Male C57/BL6 mice, 3 groups	Antibacterial property	CS-PVA-GF showed an accelerated wound healing compared to CS-PVA and control groups after 15 days of treatment.	Lu et al. (2012)
		Excision (normal)	van Beveren rabbits, 3 groups	Antibacterial property	After 10 days of treatment, the CS-PVA-graphene group witnessed a completer disappearance of the skin wound while the remaining groups had reduced wound area.	
GF	Foam (GF + MSCs)	Excision (normal)	Rats, 3 groups	Monolith of continuous and porous structure and water retention	On 14 days of post-surgery, GF + MSCs treated rat groups had a complete re-epithelization of the wound with a thickened dermis layer with significant difference compared to GF and control groups.	Li et al. (2015)
GO NSs	Electrospun nanofiber (GO + chitosan + PVP)	Excision (normal)	Male Sprague Dawley rats, 6 groups	High specific surface area, mechanical strength, and cells adherence	The nanofiber containing 1.5 and 2 wt% GO improved wound contraction without scar formation than control on 21 days post-operation.	Mahmoudi et al. (2017)

GO NSs	Film, CFGO (GO+ collagen + fibrin)	Excision (normal)	Male albino Wistar rats, 3 groups (n = 3)	Mechanical property	On 12th of the animal studies, CFGO accelerated wound healing process compared to other treatments such as CF and control due to increased collagen deposition and faster epithelial formation.	Deepachitra et al. (2014)
GO NSs	Electrospun nanofiber (GO+ chitosan + L-polylactic acid (PLLA))	Excision (normal)	Female Sprague-Dawley rats, 4 groups	Increasing surface roughness, hydrophilicity, and antimicrobial property	During the treatment period of 21 days, GO-coated CS/PLLA nanofibrous scaffold showed good potential in wound healing than CS/PLLA scaffold and the commercial wound plast.	Yang et al. (2020)
GO nanoplatelets	Electrospun nanofiber, (GO + polyurethane (PU) + siloxane (XSi))	Excision (normal)	Wistar rats, 3 groups	Enhancement in vascularization, re-epithelization, and collagen deposition	XSi-PU/GO5% promoted tissue regeneration with a higher quality of new skin after 20 days of treatment with significant difference to XSi-PU and control groups.	Shams et al. (2017)
GONSs (mean thickness 2 nm, lateral dimension 70–700 nm)	Electrospun nanofiber (GO + PCL + silver (Ag) + arginine (Arg))	Excision (normal)	Mice, 3 groups	Angiogenesis effect	PCL-GO/Ag/Arg nanofiber with 1.0 wt % GO/Ag/Arg increased tissue regeneration process significantly as compared with PCL/Ag and PCL alone treated groups.	Shahmoradi et al. (2018)

(continued)

Table 1 (continued)

GFNs and its dimension	3-D composite of GFNs	Wound model	Animal model	Features of GFNs	Main outcomes	References
GO	Hydrogel (GO + Ag + acrylic acid (AA) + N, N′-methylene bisacrylamide)	Excision (normal)	Male Sprague Dawley rats, 4 groups	Mechanical strength and antibacterial property	The optimized hydrogel group (Ag5G1) exhibited a larger wound healing ratio of 98% than that of Ag1G1 hydrogel (85%) and the control group (53%) after 15-day treatment.	Fan et al. (2014)
Few layered rGO NSs (hydrodynamic size 531 nm, polydispersity index 0.489)	Sponge (rGO + isabgol)	Excision (normal)	Male Wistar rats, 3 groups (n = 5)	Stimulation of collagen synthesis, collagen crosslinking, and wound contraction	On day 16 of treatment, rGO + isabgol scaffold enhanced wound healing with increased blood vessel formation and dense collagen bundles with a visible epithelial layer than isabgol or control treated groups.	Thangavel et al. (2018)
GO nanoflakes	Sponge (GO + decellularized rabbit adipose- derived ECM, genipin crosslinking)	Incision (normal)	Wistar rats, 7 groups (n = 3)	Activation of macrophages	The ECM sponges containing 20 μg/mL of GO showed betterment in wound healing than ECM sponges containing 10 or 50 μg/mL of GO after 4 weeks of treatment.	Nyambat et al. (2018)
Few layered rGO NSs (hydrodynamic size 531 nm, polydispersity index 0.489)	Sponge (rGO + isabgol)	Excision (diabetic)	Male Wistar rats, 3 groups (n = 5)	Collagen synthesis and wound contraction	The wound healing in diabetic rats was remarkably faster in rGO + isabgol treated rat groups compared to control and isabgol alone after 20 days of treatment.	Thangavel et al. (2018)

rGO (mean thickness 4.53 ± 2.31 nm; average lateral dimension 0.79 ± 0.37 μm)	Sponge (rGO + acellular dermal matrix + MSCs)	Excision (diabetic)	Male ICR mice, 4 groups. (n = 6)	Stability and mechanical strength	The ADM-rGO-MSCs composite scaffold supported robust vascularization, collagen deposition, and rapid re-epithelialization during diabetic wound healing.	Fu et al. (2019)
GO few- layered nanoflakes	GO + near-infrared (NIR) laser	Excision (infection by *S. aureus*)	Male albino mice, 3 groups, NIR: 1064 nm, 180 s	Laser mediated antifungal and antibacterial property	GO+laser group enhanced wound healing along with a high level of destruction of microbes than laser alone and control groups on the 12th day of treatment.	Shahnawaz Khan et al. (2015)

and promoted cell growth and proliferation. The GFs loaded with MSCs provided the necessary factors for enhanced vascularization resulting in a more complex ECM deposition. They facilitated wound closure with a thickened dermis layer at 14 days of post-surgery. An anti-scarring effect was evidenced in the scaffold treated rat groups by downregulation of transforming growth factors TGF-β1 and TGF-β3 and alpha-smooth muscle actin. The neo-skin formed in the GF + MSCs group was similar to normal skin with an organized dermal matrix comprising sebaceous glands.

Mahmoudi et al. prepared electrospun nanofibrous matrix from the composite of chitosan, polyvinylpyrrolidone, and graphene oxide nanosheets crosslinked by genipin with physicomechanical properties close to natural skins (Mahmoudi et al. 2017). MTT results using human skin fibroblast cells demonstrated that GO promoted cell viability with enhanced bactericidal capacity. During in vivo studies evaluating the superior acute-wound healing effect of graphene oxide nanosheets (0, 0.5, 1, 1.5, 2 wt. % GO) embedded in the nanofibrous scaffold, it was found that the scaffolds containing 1.5 and 2 wt% GO had the greatest improvement in wound healing as compared with the other scaffolds including control (sterile gauze). Pathological studies reported that the scaffold containing 1.5 wt% GO promoted thick dermis formation and complete epithelialization at 21 days of post-surgery. The wound healing rate was accelerated by 92% without scar formation after 14–21 days post-operation.

Deepachitra et al. studied collagen-fibrin-GO composite film (CFGO) as a wound dressing material through in vitro and in vivo biological measurements (Deepachitra et al. 2014). The presence of GO increased the mechanical strength of CFGO film. MTT assay results revealed that GO treated fibroblast cells (NIH 3T3) showed a significant decrease in cell viability while the composite CFGO improved the cell viability percentage. Histopathological and biochemical analyses showed that the CFGO treated wounds healed faster than control and collagen-fibrin (CF) treated wounds due to the presence of GO. On the 12th day of in vivo animal studies, it was observed that 97% of the wound was closed in the CFGO experimental groups. In comparison, control (untreated) and CF experimental groups had wound closure of up to 72% and 82%.

Yang et al. coated GO nanosheets on shell (chitosan, CS)-core (L-polylactic acid, PLLA) structured nanofibrous scaffolds to create a microenvironment for wound healing (Yang et al. 2020). The coatings of GO nanosheets significantly improved the hydrophilicity of CS/PLLA nanofibrous scaffolds. They contributed to excellent antimicrobial activity to Gram-negative *Escherichia coli (E. coli)* and Gram-positive *Staphylococcus aureus (S. aureus)*. The in vitro results demonstrated that the presence of GO in the composite influenced the proliferation of pig iliac endothelial cells. When the experimental rats were treated with GO-CS/PLLA nanofibrous scaffolds, wound healing was significantly better than that of treatment with other groups such as CS/PLLA and control.

Shams et al. evaluated polyurethane/siloxane network containing graphene oxide nanoplatelets (XSi-PU/GO) as wound dressing material (Shams et al. 2017). Sol-gel hydrolysis/condensation procedure was applied to distribute GO into the membrane.

In vivo assay of the XSi-PU/GO5% membrane revealed that the dressing enhanced wound healing involving in a complete re-epithelization, augmented vascularization, and collagen deposition on repaired tissue. The membrane maintained the moist environment around the wound site to enhance the migration of epithelial cells. Due to GO's antimicrobial property, the membrane decreased the number of pathogenic microorganisms and thereby reduced inflammatory response at the wound site.

Shahmoradi et al. prepared electrospun nanofiber of polycaprolactone (PCL)-GO/Ag/Arg nanocomposite with angiogenesis promoting ability and antibacterial nature for enhancing the wound healing process (Shahmoradi et al. 2018). The scaffold demonstrated a large potential to destroy *E. coli* and *S. aureus* bacterial species and meanwhile showed biocompatibility against L929 fibroblastic cell line. It was reported that the incorporation of GO and Arg into the nanocomposite increased the tensile strength, elongation at break, and hydrophilicity of nanocomposite. The in vitro and in vivo results demonstrated that the PCL-GO/Ag/Arg nanocomposite at the optimized concentration of 1.0 wt % GO/Ag/Arg possessed remarkable biological features and let to a significant increase in tissue regeneration process via reconstruction of a thickened epidermis layer on the wound surface.

A series of hydrogels were prepared with desired antibacterial performance and good water retaining ability by crosslinking Ag/graphene composites with acrylic acid and N, N′-methylene bisacrylamide at different mass ratios to promote the development of wound dressing (Fan et al. 2014). Glucose was used as a green reducing agent to minimize the toxicity effects triggered by the scaffold. In the management of an infected wound, the scaffold must maintain an appropriate level of moisture, avoiding wound infection. It was found that the hydrogel with the optimal Ag to graphene mass ratio of 5:1 (Ag5G1) exhibited more substantial antibacterial capacity against both *E. coli* and *S. aureus* than other hydrogels. The presence of graphene in hydrogel enhanced the tensile strength and elongation at break, and hydrophilic polyacrylic acid accounted for a higher swelling ratio. The in vivo experiments revealed that Ag5G1 hydrogel significantly increased the wound healing rate in rats. The histological studies showed that the hydrogel successfully reconstructed intact and thickened epidermis on 15 days of post healing of impaired wounds.

Thangavel et al. developed a composite of isabgol and reduced graphene oxide to treat normal and diabetic wounds (Thangavel et al. 2018). Graphene oxide was reduced to reduced graphene oxide under focused solar radiation. The results revealed that rGO supported collagen synthesis, collagen crosslinking, and wound contraction and reduced the wound re-epithelialization period significantly compared to control. Histology and immunohistochemistry analyses showed that isabgol + rGO scaffold treatment shortened the inflammation phase recruiting macrophages and thereby accelerated wound healing.

Nyambat et al. developed biodegradable, low immunogenic, and genipin-crosslinked adipose stem cell (ASC)-derived ECM sponges containing different amounts (10, 20, 50, and 100 μg/mL) of graphene oxide (GO) for skin-tissue regeneration (Nyambat et al. 2018). The presence of GO and genipin-crosslinking improved the mechanical properties of ECM sponges. SEM observation of sponges showed a highly porous microstructure with a pore size of 71.22 ± 19.52 μm in the scaffolds. It was found that either non-crosslinked or crosslinked ECM sponges with a medium amount (20 μg/mL) of GO triggered no severe toxic effects. Besides, sponges with a medium amount of GO showed significantly higher cellular mitochondrial activity compared to those sponges with a higher amount of GO (50 and 100 μg/mL). After subcutaneous implantation in a rat model for 4 weeks, the ECM sponges containing 20 μg/mL of GO proved as a suitable xenogenous skin substitute for full-thickness skin defects showing appropriate biodegradation and a lower inflammatory reaction.

2.2 Diabetic and Infection Type of Skin Wound Healing

Thangavel et al. reported that the composite of isabgol and reduced graphene oxide (Isabgol + rGO) enhanced diabetic wound healing in experimental Wistar rats (Thangavel et al. 2018). Isabgol + rGO treated diabetic wounds consumed 20 days for complete re-epithelialization, whereas diabetic control and isab treatments took 26 days and 23 days, respectively. Collagen crosslinking was higher with 30% higher shrinkage temperature in isabgol + rGO treated wound tissues compared to control. Also, the composite increased collagen synthesis from 8 to 12 days compared to control and isabgol treatments due to the influence of rGO.

Fu et al. fabricated a highly efficient local transplantation system for MSCs by incorporating reduced graphene oxide (rGO) nanoparticles with an acellular dermal matrix (ADM) for healing diabetic wounds (Fu et al. 2019). Due to the presence of rGO, the composite scaffolds (ADM-rGO) achieved strong mechanical behavior and high stability. rGO was produced by reducing GO in the presence of L-ascorbic acid. rGO imparted mechanical strength in the composite scaffold and delayed enzymatic biodegradation and helped for cell implantation and multiplication. After MSCs were cocultured, the wound healing efficacy of the ADM-rGO-MSC composite scaffold was investigated in a diabetic cutaneous wound model that was induced by streptozotocin (STZ). The conductive ADM-rGO-MSC composite scaffold promoted diabetic wound closure with robust vascularization and collagen deposition within the wound. A rapid re-epithelialization of newborn skin was also observed.

The wound infection in artificial wounds created on the dorsal surface of mice was controlled using near-infrared (NIR laser, $\lambda = 1064$ nm) photothermal responsive GO (Shahnawaz Khan et al. 2015). They suggested that this photothermal method would serve as a cheap alternative to antibiotics with higher efficiency. Generally, patients with compromised immune systems and severe burns are more susceptible to different bacterial and fungal diseases. The in vitro antimicrobial studies of GO were studied in different bacterial strains (*Pseudomonas aeruginosa* and *Staphylococcus aureus*) and fungi (*Saccharomyces cerevisiae* and *Candida utilis*). For in vivo studies, three wounds were created on each mouse's dorsal side and then infected by *S. aureus*. After the bacterial infection, the mice were treated with NIR laser for 180 s. The three wounds were assigned for the study as control, laser only, GO+laser groups. It was found that the GO+laser group destroyed *S. aureus* abundantly and helped for faster wound healing than the remaining groups.

3 Conclusions and Perspectives

The present review demonstrates the contribution of graphene family nanomaterials in skin tissue regeneration with different wound models such as a normal excisional or incisional wound, diabetic, and infection wound. Though GFNs have proved as better candidates in skin wound healing, there are some issues to be addressed. (1) GFNs have been widely applied for neuronal, bone, and cardiac tissue regeneration with remarkable physicochemical and biological properties; however, few kinds of literature are available to study its potential application in skin tissue regeneration. Of them, very few contributions of GFNs are reported to investigate their skin tissue regeneration ability in incisional, diabetic, and infection wound models. (2) Apart from nanosheets, other 2D GFNs are less documented. (2) In many studies, the particle properties, including size and lateral dimension of GFNs were not sufficient. (3) The demonstration of the in vivo clearance of GFNs from the body tissues is not adequate, restricting their use in human clinical trials. The following are proposals for the future of this field: (1) An extensive study to show the influence of GFNs in all types of skin wound model. (2) Sophisticated techniques and methodology to measure the particle dimension and morphology of GFNs and their composites in different forms such as a hydrogel, bioink, nanofiber, film, and sol. (3) A comprehensive study to investigate biodistribution, metabolism, and renal clearance of GFNs absorbed onto the wound tissue. We expect that the researchers with interdisciplinary backgrounds will develop the field of skin tissue regeneration using GFNs by contemplating the challenges and the concerning suggestions.

Acknowledgments This research was supported by National Research Foundation of Korea (NRF) funded by the Ministry of Science (NRF-2019R1A4A1024116) and by Korea Environment Industry & Technology Institute (KEITI) through project to develop eco-friendly new materials and processing technology derived from wildlife, funded by Korea Ministry of Environment (MOE) (2021003270006).

References

Alavi A, Sibbald RG, Mayer D, Goodman L, Botros M, Armstrong DG, Woo K, Boeni T, Ayello EA, Kirsner RS (2014) Diabetic foot ulcers: part I. pathophysiology and prevention. J Am Acad Dermatol 70(1):1. e1–18; quiz 19–20

Anisha BS, Biswas R, Chennazhi KP, Jayakumar R (2013) Chitosan-hyaluronic acid/nano silver composite sponges for drug resistant bacteria infected diabetic wounds. Int J Biol Macromol 62: 310–320

Ansell DM, Campbell L, Thomason HA, Brass A, Hardman MJ (2014) A statistical analysis of murine incisional and excisional acute wound models. Wound Repair Regen 22(2):281–287

Arya AK, Tripathi R, Kumar S, Tripathi K (2014) Recent advances on the association of apoptosis in chronic non healing diabetic wound. World J Diabetes 5(6):756–762

Balañá ME, Charreau HE, Leirós GJ (2015) Epidermal stem cells and skin tissue engineering in hair follicle regeneration. World J Stem Cells 7(4):711–727

Bao R, Tan B, Liang S, Zhang N, Wang W, Liu W (2017) A π-π conjugation-containing soft and conductive injectable polymer hydrogel highly efficiently rebuilds cardiac function after myocardial infarction. Biomaterials 122:63–71

Bramini M, Sacchetti S, Armirotti A, Rocchi A, Vázquez E, León Castellanos V, Bandiera T, Cesca F, Benfenati F (2016) Graphene oxide nanosheets disrupt lipid composition, Ca^{2+} homeostasis, and synaptic transmission in primary cortical neurons. ACS Nano 10 (7):7154–7171

Cha C, Shin SR, Gao X, Annabi N, Dokmeci MR, Tang XS, Khademhosseini A (2014) Controlling mechanical properties of cell-laden hydrogels by covalent incorporation of graphene oxide. Small 10(3):514–523

Costa PZ, Soares R (2013) Neovascularization in diabetes and its complications. Unraveling the angiogenic paradox. Life Sci 92(22):1037–1045

Daisy ERAC, Rajendran NK, Houreld NN, Marraiki N, Elgorban AM, Rajan M (2020) Curcumin and Gymnema sylvestre extract loaded graphene oxide-polyhydroxybutyrate-sodium alginate composite for diabetic wound regeneration. React Funct Polym 154:104671

Deepachitra R, Ramnath V, Sastry TP (2014) Graphene oxide incorporated collagen–fibrin biofilm as a wound dressing material. RSC Adv 4(107):62717–62727

Eda G, Fanchini G, Chhowalla M (2008) Large-area ultrathin films of reduced graphene oxide as a transparent and flexible electronic material. Nat Nanotechnol 3(5):270–274

Fan Z, Liu B, Wang J, Zhang S, Lin Q, Gong P, Ma L, Yang S (2014) A novel wound dressing based on Ag/graphene polymer hydrogel: effectively kill bacteria and accelerate wound healing. Adv Funct Mater 24(25):3933–3943

Frontiñán-Rubio J, Gómez MV, Martín C, González-Domínguez JM, Durán-Prado M, Vázquez E (2018) Differential effects of graphene materials on the metabolism and function of human skin cells. Nanoscale 10(24):11604–11615

Fu J, Zhang Y, Chu J, Wang X, Yan W, Zhang Q, Liu H (2019) Reduced graphene oxide incorporated acellular dermal composite scaffold enables efficient local delivery of mesenchymal stem cells for accelerating diabetic wound healing. ACS Biomater Sci Eng 5(8):4054–4066

Galkowska H, Wojewodzka U, Olszewski WL (2006) Chemokines, cytokines, and growth factors in keratinocytes and dermal endothelial cells in the margin of chronic diabetic foot ulcers. Wound Repair Regen 14(5):558–565

Geim AK (2009) Graphene: status and prospects. Science 324(5934):1530–1534

Geim AK, Novoselov KS (2007) The rise of graphene. Nat Mater 6(3):183–191

Goenka S, Sant V, Sant S (2014) Graphene-based nanomaterials for drug delivery and tissue engineering. J Control Release 173(1):75–88

Guo F, Kim F, Han TH, Shenoy VB, Huang J, Hurt RH (2011) Hydration-responsive folding and unfolding in graphene oxide liquid crystal phases. ACS Nano 5(10):8019–8025

Hong SW, Lee JH, Kang SH, Hwang EY, Hwang Y-S, Lee MH, Han D-W, Park J-C (2014) Enhanced neural cell adhesion and neurite outgrowth on graphene-based biomimetic substrates. Biomed Res Int 2014(1):212149

Jakus AE, Shah RN (2017) Multi and mixed 3D-printing of graphene-hydroxyapatite hybrid materials for complex tissue engineering. J Biomed Mater Res A 105(1):274–283

Kakran M, Sahoo NG, Bao H, Pan Y, Li L (2011) Functionalized graphene oxide as nanocarrier for loading and delivery of ellagic acid. Curr Med Chem 18(29):4503–4512

Kim F, Cote LJ, Huang J (2010a) Graphene oxide: surface activity and two-dimensional assembly. Adv Mater 22(17):1954–1958

Kim J, Cote LJ, Kim F, Yuan W, Shull KR, Huang J (2010b) Graphene oxide sheets at interfaces. J Am Chem Soc 132(23):8180–8186

Komarcević A (2000) The modern approach to wound treatment. Med Pregl 53(7–8):363–368

Lakshmanan R, Maulik N (2018) Graphene-based drug delivery systems in tissue engineering and nanomedicine. Can J Physiol Pharmacol 96(9):869–878

Lee EJ, Lee JH, Shin YC, Hwang D-G, Kim JS, Jin OS, Jin L, Hong SW, Han D-W (2014) Graphene oxide-decorated PLGA/collagen hybrid fiber sheets for application to tissue engineering scaffolds. Biomater Res 18(1):18–24

Lee JH, Shin YC, Jin OS, Kang SH, Hwang Y-S, Park J-C, Hong SW, Han D-W (2015a) Reduced graphene oxide-coated hydroxyapatite composites stimulate spontaneous osteogenic differentiation of human mesenchymal stem cells. Nanoscale 7(27):11642–11651

Lee JH, Shin YC, Lee S-M, Jin OS, Kang SH, Hong SW, Jeong C-M, Huh JB, Han D-W (2015b) Enhanced osteogenesis by reduced graphene oxide/hydroxyapatite nanocomposites. Sci Rep 5 (1):18833

Lee H, Paeng K, Kim IS (2018) A review of doping modulation in graphene. Synth Met 244:36–47

Li Y, Liu Y, Fu Y, Wei T, Le Guyader L, Gao G, Liu RS, Chang YZ, Chen C (2012) The triggering of apoptosis in macrophages by pristine graphene through the MAPK and TGF-beta signaling pathways. Biomaterials 33(2):402–411

Li Z, Wang H, Yang B, Sun Y, Huo R (2015) Three-dimensional graphene foams loaded with bone marrow derived mesenchymal stem cells promote skin wound healing with reduced scarring. Mater Sci Eng C Mater Biol Appl 57:181–188

Linares J, Matesanz MC, Vila M, Feito MJ, Gonçalves G, Vallet-Regí M, Marques PA, Portolés MT (2014) Endocytic mechanisms of graphene oxide nanosheets in osteoblasts, hepatocytes and macrophages. ACS Appl Mater Interfaces 6(16):13697–13706

Liu N, Luo F, Wu H, Liu Y, Zhang C, Chen J (2008) One-step ionic-liquid-assisted electrochemical synthesis of ionic-liquid-functionalized graphene sheets directly from graphite. Adv Funct Mater 18(10):1518–1525

Lu B, Li T, Zhao H, Li X, Gao C, Zhang S, Xie E (2012) Graphene-based composite materials beneficial to wound healing. Nanoscale 4(9):2978–2982

Mahmoudi N, Eslahi N, Mehdipour A, Mohammadi M, Akbari M, Samadikuchaksaraei A, Simchi A (2017) Temporary skin grafts based on hybrid graphene oxide-natural biopolymer nanofibers as effective wound healing substitutes: pre-clinical and pathological studies in animal models. J Mater Sci Mater Med 28(5):73

Mohandas A, Anisha BS, Chennazhi KP, Jayakumar R (2015) Chitosan-hyaluronic acid/VEGF loaded fibrin nanoparticles composite sponges for enhancing angiogenesis in wounds. Colloids Surf B Biointerfaces 127:105–113

Novoselov KS, Geim AK, Morozov SV, Jiang D, Zhang Y, Dubonos SV, Grigorieva IV, Firsov AA (2004) Electric field effect in atomically thin carbon films. Science 306(5696):666–669

Nyambat B, Chen C-H, Wong P-C, Chiang C-W, Satapathy MK, Chuang E-Y (2018) Genipin-crosslinked adipose stem cell derived extracellular matrix-nano graphene oxide composite sponge for skin tissue engineering. J Mater Chem B 6(6):979–990

Park S, An J, Jung I, Piner RD, An SJ, Li X, Velamakanni A, Ruoff RS (2009) Colloidal suspensions of highly reduced graphene oxide in a wide variety of organic solvents. Nano Lett 9(4):1593–1597

Park J, Kim YS, Ryu S, Kang WS, Park S, Han J, Jeong HC, Hong BH, Ahn Y, Kim B-S (2015) Graphene potentiates the myocardial repair efficacy of mesenchymal stem cells by stimulating the expression of angiogenic growth factors and gap junction protein. Adv Funct Mater 25 (17):2590–2600

Patel S, Srivastava S, Singh MR, Singh D (2019) Mechanistic insight into diabetic wounds: pathogenesis, molecular targets and treatment strategies to pace wound healing. Biomed Pharmacother 112:108615

Pinto AM, Gonçalves IC, Magalhães FD (2013) Graphene-based materials biocompatibility: a review. Colloids Surf B Biointerfaces 111C:188–202

Pumera M (2011) Graphene-based nanomaterials for energy storage. Energy Environ Sci 4 (3):668–674

Qiu Z, Kwon AH, Kamiyama Y (2007) Effects of plasma fibronectin on the healing of full-thickness skin wounds in streptozotocin-induced diabetic rats. J Surg Res 138(1):64–70

Qiu Y, Wang Z, Owens ACE, Kulaots I, Chen Y, Kane AB, Hurt RH (2014) Antioxidant chemistry of graphene-based materials and its role in oxidation protection technology. Nanoscale 6 (20):11744–11755
Qu M, Li Y, Wu Q, Xia Y, Wang D (2017) Neuronal ERK signaling in response to graphene oxide in nematode Caenorhabditis elegans. Nanotoxicology 11(4):520–533
Raja IS, Kang MS, Kim KS, Jung YJ, Han D-W (2020) Two-dimensional theranostic nanomaterials in cancer treatment: state of the art and perspectives. Cancers 12(6):1657
Saravanan S, Chawla A, Vairamani M, Sastry TP, Subramanian KS, Selvamurugan N (2017) Scaffolds containing chitosan, gelatin and graphene oxide for bone tissue regeneration in vitro and in vivo. Int J Biol Macromol 104(Pt B):1975–1985
Sasidharan A, Panchakarla LS, Chandran P, Menon D, Nair S, Rao CNR, Koyakutty M (2011) Differential nano-bio interactions and toxicity effects of pristine versus functionalized graphene. Nanoscale 3(6):2461–2464
Sasidharan A, Panchakarla LS, Sadanandan AR, Ashokan A, Chandran P, Girish CM, Menon D, Nair SV, Rao CN, Koyakutty M (2012) Hemocompatibility and macrophage response of pristine and functionalized graphene. Small 8(8):1251–1263
Shahmoradi S, Golzar H, Hashemi M, Mansouri V, Omidi M, Yazdian F, Yadegari A, Tayebi L (2018) Optimizing the nanostructure of graphene oxide/silver/arginine for effective wound healing. Nanotechnology 29(47):475101
Shahnawaz Khan M, Abdelhamid HN, Wu H-F (2015) Near infrared (NIR) laser mediated surface activation of graphene oxide nanoflakes for efficient antibacterial, antifungal and wound healing treatment. Colloids Surf B Biointerfaces 127:281–291
Shams E, Yeganeh H, Naderi-Manesh H, Gharibi R, Mohammad Hassan Z (2017) Polyurethane/siloxane membranes containing graphene oxide nanoplatelets as antimicrobial wound dressings: in vitro and in vivo evaluations. J Mater Sci Mater Med 28(5):75
Shang NG, Papakonstantinou P, McMullan M, Chu M, Stamboulis A, Potenza A, Dhesi SS, Marchetto H (2008) Catalyst-free efficient growth, orientation and biosensing properties of multilayer graphene nanoflake films with sharp edge planes. Adv Funct Mater 18 (21):3506–3514
Shang L, Qi Y, Lu H, Pei H, Li Y, Qu L, Wu Z, Zhang W (2019) 7 – Graphene and graphene oxide for tissue engineering and regeneration. In: Cui W, Zhao X (eds) Theranostic bionanomaterials. Elsevier, Amsterdam, pp 165–185
Shao Y, Wang J, Wu H, Liu J, Aksay IA, Lin Y (2010) Graphene based electrochemical sensors and biosensors: a review. Electroanalysis 22(10):1027–1036
Sharp A, Clark J (2011) Diabetes and its effects on wound healing. Nurs Stand 25(45):41–47
Shin SR, Aghaei-Ghareh-Bolagh B, Dang TT, Topkaya SN, Gao X, Yang SY, Jung SM, Oh JH, Dokmeci MR, Tang X, Khademhosseini A (2013) Cell-laden microengineered and mechanically tunable hybrid hydrogels of gelatin and graphene oxide. Adv Mater 25(44):6385–6391
Shin YC, Lee JH, Jin L, Kim MJ, Kim Y-J, Hyun JK, Jung T-G, Hong SW, Han D-W (2015) Stimulated myoblast differentiation on graphene oxide-impregnated PLGA-collagen hybrid fibre matrices. J Nanobiotechnol 13(1):21
Shin SR, Li Y-C, Jang HL, Khoshakhlagh P, Akbari M, Nasajpour A, Zhang YS, Tamayol A, Khademhosseini A (2016) Graphene-based materials for tissue engineering. Adv Drug Deliv Rev 105(Pt B):255–274
Sreenivasulu B, Ramji BR, Nagaral M (2018) A review on graphene reinforced polymer matrix composites. Mater Today Proc 5(1):2419–2428
Takeo M, Lee W, Ito M (2015) Wound healing and skin regeneration. Cold Spring Harb Perspect Med 5(1):a023267
Tamayol A, Akbari M, Annabi N, Paul A, Khademhosseini A, Juncker D (2013) Fiber-based tissue engineering: Progress, challenges, and opportunities. Biotechnol Adv 31(5):669–687
Tan WS, Arulselvan P, Ng SF, Mat Taib CN, Sarian MN, Fakurazi S (2019) Improvement of diabetic wound healing by topical application of Vicenin-2 hydrocolloid film on Sprague Dawley rats. BMC Complement Altern Med 19(1):20

Thangavel P, Kannan R, Ramachandran B, Moorthy G, Suguna L, Muthuvijayan V (2018) Development of reduced graphene oxide (rGO)-isabgol nanocomposite dressings for enhanced vascularization and accelerated wound healing in normal and diabetic rats. J Colloid Interface Sci 517:251–264

Thu HE, Zulfakar MH, Ng SF (2012) Alginate based bilayer hydrocolloid films as potential slow-release modern wound dressing. Int J Pharm 434(1–2):375–383

Tran TH, Nguyen HT, Pham TT, Choi JY, Choi HG, Yong CS, Kim JO (2015) Development of a graphene oxide nanocarrier for dual-drug chemo-phototherapy to overcome drug resistance in cancer. ACS Appl Mater Interfaces 7(51):28647–28655

Weaver CL, Cui XT (2015) Directed neural stem cell differentiation with a functionalized graphene oxide nanocomposite. Adv Healthc Mater 4(9):1408–1416

Yang X, Qiu L, Cheng C, Wu Y, Ma Z-F, Li D (2011) Ordered gelation of chemically converted graphene for next-generation electroconductive hydrogel films. Angew Chem Int Ed 50 (32):7325–7328

Yang C, Yan Z, Lian Y, Wang J, Zhang K (2020) Graphene oxide coated shell-core structured chitosan/PLLA nanofibrous scaffolds for wound dressing. J Biomaert Sci Polym Ed 31 (5):622–641

Zhao H, Ding R, Zhao X, Li Y, Qu L, Pei H, Yildirimer L, Wu Z, Zhang W (2017) Graphene-based nanomaterials for drug and/or gene delivery, bioimaging, and tissue engineering. Drug Discov Today 22(9):1302–1317

Part III

Graphene as Agents for Drug Delivery, Bioimaging, Theranostics, and Therapeutics

Graphene-Based Nanomaterials as Drug Delivery Carriers

Woo Yeup Jeong, Hye Eun Choi, and Ki Su Kim

Abstract

Graphene and graphene-based materials have been attracted in the past few years for biomedical applications due to their physicochemical and biological properties such as large surface area, chemical and mechanical stability, excellent conductivity, and good biocompatibility. Graphene-based materials not only surface modified graphene-based materials like graphene oxide (GO) or reduced graphene oxide (rGO) but also other structural forms like fullerene, carbon nanotubes, and graphite have been applied to advanced drug delivery systems. In this chapter, we review on the application of graphene-based materials in the drug delivery system with their physicochemical properties, methods for the preparation of graphene-based carriers, followed by analysis about their biodistribution and biosafety whether they are suitable as drug delivery carriers.

Keywords

Graphene · Graphene-based materials · Drug delivery system

Woo Yeup Jeong and Hye Eun Choi equally contributed to this work.

W. Y. Jeong · H. E. Choi
School of Chemical Engineering, College of Engineering, Pusan National University, Busan, South Korea

K. S. Kim (✉)
School of Chemical Engineering, College of Engineering, Pusan National University, Busan, South Korea

Institute of Advanced Organic Materials, Pusan National University, Busan, South Korea
e-mail: kisukim@pusan.ac.kr

D.-W. Han, S. W. Hong (eds.), *Multifaceted Biomedical Applications of Graphene*, Advances in Experimental Medicine and Biology 1351,
https://doi.org/10.1007/978-981-16-4923-3_6

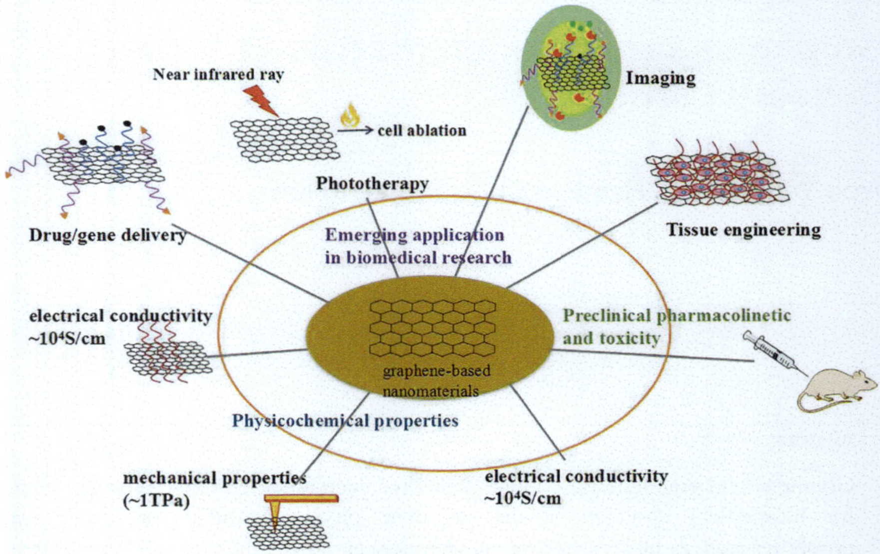

Fig. 1 The various biomedical applications of graphene-based nanomaterials (Copyright by Qu et al. 2018)

1 Introduction

Graphene, a two-dimensional (2D) material with honeycomb structure consisting of carbons and its derivatives, are classified based on either shape or number of layers, or their chemical modification. Depending on the structure forms, it is called with various names, namely spherical structure, fullerenes (0D), rolled structure with few layers, carbon nanotubes (1D), and stacked structure with multi-layers, graphite (3D). In addition, depending on oxygen content and their chemical composition, it exists in three forms, namely just exfoliated forms of graphite, graphene (G), exfoliated forms of oxidized graphite, graphene oxide (GO), and chemically reduced form of graphene oxide, reduced graphene oxide (rGO). Among them, chemically modified graphene-based materials can more offer an excellent capability to immobilize a large number of substances, including drugs, biomolecules fluorescent probes, and cells, compared to non-modified graphene because the presence of functional groups on the surface provides more possible interactions. Since these graphene-based materials are discovered, numerous researches have been carried out to confirm their properties and application in various fields from electronic devices to biomedical applications (Fig. 1).

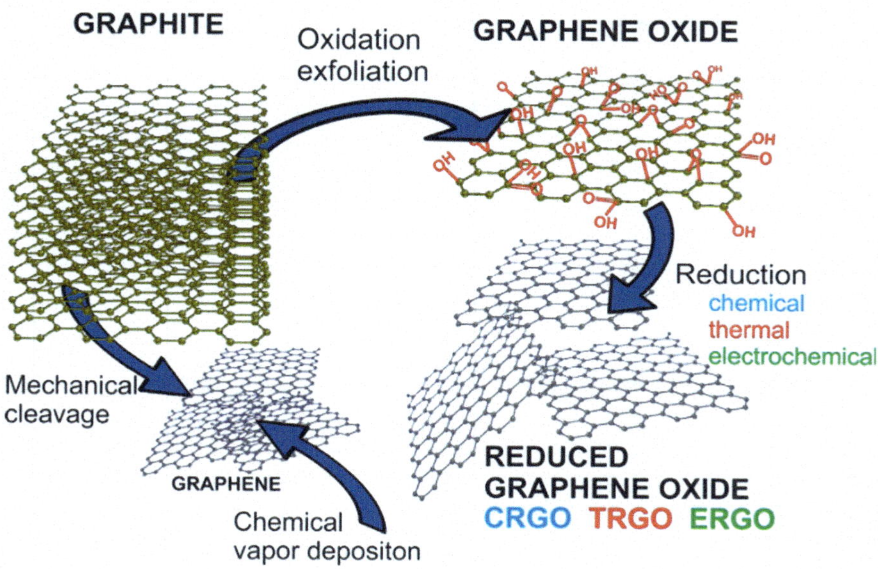

Fig. 2 A schematic illustration of the synthesis of graphene, graphene oxide (GO), and reduced graphene oxide (rGO) (Image from Neale et al. 2018 with open access under Creative Commons (Elsevier))

Graphene oxide (GO) is the chemically modified planar form of oxidized graphite that consists of a single thick layer of graphene sheets with carboxylic acid (-COOH), epoxide (-O-), and hydroxyl (-OH) groups on the surface. The carboxyl groups introduced on the surface show a pH-dependent negative surface charge. And then, the electrical repulsion generated by this negative charge gives colloidal stability (Park and Ruoff 2009). The epoxy group (-O-) and hydroxyl group (-OH) present hydrogen bonding, which is weak polar interactions and other surface interactions (Kim et al. 2010). The surface also contains free π electrons from non-oxidized areas of graphene, which are hydrophobic and capable of π–π interactions for drug loading and non-covalent functionalization. Thus, GO shows an amphiphilic, which can be used as a surfactant to stabilize hydrophobic molecules in a solution (Kudin et al. 2001). GO can be commonly synthesized with Hummer's method through oxidative exfoliation of graphite. Multilayered GO is produced by rough oxidation of graphite followed by dispersion in the aqueous medium through sonication, and by repeating this treatment, monolayer GO is produced. And GO can be also synthesized with an ultrasonication method that allows more homogeneous size comparing exfoliation method (Muthoosamy and Manickam 2017).

As shown in Fig. 2, reduced graphene oxide (rGO) can be synthesized from GO by chemically, thermally, or electrochemically reducing process (Park et al. 2009; Neale et al. 2018). In this process, various carbon-oxygen ratio and chemical compositions in rGO can be obtained by using different reducing agents. Therefore, the controllable ratio of oxygen functional group makes rGO very attractive material in biomedical applications (Pei and Cheng 2012). rGO is mainly produced to enhance the electrical conductivity and optical absorbance in GO while reducing the oxygen content, surface charge, and hydrophilicity (Bagri et al. 2010). This synthesized rGO is highly biocompatible towards normal cells and suitable for drug delivery (Muthoosamy et al. 2016).

Herein, we provide an overview of recent findings in the fields of drug delivery system concerning the surface properties, biodistribution and biosafety of graphene-based materials and preparation of graphene-based drug delivery carriers, and to highlight the promising perspectives for the possible applications of graphene-based materials.

2 Properties of Graphene-Based Nanomaterials

Graphene-based materials have been developed for use in various applications. These nanomaterials have interesting properties which include high thermal conductivity, high charge carrier mobility, large surface area, and mechanical properties, and we focus on these properties in detail in association with their biomedical applications.

2.1 Surface Properties of Graphene-Based Nanomaterials

The surface properties of graphene are related to the characteristic of the graphite. Theoretically, the graphene's two-dimensional honeycomb lattice is composed of sp^2 hybridization carbon atoms bonded together with σ bonds. The remaining ñ orbitals on each carbon atom overlaps with the three neighboring carbon atoms to form a π orbital that contributes to a delocalized network of electrons. Therefore, the electrons can freely move in a graphene structure. This is responsible for the strong electrical conductivity (ballistic electron transport with carrier mobility of 10,000 $cm^2\ V^{-1}\ S^{-1}$) of graphene (Geim and Novoselov 2007). The hardness of single layer is higher than diamond, consisted of only carbon atom like graphene, because of strong C-C bonding, bonded interaction, in the same layer, while diamond has interlayer bonding via weak Van der Waals forces, non-bonded interaction, make it soft. In addition, graphene-based materials have a large surface area that makes graphene-based materials absorb or bind more cargos (Cornelissen et al. 2013). Compared with the hydrophobic graphene consisting of only carbon, GO has both hydrophobic plane region and hydrophilic functional groups region that make it show an amphiphilic characteristic. On the hydrophobic plane region, the π–π stacking interaction of its surface can load chemical materials like aromatic materials. On the other hydrophilic functional group region, the functional groups can conjugate with other groups of other materials like polymer and drugs. Of course, though the presence of the functional groups creates high defect density resulting in reduction of π–π interaction, it can give a variety of properties through conjugation with other materials.

2.2 Photothermal Properties of Graphene-Based Nanomaterials

The photothermal effect is absorption and conversion of light into thermal energy. Graphene-based materials have this valuable property promising various applications. This photo-induced heating property is useful for applications like photothermal therapy and photothermal imaging. The near-infrared (750–1000 nm, NIR) is commonly used for photothermal effect due to its high penetration into tissue. When graphene-based materials are irradiated with NIR laser light, they can produce a temperature increasing through interaction with the light. The increased temperature variation is dependent on the concentration of graphene-based materials and on the laser intensity and duration time. Compared with other inorganic materials like gold, having the photothermal property, graphene-based materials are suitable to synthesize and modify and absorb a broad spectrum of light. Among the graphene-based materials, rGO can display a 3–4-fold higher NIR absorption than GO (Yang et al. 2012). In NIR irradiation, the GO can increase temperature to about 44 °C, while the rGO can increase to about 58 °C. However, because rGO can aggregate during synthesis, it needs to progress additional processing. Due to this problem, GO is still widely used in photothermal applications.

2.3 Other Properties of Graphene-Based Nanomaterials

The mechanical properties of graphene have been determined by several methods like numerical simulations, force displacement, force volume and nano-indentation atomic force microscopy (AFM). Strength of graphene is approximately 200 times higher than steel and it makes graphene one of the strongest materials (Luila et al. 2012). Young's modulus, Poisson's ratio, and fracture strength for graphene are 1 TPa, 0.149 GPa, and 130 GPa, respectively (Li et al. 2006). Due to its high strength, graphene has been explored for enhancing the mechanical properties of polymeric materials. Graphene reinforcement into polymethyl methacrylate (PMMA) and polyvinyl alcohol (PVA) significantly increased the modulus and hardness of these composites for biological applications (Barun et al. 2009). Interestingly, when graphene is used along with other carbon family materials, the strength, stiffness, and hardness of polymer composites increased up compared to composited prepared with individual nanomaterials alone (Chen et al. 2008). High strength and ability to tune the mechanical properties using various forms/ functionalization strategies imply potential applications of graphene as fillers or reinforcements in medical implants, hydrogels, and scaffolds used in tissue engineering.

Graphene is an electrically conductive material with high electron mobility (25 $m^2\ V^{-1}\ s^{-1}$) (Novoselov et al. 2012) and electrical conductivity (6500 Sm^{-1}) (Park et al. 2009) consisting of 2D layers of sp^2 carbon one atom thick. Graphene has been shown to greatly enhance the electrical conductivity of polymers at low filler contents (Stankovich et al. 2006). In the general fabrication of GO, the process results in disruption of the sp^2 bonding orbitals of graphene and the addition of abundant surface groups that inhibit its electrical conductivity, making GO electrically resistive (1.64×10^4 Ωm) (Wang et al. 2012). As the results of this high resistivity, researchers have explored reduction techniques of GO to form rGO. Upon reduction, the electrical conductivity of GO can be greatly improved and can be tuned over several orders of magnitude with conductivities ranging from (~0.1 Sm^{-1}) (Eda et al. 2008) to (2.98×10^4 S^{-1}) (Pei et al. 2010). Even after reduction, the rGO contains residual sp^3 bonded carbon to oxygen, which disturbs the movement of charge carriers through the rest of the sp^2 clusters. Electrical transport in rGO occurs primarily by hopping, which differs from that of mechanically exfoliated graphene (Bagri et al. 2010).

3 Synthesis and Preparation of Graphene-Based Drug Delivery Carriers

As described earlier, graphene and its derivatives can be conjugated with other aromatic materials on the surface through the π–π stacking, and the large surface area of graphene increased opportunities for multi-drug delivery to the target site from the site of administration. Unlike graphene, which is consisted of simply carbon atoms, the derivatives, GO and reduce rGO, have a large quantity of carboxyl group and hydroxyl group that can form various chemical interactions. Therefore, graphene-based materials can be exploited as drug delivery carriers in combination with other substances to demonstrate good biocompatibility and drug delivery efficiency. The graphene-based drug delivery carriers exist in three types, a drug-loaded in the only graphene-based materials, a drug-loaded in the composite of graphene-based materials and other substances, and drug-loaded in a carrier with graphene-based materials. However, the first form can only passive targeting that leads to low concentration in targeting cells because of the leakage in the activation of a drug during the prolonged circulation. So, to enhance the efficiency and safety of drug delivery carriers synthesizing with graphene-based materials, the other two forms have been studied a lot. Table 1 summarizes most of the drugs or genes that have been loaded onto graphene-based materials. As can be seen, the majority of drugs, such as doxorubicin (DOX), have been stably complexed with the surface of graphene-based materials avoiding chemical conjugation.

Generally, nanocarriers are delivered and interact with the cell membrane and uptake into the cells by endocytosis. For drug delivery into cells, it is important that the drug delivery carriers resist the endosomal degradation until loaded drug is released in cytoplasma. Functionalized graphene derivatives have been successfully applied to develop stimuli-responsive nanocarriers that release the drug in the cytosol. Polymeric modification and conjugation strategies also enhance biocompatibility and circulation times in vivo (Nurunnabi et al. 2013). For example, Fig. 3 describes the manipulation of the hydrophilic-lipophilic properties of graphene through chemical modification, which would allow interactions with biological membranes for drug delivery into the interior of a cell (Novoselov et al. 2012).

Table 1 Therapeutic molecules loaded onto graphene-based materials

Materials	Delivered Molecule	Functionlization	Target	References
G	Heparin	–	Red blood cells	Lee et al. (2011)
G	Tamoxifen	PY(+) -Chol	MDA-MB231	Misra et al. (2012)
G	Doxorubicin	F127	MCF-7 breast cancer	Hu et al. (2012)
GO	Doxorubicin	–	–	Yang et al. (2008)
GO	Doxorubicin	PEG	B-cell lymphoma	Sun et al. (2008)
GO	Camptothecin	PEG	7 breast cancer	Liu et al. (2008)
GO	Camptothecin	PEG	HCT-116 cells	Liu et al. (2008)
GO	Rituxan	PEG	CEM NK T-cell	Sun et al. (2008)
GO	Doxorubicin	FA	7 breast cancer	Zhang et al. (2010b)
GO	5-fluorouracil	Chitosan	Lymphoblastic leukemia	Rana et al. (2010)
GO	DNA	Aptamer, carboxyfluorescein	JB6 cells	Wang et al. (2010)
GO	DNA	–	HeLa cells	Lu et al. (2010)
GO	Chlorin e6	PEG	Nasopharyngeal cancer	Tian et al. (2011)
GO	Doxorubicin	PEG	EMT6 murine tumor	Zhang et al. (2011)
GO	^{125}I	PEG	Liver and spleen	Yang et al. (2011)
GO	pDNA	Chitosan		Bao et al. (2011)
GO	pDNA	PEI	HeLa cells	Feng et al. (2011)
GO	siRNA	PEG, FA	Nasopharyngeal cancer	Tao et al. (2012)
GO	DOX	PEG, PLGA, CYS, FA	MCF-7 breast cancer	Huang et al. (2018)
GO	DOX	Galactosylated chitosan	HepG2 and SMMC-7721 cells	Wang et al. (2018)
GO	DOX	Alginate	A549 cells	Xu et al. (2018)
GO	DOX	Alendronate trihydrate	MDA-MB-231 cells	Pham et al. (2019)
GO	Camptothecin	Poly (N-isopropylacrylamide)	A-5RT3 cells	Pan et al. (2011)
GO	Camptothecin	Poly(vinyl alcohol)	MDA-MB-231 cells	Sahoo et al. (2011)

(continued)

Table 1 (continued)

Materials	Delivered Molecule	Functionlization	Target	References
GO	Paclitaxel	Poly(lactide), PEG	A549 cells	Angelopoulou et al. (2015)
GO	Ellagic acid	Poloxamer 108	MCF-7, HT-29 cells	Kakran et al. (2011)
GO	Methotrexate	Gelatin	A549 cells	An et al. (2013)
GO	Cisplatin analog	PEG	4 T1-bearing mice	Li et al. (2015)
GO	DOX, PTX	HA	MDA-MB-231 cells	
rGO	β-Lapachone	Fe3O4	MCF-7 cells	Fan et al. (2013)
rGO	Doxorubicin	PEG, PEI	–	Kim et al. (2013)
rGO	Doxorubicin	FA	MDA MB 231 cells	Park et al. (2015)

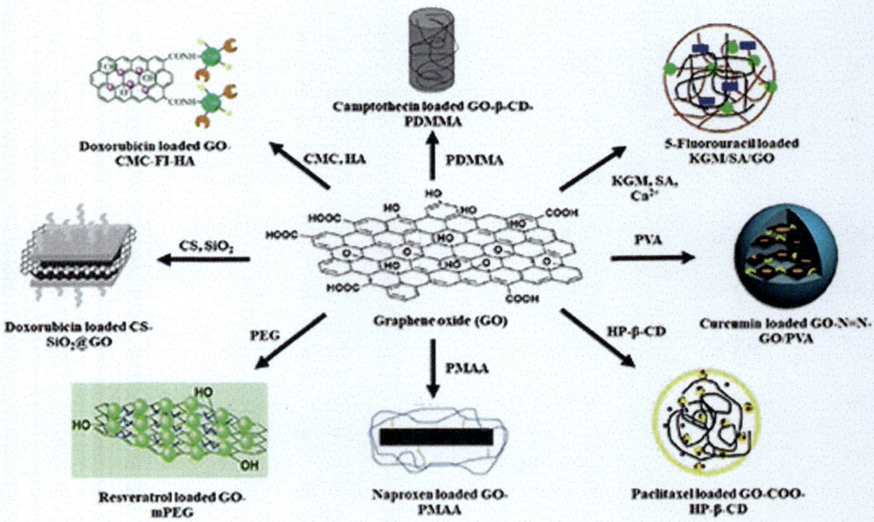

Fig. 3 The various drug delivery carriers using GO (Copyright by Ghawanmeh et al. 2019)

Nano-graphene oxide (NGO) is used as an efficient nanocarrier for the delivery of water-insoluble aromatic anticancer drugs into cells (Kim et al. 2013; Liu et al. 2008; Dembereldorj et al. 2012; Novoselov et al. 2012). Kim et al. studied near-infrared

(NIR), acidic pH and high intracellular levels glutathione (GSH) for delivery of DOX with PEG and branched polyethylenimine (bPEI)-functionalized rGO (PEG-bPEI-rGO) nanocarriers to kill the cancer cells. The nanocarriers were exposed to NIR irradiation to induce endosomal disruption and loaded DOX was released. This was further enhanced by the presence of GSH resulting in much cancer cell death than those without irradiation. In addition, DOX releasing from the PEG-GO nanocomposites was confirmed in *in vitro* cell experiments with GSH condition. Liu et al. carried out PEG-functionalized nanographene oxide (NGO) sheets loaded with camptothecin (CPT) analog, SN38. NGO-PEG-SN38 complexes exhibited good water solubility and high loading efficiency of SN38 and showed higher cytotoxicity to HCT-116 cells compared to free SN 38. After then, the same group has studied the targeted delivery of rituxan conjugated PEG-NGO (Sun et al. 2008). The π–π stacking was used to load DOX onto PEG-NGO conjugate and pH-dependent drug release was studied in *vitro.* Joo et al. reported that PEGylated GO loaded DOX via π–π interactions shows a promising real-time release of DOX from PEGylated GO at a specific tissue after an external triggering by GSH.

Chen et al. studied superparamagnetic GO–Fe_3O_4 nanohybrid that is used as carrier for targeting agents towards tumor cells, whereas the drug release is controlled by pH values. The carrier is prepared by the chemical precipitation method, followed by conjugation with folic acid (FA) onto Fe_3O_4 nanoparticles via imide linkage with amino groups of 3-aminopropyl triethoxysilane modified GO–Fe_3O_4 nanohybrid. Doxorubicin (DOX) as an anti-tumor drug model is then loaded onto the surface of the functionalized GO via π–π stacking. Furthermore, the release of DOX has been exhibiting pH dependence because of carboxylic acid groups on surface of GO. The surface-functionalized exfoliated graphene is also used as carrier to release tamoxifen citrate in a time-dependent manner to selectively enhance the death of transformed cancer cells compared to normal cells (Misra et al. 2012). Folic acid (FA) conjugated rGO showed good dispersion stability and DOX loading efficiency (Park et al. 2015). The in vivo delivery experiments indicated that DOX loaded rGO/FA could be specifically delivered to MDA MB 231 cells, with an excellent drug release efficiency.

In addition, graphene-based materials show the feasibility *in vitro* gene transfection, which is important in the treatment of various diseases caused by genetic disorders. The overall negatively charged graphene-based materials have been utilized to form electrostatic interactions with positively charged polymers such as polyethylenimine (PEI) (Feng et al. 2011). PEI was complexed at the surface of graphene-based materials and also complexed with negatively charged RNA or DNA molecules for gene transfer purposes. The PEI-graphene-nucleic acid complex has shown to offer better protection from enzymes. Silencing RNA (siRNA), specific DNA segments and plasmid DNA (pDNA), etc., are loaded in functionalized graphene-based materials for intracellular transfection into cancer cells with improved cell killing efficacy (Feng et al. 2011; Bao et al. 2011; Varghese et al. 2009). Liu et al. loaded pDNA in PEI modified GO with covalent interaction and electrostatic interaction. In this study, PEI is thus used for high-quality transfection effectiveness containing low cytotoxicity than pDNA/PEI complexes. In addition, chitosan modified GO (CS-GO) has been integrated. This complex is used for the delivery of anticancer medicine and pDNA stacked individually by electrostatic and π–π interaction (Bao et al. 2011).

Graphene-based materials have also been used as a multi-functional platform; they are not only drug delivery carriers but also tools for imaging purposes (Wang et al. 2010; Hong et al. 2012). Wang et al. studied the ability of GO to provide real-time, in situ monitoring of living cells using apatamer–carboxyfluorescein GO complex. And Hong et al. studied a graphene oxide-based tracking method that was covalently conjugated to amine-terminated PEG molecules that conjugated with Gallium 66 (^{66}Ga) labeled triacetic acid. An antibody with affinity to tumor vascular structure was also conjugated onto GO and positron emission tomography (PET) imaging was used to analyze. As the results, there was significant uptake in the liver and spleen. Graphene materials were also explored for imaging purposes which involved the non-covalent conjugation of DOX and iron oxide particles on covalently PEGylated GO sheets (Ma et al. 2012). More particularly, this work illustrated how GO could be used for theranostic applications, a carrier for DOX.

4 Biodistribution and Biosafety of Graphene-Based Nanomaterials

The graphene-based materials are used in a broad range due to mechanical, electrical, thermal, and chemical properties that have been explored for biomedical applications. However, that is why more accurate analysis of the biological and environmental hazards of graphene-based materials is needed to be understood and mitigated. The assessment of cytotoxicity is the initial step towards significantly expensive and elaborate *in vivo* studies. Extensive studies have been conducted to illuminate the preclinical biocompatibility of graphene-based nanomaterials in vitro and *in vivo*, which is critical to identify suitable candidates for clinical applications.

4.1 Biodistribution of Graphene-Based Nanomaterials

Biodistribution is a method of tracking where specific materials are delivered in animals or humans. When external materials are absorbed with the various routes such as inhalation, oral feeding, intravenous (*i.v.)* injection, etc., the materials distribute and affect all parts of the body. In this process, they can be degraded or accumulated in certain tissues. Finally, they are excreted from the body. This whole process is simply described in order of absorption, distribution, metabolism, and excretion. In the whole process, methods for identifying the materials are conjugating or loading certain substances that emit a specific wavelength and analyzing changes in the body that occurred by the materials.

Yang et al. used PEGylated graphene to study the long-term distribution in mice using radiolabeled substrate (Fig. 4). Their results demonstrated that ^{125}I loaded NGS-PEG mainly accumulate in the RES after *i.v.* administration, such as liver and

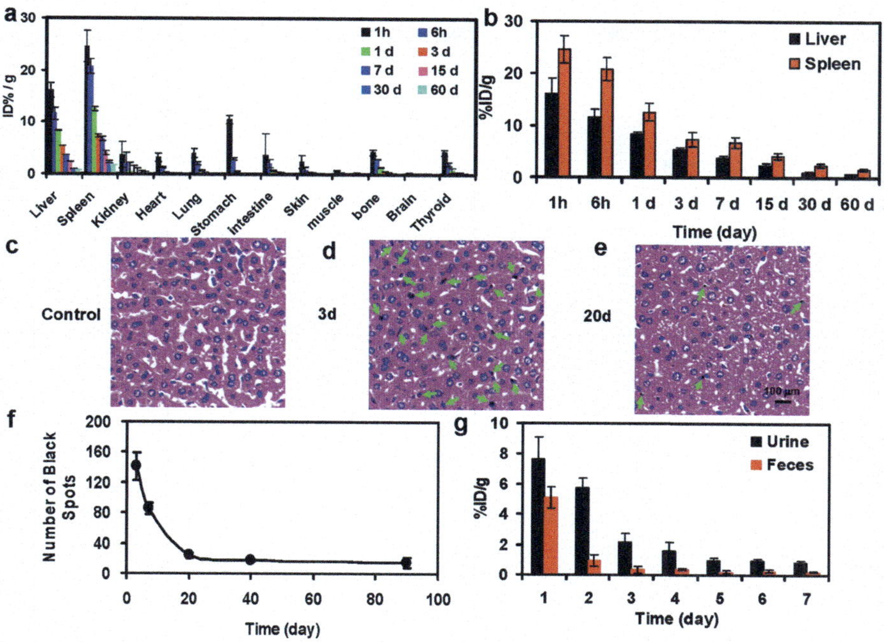

Fig. 4 Biodistribution and clearance of NGO-PEG nanoparticles over time (Copyright by Yang et al. 2011)

spleen followed by clearance through renal and fecal excretion. As a result, the distribution based on radioactivity measurements was in agreement with the immunohistochemistry (IHC) studies using hematoxylin and eosin (H&E) staining.

4.2 Biosafety of Graphene-Based Nanomaterials

The most significant factor on biosafety is cytotoxicity, which indicates how toxic the material is to the human body and some of the things that affect the cytotoxicity of materials include physical factors such as dose, morphology, and size, and chemical, the functionalization of the surface, and biological factors, the type of cells that act are studied to use graphene-based materials more safely. Zhang et al. studied the dose-dependent cytotoxicity of graphene using MTT assay compared to the results with single-walled carbon nanotubes (SWCNT). Figure 5 shows that more than 70% cell death was observed at 100 μg/ml of concentration condition of SWCNTs. Besides, no cell death was observed for 0.01–10 μg/ml concentration conditions of graphene and nearly 15–20% cell death was observed for graphene treatment at 100 μg/ml condition. These results showed the dependence of a dose as well as the morphology of the nanomaterial, with graphene exhibiting overall lower cytotoxicity compared to SWCNTs.

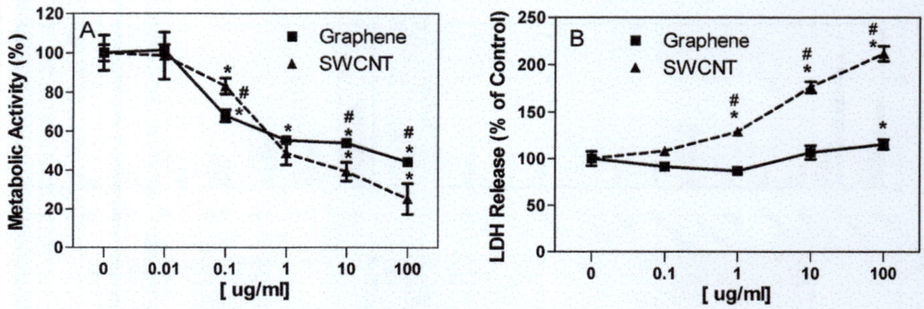

Fig. 5 Effect of Graphene or SWCNT on (**a**) mitochondrial toxicity and (**b**) LDH release (cell membrane damage marker) (Copyright by Zhang et al. 2010a)

Talukdar et al. studied the morphology dependent cytotoxicity of graphene nanostructures of various morphologies such as oxidized-graphene nanoribbons (GONRs), oxidized-graphene nanosheets (GONSs), and graphene nanospheres (GNSs) on the toxicity and stem cell differentiation potential of human mesenchymal stem cells (hMSCs). hMSCs were treated with 5–300 μg/ml of various concentrations conditions of GONRs, GONSs and GNSs for 24 or 72 h and cytotoxicity was evaluated using Alamar blue and CalceinAM assays. The results showed dose-dependent cytotoxicity of various 2D graphene nanostructures with concentrations >50 μg/ml showing no cytotoxicity. TEM imaging showed cellular and nuclear uptake of GNSs and GONSs. Das et al. studied the functionalization dependent cytotoxicity of graphene-based materials. They reported higher cytotoxicity of GO compared to reduced graphene oxide. HUVEC cells treated with GO and rGO showed a functionalization dependent cytotoxicity. Furthermore, size-dependent cytotoxicity was also observed for both GO and rGO in this study. Upon a 10-fold reduction in the size of oxidized and reduced graphene sheets, smaller graphene nanosheets showed higher toxicity compared to the larger one. Chowdhury et al. reported the cell type-dependent cytotoxicity using graphene oxide nanoribbons (GONRs) dispersed in DSPEPEG (1,2-distearoyl-sn-glycero-3-phosphoethanolamine-N-[amino(polyethylene glycol)]) in four cell lines: NIH-3T3 cells, HeLa cells, MCF7 cells, and SKBR3 cells. As a result, HeLa cells showed the lowest cell viability with 5–25% compared to other cell types with 78–100%.

5 Conclusion and Perspective

Since the development of various types of graphene-based materials, they have been widely applied in the biomedical field due to their unique structures and interesting properties such as a high surface area, high mechanical strength, easy modification, and good biocompatibility (Bitounis et al. 2013). In this chapter, we have described graphene-based materials used in specific applications, drug delivery systems, with different nanomaterials and recent studies. Graphene-based materials were often used as drug themselves, but they were mainly used as drug delivery carriers and various studies have been conducted for more efficient delivery. However, there were several limitations to being used, such as biosafety issues and limited applicable organs. First, yet graphene-based materials have dose-dependent cytotoxicity, in vivo aggregation and excretion. Thus, the biological interactions between graphene-based materials and cells, tissues, and organs should be studied in depth to understand the *in vivo* activation mechanisms. Second, most of the graphene-based materials were focused on applications in lungs, liver, or cancer cells. It is possible to apply graphene-based materials for other applications and more various organs through more studies. To solve these problems, graphene-based materials have been modified with other substances including small molecules, nanoparticles, and polymers. By forming composites with these substances, graphene-based materials can have improved biocompatibility and targeting effects, which can result in higher biosafety and delivery efficiency. As these various studies are progressed, graphene-based materials are expected to be used in wide fields in the future.

Acknowledgments This research was supported by National Research Foundation of Korea (NRF) funded by the Ministry of Science (NRF-2019R1A4A1024116) and Korea Environment Industry & Technology Institute (KEITI) through project to develop eco-friendly new materials and processing technology derived from wildlife, funded by Korea Ministry of Environment (MOE) (2021003270006).

References

An J, Gou Y, Yang C, Hu F, Wang C (2013) Synthesis of a biocompatible gelatin functionalized graphene nanosheets and its application for drug delivery. Mater Sci Eng C 33(5):2827–2837

Angelopoulou A, Voulgari E, Diamanti EK, Gournis D, Avgoustakis K (2015) Graphene oxidestabilized by PLA–PEG copolymers for the controlled delivery of paclitaxel. Eur J Pharm Biopharm 93:18–26

Bagri A, Mattevi C, Acik M, Chabal Y, Chhowalla M, Shenoy V (2010) Structural evolution during the reduction of chemically derived graphene oxide. Nat Chem 2:581–587

Bao H, Pan Y, Ping Y, Sahoo NG, Wu T, Li L, Li J, Gan LH (2011) Chitosan-functionalized grapheneoxide as a nanocarrier for drug and gene delivery. Small 7:1569–1578

Barun D, Prasad KE, Ramamurty U, Rao C (2009) Nano-indentation studies on polymer matrix composites reinforced by few-layer graphene. Nanotechnology 20:125705

Bitounis D, Boucetta H, Hong BH, Min DH, Kostarelos K (2013) Prospects and challenges of graphene in biomedical applications. Adv Mater 25:2258–2268

Chen H, Müller M, Gilmore K, Wallace G, Li D (2008) Mechanically strong, electrically conductive, and biocompatible graphene paper. Adv Mater 20:3557–3561

Chowdhury S, Lalwani G, Zhang K, Yang J, Neville K, Sitharaman B (2013) Cell specific cytotoxicity and uptake of graphene nanoribbons. Biomaterials 34:283–293

Cornelissen B, Able S, Lersemans V, Waghorn P, Myhra S, Jurkshat K, Crossley A, Vallis K (2013) Nanographene oxide-based radioimmunoconstructs for *in vivo* targeting and SPECT imaging of HER@-positive tumors. Biomaterials 34:1146–1154

Das S, Singh S, Sinh V, Joung DH, Dowding J, Reif D, Anderson J, Zhai L, Khondaker S, Self W, Seal S (2013) Oxygenated functional group density on graphene oxide: its effect on cell toxicity. Part Part Syst Charact 30:148–157

Dembereldorj U, Kim MR, Kim SM, Ganbold E, Lee SY, Joo SW (2012) A spatiotemporal anticancer drug release platform of PEGylated graphene oxide triggered by glutathione *in vitro* and *in vivo*. J Mater Chem 22:23845–23851

Eda G, Fanchini G, Chhowalla M (2008) Large-area ultrathin films of reduced graphene oxide as a transparent and flexible electronics materials. Nat Nanotechnol 3:270

Fan X, Jiao G, Zhao W, Jin P, Li X (2013) Magnetic Fe3O4–graphene composites as targeted drug nanocarriers for ph-activated release. Nanoscale 5(3):1143–1152

Feng L, Zhang S, Liu Z (2011) Graphene based gene transfection. Nanoscale 3(3):1252–1257

Geim A, Novoselov K (2007) The rise of graphene. Nat Mater 6:183–191

Ghawanmeh AA, Ali GA, Algarni H, Sarkar SM, Chong KF (2019) Graphene oxide-based hydrogels as a nanocarrier for anticancer drug delivery. Nano Res 12:973–990

Hong H, Zhang Y, Engle J, Nayak T, Theuer C, Nickles R, Barnhart T, Cai W (2012) *In vivo* targeting and positron emission tomography imaging of tumor vasculature with 66Ga-labeled nano-graphene. Biomaterials 33(16):4147–4156

Hu H, Yu J, Li Y, Zhao J, Dong H (2012) Engineering of a novel pluronic F127/graphene nanohybrid for pH responsive drug delivery. J Biomed Res A 100(1):141–148

Huang C, Wu J, Jiang W, Liu R, Li Z, Luan Y (2018) Amphiphilic prodrug-decorated graphene oxide as a multi-functional drug delivery system for efficient cancer therapy. Mater Sci Eng C 89:15–24

Kakran M, Sahoo NG, Bao H, Pan Y, Li L (2011) Functionalized graphene oxide as nanocarrier for loading and delivery of ellagic acid. Curr Med Chem 18(29):4503–4512

Kim JY, Cote L, Kim F, Yuan W, Shull K, Huang J (2010) Graphene oxide sheets at interfaces. J Am Chem Soc 132(23):8180–8186

Kim HW, Lee DH, Kim JH, Kim TI, Kim WJ (2013) Photothermally triggered cytosolic drug delivery via endosome disruption using a functionalized reduced graphene oxide. ACS Nano 7: 6735–6746

Kudin K, Scuseria G, Yakobson B (2001) C_2F, BN, and C nanoshell elasticity from *ab initio* computations. Phys Rev B 64:235406

Lee DY, Khatun Z, Lee JH, Lee YK, In I (2011) Blood compatible graphene/heparin conjugate through noncovalent chemistry. Biomacromolecules 12(2):336–341

Li J, Kudin K, McAllister MJ, Prud'homme R, Aksay I, Car R (2006) Oxigen-driven unzipping of graphitic materials. Phys Rev Lett 96:176101

Li J, Lyv Z, Li Y, Liu H, Wang J, Zhan W, Chen H, Li X (2015) A theranostic prodrug delivery system based on Pt(IV) conjugated nano-graphene oxide with synergistic effect to enhance the therapeutic efficacy of Pt drug. Biomaterials 51, microwave method. Appl Catal B-Environ 176–7:500–512

Liu Z, Robinson J, Sun X, Dai H (2008) PEGylated nanographene oxide for delivery of water-insoluble cancer drugs. J Am Chem Soc 130(33):10876–10877

Lu C, Zhu C, Li J, Liu J, Chen X, Yang H (2010) Using graphene to protect DNA from cleavage during cellular delivery. Chem Commun 46:3116–3118

Luila T, Bose S, Mishra AK, Kim NH, Lee JH (2012) Chemical functionalization of graphene and its applications. Prog Mater Sci 57:1061–1105

Ma X, Tao H, Feng L, Cheng L, Shi X, Li Y (2012) A functionalized graphene oxide-iron oxide nanocomposite for magnetically targeted drug delivery, photothermal therapy, and magnetic resonance imaging. Nano Res 5(3):199–212

Misra SK, Kondaiah P, Bhattacharya S, Rao CNR (2012) Graphene as a nanocarrier for tamoxifen induces apoptosis in transformed cancer cell lines of different origins. Small 8:131–143

Muthoosamy K, Manickam S (2017) State of the art and recent advances in the ultrasound-assisted synthesis, exfoliation and functionalization of graphene derivatives. Ultrason Sonochem 39: 478–493

Muthoosamy K, Abubakar I, Bai R, Loh HS, Manickam S (2016) Exceedingly higher co-loading of curcumin and paclitaxel onto polymer-functionalized reduced graphene oxide for highly potent synergistic anticancer treatment. Sci Rep 6:32808

Neale SJ, Randviir EP, Dena AS, Banks CE (2018) An overview of recent applications of reduced graphene oxide as a basis ofelectroanalytical sensing platforms. Appl Mater Today 10:218–226
Novoselov K, Fal'ko V, Colombo L, Gellert P, Schwab M, Kim K (2012) A roadmap for graphene. Nature 490:192–200
Nurunnabi M, Khatun Z, Huh KM, Park SY, Lee DY, Cho KJ, Lee YK (2013) *In vivo* biodistribution and toxicology of carboxylated graphene quantum dots. ACS Nano 7(8): 6858–6867
Pan Y, Bao H, Sahoo NG, Wu Y, Li L (2011) Water-soluble poly(n-isopropylacrylamide)-graphene sheets synthesized via click chemistry for drug delivery. Adv Funct Mater 21(14):2754–2763
Park S, Ruoff R (2009) Chemical methods for the production of graphenes. Nat Nanotechnol 4:217
Park SJ, Jung IH, Piner R, An SJ, Li X, Velamakanni A, Ruoff R (2009) Colloidal suspensions of highly reduced graphene oxide in a wide variety of organic solvents. Nano Lett 4:1593–1597
Park YH, Park SY, In I (2015) Direct noncovalent conjugation of folic acid on reduced graphene oxide as anticancer drug carrier. J Ind Eng Chem 30:190–196
Pei S, Cheng HM (2012) The reduction of graphene oxide. Carbon 50:3210–3228
Pei S, Zhao J, Du J, Ren W, Cheng H (2010) Direct reduction of graphene oxide films into highly conductive and flexible graphene films by hydrohalic acids. Carbon 48(15):4466–4474
Pham TT, Nguyen HT, Phung CD, Pathak S, Regmi S, Ha DH, Kim JO, Yong CS, Kim SK, Choi JE (2019) Targeted delivery of doxorubicin for the treatment of bone metastasis from breast cancer using alendronate-functionalized graphene oxide nanosheets. J Ind Eng Chem 76:310–317
Qu Y, He F, Yu C, Liang X, Liang D, Ma L, Zhang Q, Lv J, Wu J (2018) Advances on graphene-based nanomaterials for biomedical applications. Mater Sci Eng C 90:764–780
Rana V, Choi MC, Kong JY, Kim GY, Kim MJ, Kim SH, Mishra S, Singh R, Ha CS (2010) Synthesis and drug-delivery behavior of chitosan-functionalized graphene oxide hybrid nanosheets. Macromol Mater Eng 296:131–140
Sahoo NG, Bao H, Pan Y, Pal M, Kakran M, Cheng HKF, Li L, Tan LP (2011) Functionalized carbon nanomaterials as nanocarriers for loading and delivery of a poorly water-soluble anticancer drug: a comparative study. Chem Commun 47(18):5235–5237
Stankovich S, Dikin D, Dommett G, Kohlhaas K, Zimney E, Stach E, Piner R, Nguyen S, Ruoff R (2006) Graphene-based composite materials. Nature 442:282–286
Sun X, Liu Z, Welsher K, Robinson J, Goodwin A, Zaric S, Dai H (2008) Nano-graphene oxide for cellular imaging and drug delivery. Nano Res 1:203–212
Tao C, Wang J, Qin S, Long Y, Zhu H, Jiang Z (2012) Fabrication of pH-sensitive graphene oxide–drug supramolecular hydrogels as controlled release systems. J Mater Chem 22:24856–24861
Tian B, Wang C, Zhang S, Feng L, Liu Z (2011) Photothermally enhanced photodynamic therapy delivered by nano-graphene oxide. ACS Nano 5(9):7000–7009
Varghese N, Mogera U, Govindaraj A, Das A, Maiti PK, Sood AK, Rao CNR (2009) Binding of DNA nucleobases and nucleosides with graphene. ChemPhysChem 10:206–210
Wang Y, Li Z, Hu D, Lin C, Li J, Lin Y (2010) Aptamer/graphene oxide nanocomplex for in situ molecular probing in living cells. J Am Chem Soc 132(27):9274–9276
Wang Z, Nelson J, Hillborg H, Zhao S, Schadler L (2012) Graphene oxide filled nanocomposite with novel electrical and dielectric properties. Adv Mater 24:3134–3137
Wang C, Zhang Z, Chen B, Gu L, Li Y, Yu S (2018) Design and evaluation of galactosylated chitosan/graphene oxide nanoparticles as a drug delivery system. J Colloid Interface Sci 516: 332–341
Xu X, Wang J, Wang Y, Zhao L, Li Y, Liu C (2018) Formation of graphene oxide-hybridized nanogels for combinative anticancer therapy. Nanomedicine 14:2387–2395
Yang X, Zhang X, Liu Z, Ma Y, Huang Y, Chen Y (2008) High-efficiency loading and controlled release of doxorubicin hydrochloride on graphene oxide. J Phys Chem C 112(45):17554–17558
Yang K, Wan J, Zhang S, Zhang Y, Lee S, Liu Z (2011) *In vivo* pharmacokinetics, long-term biodistribution, and toxicology of PEGylated graphene in mice. ACS Nano 5(1):516–522

Yang K, Wan J, Zhang S, Tian B, Zhang Y, Liu Z (2012) The influence of surface chemistry and size of nanoscale graphene oxide on photothermal therapy of cancer using ultra-low laser power. Biomaterials 33:2206–2214

Zhang Y, Ali SF, Dervishi E, Xu Y, Li Z, Casciano D, Biris AS (2010a) Cytotoxicity effects of graphene and single-wall carbon nanotubes in neural phaeochromocytoma-derived PC12 cells. ACS Nano 4(6):3181–3186

Zhang L, Xia J, Zhao Q, Liu L, Zhang Z (2010b) Functional graphene oxide as a nanocarrier for controlled loading and targeted delivery of mixed anticancer drugs. Small 6(4):537–544

Zhang W, Guo Z, Huang D, Liu Z, Guo X, Zhong H (2011) Synergistic effect of chemo-photothermal therapy using PEGylated graphene oxide. Biomaterials 32(33):8555–8561

Graphene-Based Nanomaterials for Biomedical Imaging

So Yun Lee, Mina Kwon, Iruthayapandi Selestin Raja, Anara Molkenova, Dong-Wook Han, and Ki Su Kim

Abstract

Graphene is sp^2-hybridized carbon structure-based two-dimensional (2D) sheet. Graphene-based nanomaterials possess several features such as unique mechanical, electronic, thermal, and optical properties, high specific surface area, versatile surface functionalization, and biocompatibility, which attracted researcher's interests in various fields including biomedicine. In this chapter, we particularly focused on the biomedical imaging applications of graphene-based nanomaterials like graphene oxide (GO), reduced graphene oxide (rGO), graphene quantum dots (GQDs), graphene oxide quantum dots (GOQDs), and other derivatives, which utilize their outstanding optical properties. There are some biomedical

So Yun Lee and Mina Kwon equally contributed to this work.

S. Y. Lee · M. Kwon
School of Chemical Engineering, College of Engineering, Pusan National University, Busan, South Korea
e-mail: kmn1126111@pusan.ac.kr

I. S. Raja · A. Molkenova
BIO-IT Fusion Technology Research Institute, Pusan National University, Busan, South Korea

D.-W. Han
Department of Cogno-Mechatronics Engineering, College of Nanoscience and Nanotechnology, Pusan National University, Busan, South Korea
e-mail: nanohan@pusan.ac.kr

K. S. Kim (✉)
School of Chemical Engineering, College of Engineering, Pusan National University, Busan, South Korea

Institute of Advanced Organic Materials, Pusan National University, Busan, South Korea
e-mail: kisukim@pusan.ac.kr

D.-W. Han, S. W. Hong (eds.), *Multifaceted Biomedical Applications of Graphene*, Advances in Experimental Medicine and Biology 1351,
https://doi.org/10.1007/978-981-16-4923-3_7

imaging modalities using Graphene-based Nanomaterials, among which we will highlight fluorescence imaging, Raman imaging, magnetic resonance imaging, and photoacoustic imaging. We also discussed the brief perspectives and future application related to them.

Keywords

Graphene · Graphene oxide · Reduced graphene oxide · Graphene quantum dots · Graphene oxide quantum dots · Biomedical imaging

1 Introduction

Graphene is the sp^2 -hybridized carbon structure-based two-dimensional (2D) sheet isolated from multiple-layered graphite. It and its derivatives such as graphene oxide (GO), reduced graphene oxide (rGO), graphene quantum dots (GQDs), graphene oxide quantum dots (GOQDs), and graphene nanoribbons (GNR) show potential in the fields of physics, materials science, biology, and medicine, since its discovery in 2004 (Novoselov et al. 2004; Geim and Kim 2008; Geim 2009; Rao et al. 2009a). For example, GO, an oxidized form of graphene, is water soluble and has large surface area with rich functional groups such as hydroxyl (-OH) and carboxyl acid (-COOH) group that can react with other functional groups and exposed on its surface. Moreover, since GO also has free π electrons that can be utilized in π- π stacking, various drugs and imaging agents can be loaded on it through covalent or noncovalent interactions with high stability (Sun et al. 2008; Kuila et al. 2012; Huang et al. 2011; Zhang et al. 2010a, b; Goenka et al. 2014). In addition, because of high optical absorbance of GO, rGO, and GQDs in the near-infrared region, some researchers have focused on utilizing these graphene-based nanomaterials to selective photothermal / photodynamic applications (Robinson et al. 2011; Tian et al. 2011; Zhou et al. 2014; Cheung 2020). Beside their unique characteristics discussed above, graphene-based nanomaterials are known to have interesting properties such as ultra-high surface area, good biocompatibility, and surface functionalization (Ghuge et al. 2017; Jo et al. 2012; Galiotis et al. 2015; Zhu et al. 2014; Seo et al. 2016; Novoselov et al. 2004; Geim and Kim 2008; Geim 2009; Georgakilas et al. 2012; Georgakilas et al. 2016; Kuila et al. 2012; Zhang et al. 2010c; Rao et al. 2009b; Balandin et al. 2008; Latil and Henrard 2006). Therefore, graphene-based nanomaterials have attracted significant interest in various biomedical applications including optical/electrochemical sensors, drug delivery, bioimaging, electronic devices, theranostics, and tissue engineering (Wang et al. 2011; Novoselov et al. 2012; Shao et al. 2010; Akhavan et al. 2012; Mohanty and Berry 2008; Hu et al. 2010; Ma et al. 2011; Yang et al. 2012a; Zhang et al. 2011; Liu et al. 2008a; Heo et al. 2011; Agarwal et al. 2010; Chen et al. 2012; Park et al. 2010; Palmieri et al. 2020).

Among them, biomedical imaging is the technology of examining the structures, functions, and processes of biological systems at the cellular level, such as single-photon emission computed tomography (SPECT), positron emission tomography (PET), computed tomography (CT), ultrasound imaging (US), and magnetic resonance imaging (MRI) (Xie et al. 2011; Weissleder and Pittet 2008). The key element of the bioimaging system is the markers or reporters, which can be exactly localized and accumulated in desired site. Bioimaging provides various information, such as in vivo targeting efficacy or organ toxicity of new nanodrugs for clinical use (Lin et al. 2016; Yoo et al. 2015). Generally, organic fluorescent dyes with high quantum yield, sensitivity, selectivity, and biocompatibility have been used in bioimaging (Wang et al. 2008; VanEngelenburg and Palmer 2008; Borisov and Wolfbeis 2008; Domaille et al. 2008). However, most organic dyes have the disadvantage of having fluorescence only in a dilute solution, because the fluorescence quenching can occur if concentration is relatively high or they are in the aggregation state (Hong et al. 2011). In addition, current imaging technology is very expensive to detect diseases, because of the cost of contrast agents and imaging equipment (Larson and Ghandehari 2012). Therefore, graphene and its derivatives have received increasing attention, since they are easy to fabricate and cost effective and can provide anatomical details of living systems (Kobayashi et al. 2010). The development of graphene-based nanomaterials which are biocompatible, target specific, nontoxic, and cost effective could solve various current challenges in biomedical imaging (Close et al. 2011).

In this chapter, we will review the recent researches in the field of biomedical imaging regarding the graphene-based nanomaterials and discuss the promising perspectives for the feasible applications of graphene-based nanomaterials (Fig. 1).

2 Optical Properties of Graphene-Based Nanomaterials

2.1 Photoluminescence of Graphene-Based Nanomaterials

Pristine graphene is not photoluminescent because it has a simple band structure with zero band gap energy (Kavitha and Jaiswal 2016). However, owing to its photochemical stability, biocompatibility, and low toxicity, many studies have been conducted to achieve PL graphene by cutting graphene sheets into smaller sheets or functionalization (Pan et al. 2010; Eda et al. 2010; Chien et al. 2012; Xin et al. 2012). The small fragments of graphene sheets that confined to 3D space with a size of less than 20 nm is named graphene quantum dots (GQDs) (Pan et al. 2010; Zhuo et al. 2012; Shen et al. 2012b). GQDs are semiconductor nanoparticles and show photoluminescence owing to quantum confinement and edge effects (Pan et al. 2010; Jin et al. 2013).

In 2010, Pan et al. developed the hydrothermal routes for the preparation of functionalized GQDs with ~10 nm from graphene sheets (Pan et al. 2010). In that study, in case of oxidized graphene sheets (GSs), the absorption peak at around 230 nm that assigned to the $\pi \rightarrow \pi^*$ transition of aromatic sp^2 domains was observed (Gokus et al. 2009), while that of GQDs, in addition to the strong absorption peak at 230 nm, a new absorption peak at 320 nm was also observed. And similar to luminescent carbon nanoparticles, an excitation-dependent PL behavior of the

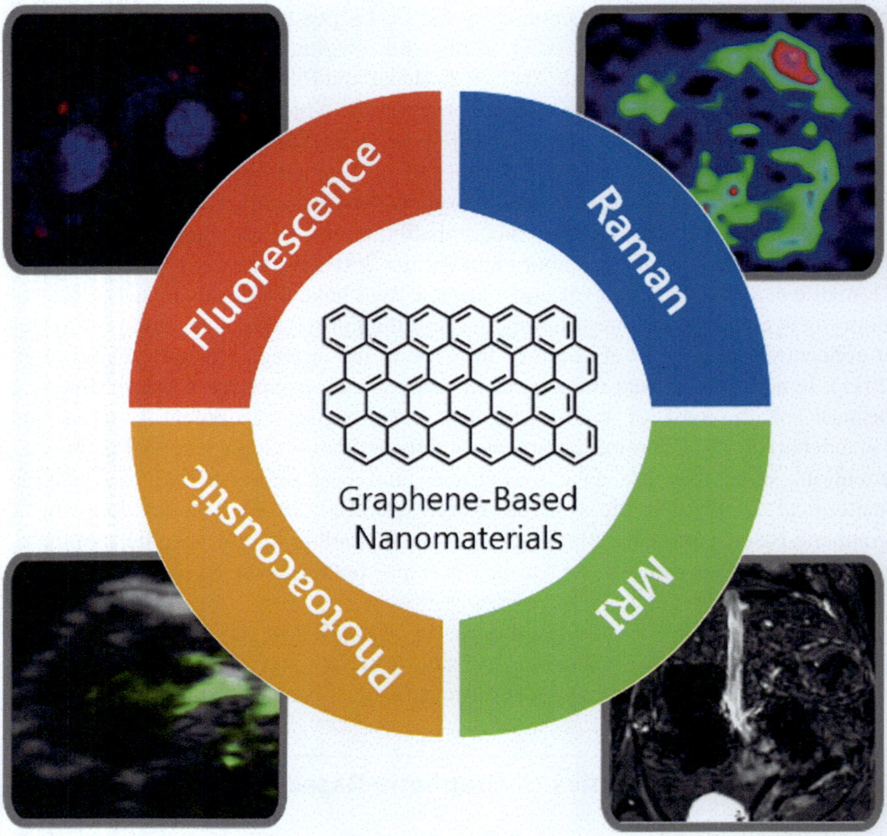

Fig. 1 The scheme represents the biomedical imaging of graphene-based nanomaterials (Reproduced from Yan et al. 2015, Zhang et al. 2016, Kanakia et al. 2015 and Sheng et al. 2013, Copyright (2013))

GQDs is accomplished in this study via changing the excitation wavelength from 320 to 420 nm. Two peaks at 257 and 320 nm are observed when the PL excitation spectrum were recorded. Furthermore, it is confirmed that the GQDs show strong blue luminescence even in neutral media, while the oxidized GSs show no photoluminescence (PL) even in alkali conditions. Therefore, the hydrothermal treatment can be considered to have significant effect on optical properties of the GQDs that exhibit strong blue photoluminescence, which was induced by the large edge effect. The mechanism of the hydrothermal cutting of oxidized GSs into GQDs is explained by the reversible structural change of GQDs depending on pH and electron transitions observed in the optical spectra.

Recently, Balaji et al. reported the one-step microwave assisted greener method for fabricating the fluorescent reduced graphene oxide quantum dots (rGOQDs) (Balaji et al. 2018). The rGOQDs solution exhibit the green fluorescent in UV-vis and it was confirmed that it was distributed mainly in the digestive system by fluorescent images of in vivo cellular uptake experiments. Besides the QDs type of graphene-base nanomaterials, graphene oxide and reduced graphene oxide also can show PL. In 2012, Chien et al. presented PL of graphene oxide (GO) depending on degree of reduction. In this work, it is confirmed that the evolution of PL was affected by reduction time and the two different types of electronically excited states are responsible for the PL features (Chien et al. 2012).

2.2 Raman Spectroscopy of Graphene-Based Nanomaterials

Graphene exhibits the characteristic peaks in the Raman spectrum which correspond to vibrational modes (Ferrari and Basko 2013; Malard et al. 2009). Figure 2 shows the Raman spectra of monolayer graphene using laser excitation at 2.41 eV (Malard et al. 2009). The D peak at ~1345 cm^{-1} that assigned to breathing mode and D' peak at ~1625 cm^{-1} could be observed in case of Raman spectrum of disordered graphene or at the edges. Therefore, the quality of edges in graphene-based nanomaterials can be evaluated using Raman spectroscopy. In 2004, CanÇado et al. first reported the on step-like graphite edges by Raman spectroscopy. In that work, the dependence of the intensity of D peak in Raman spectra on the atomic structure of edges, namely, the armchair and zigzag edges, was caused by defects (CanÇado et al. 2004). The G peak at ~1585 cm^{-1} and G' peak at ~2700 cm^{-1} are the most prominent peaks and intrinsic vibrations of defect-free graphene. The G' peak is often called 2D peak because it is located at approximately twice the frequency of the D peak and originates from a second-order process with two iTO phonons at the K point. The G peak comes from normal first-order Raman scattering process in graphene and is associated with the in-plane vibration of sp^2 hybridized carbon atoms, which is reflected by the E_{2g} symmetry phonon modes (Malard et al. 2009). Therefore, in case of the perfect graphene, the G' peak is displayed but the D peak is not (Kang et al. 2015). Because the enhanced intensity of Raman scattering is needed for the

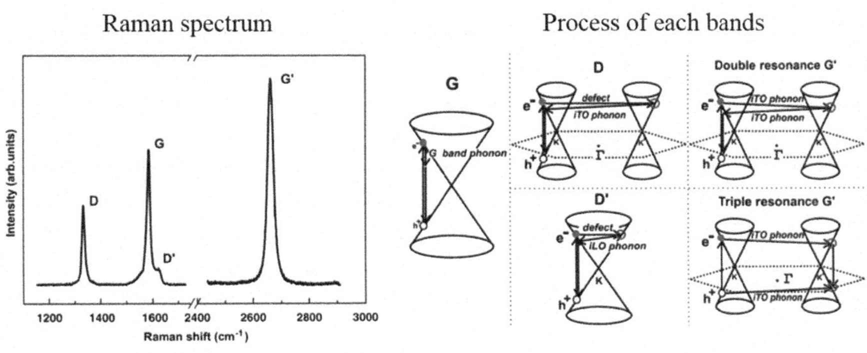

Fig. 2 Raman spectrum of a graphene edge and the process of each band (Reproduced from Malard et al. 2009)

bioimaging applications, various research using graphene-based nanomaterials with metallic nanoparticles (NPs) such as Au, Ag, Cu, and others were reported. For example, Schedin et al. in 2010 revealed that the intensity of G peak and the G' peak of graphene can be enhanced by Au NPs. The Raman enhancements of graphene are mainly caused by dipole effect of metal NPs (Schedin et al. 2010).

3 Synthesis and Preparation of Graphene-Based Nanomaterials

3.1 Graphene

Graphene can be prepared by bottom-up approach, which included chemical vapor deposition (CVD) and solvothermal synthesis, and top-down approach, which included chemical, mechanical, and physical exfoliation (Yang et al. 2013; Zhang et al. 2013; Whitener and Sheehan 2014; Yoon and Dong 2020). For instance, in 2006, Somani et al. reported the synthesis of few-layer graphene through CVD (Somani et al. 2006). After that, several groups utilized CVD to control the number of layer and the thickness on metal substrates (Cao et al. 2010; Bhaviripudi et al. 2010; Kim et al. 2009; Li et al. 2009; Chae et al. 2009; Lee et al. 2010a, b). Graphene also can be fabricated through plasma enhanced chemical vapor deposition which is a technique carried out at a relatively low temperature compared to the conventional CVD. The Wang group for the first time reported, in 2004, this technique by applying a gas mixture of CH_4 in H_2 at the power and the temperature at 680 °C for fabricating few-layer graphene sheets (Wang et al. 2004a, b).

3.2 Graphene Oxide and Reduced Graphene Oxide

Graphene oxide (GO) is monolayer of graphite oxide that can be manufactured by exfoliating the graphite oxide. The conversion of graphite into GO is attracting increasing attention as alternative of the graphene because the cost related to synthesize the graphene is expensive (Stankovich et al. 2006; Schniepp et al. 2006; Gomez-Navarro et al. 2007). Figure 3 illustrates the structure and formation of GO from graphene (upper part) and that of reduced graphene oxide (rGO) from GO (lower part) (Banerjee 2018). From the molecular structure, it can be seen that GO has several functional groups on the surface such as carbonyl and carboxyl groups on sp^2 hybridized carbon hydroxyl and epoxy groups on sp^3 hybridized carbon. Furthermore, having both aliphatic (sp^3) and aromatic (sp^2) domains can contribute to interacting with other molecules (Schniepp et al. 2006; Kuila et al. 2012; Georgakilas et al. 2016). GO is usually synthesized through the Hummers methods or its variants (Wang et al. 2012; Yang et al. 2009; Hirata et al. 2004; Olorunkosebi et al. 2021), which is basically the chemical treatment of graphite via oxidation using oxidant including $KMnO_4$, H_2SO_4, and H_3PO_4 (Hummers and Offeman 1958). Yanwu et al. described the usage of graphite salts that produced

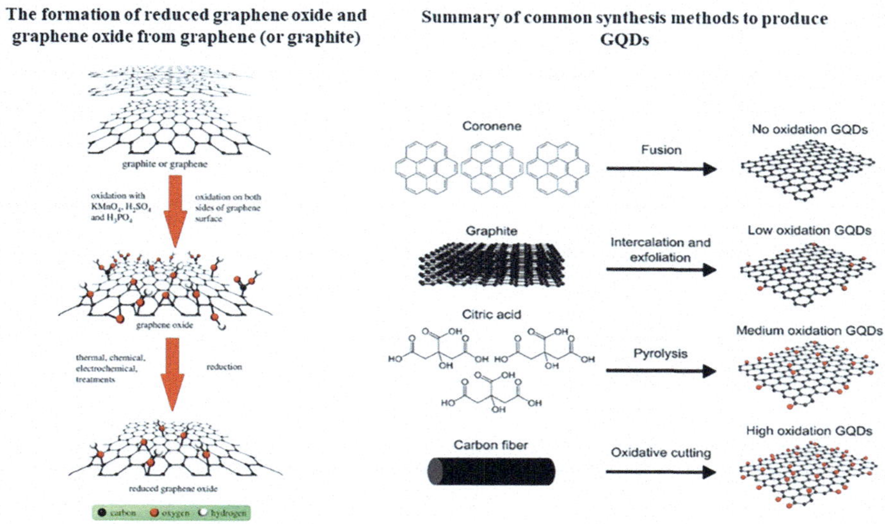

Fig. 3 Schematic of the formation of reduced graphene oxide and graphene oxide from graphene (or graphite) and summary of common synthesis methods to produce GQDs (Reproduced from Banerjee 2018and Jeong et al. 2020)

by intercalating graphite with acids like $KClO_3$, H_2SO_4 as precursors for the following oxidation to GO. Reduced graphene oxide (rGO) can be made by reducing the graphene oxide of graphite oxide chemically or thermally in the presence of reducing agents such as sodium borohydride ($NaBH_4$) (Athanasios et al. 2003; Shin et al. 2009), hydrazine (Becerril et al. 2008; Gomez-Navarro et al. 2007; Stankovich et al. 2007; Eda et al. 2008), and ascorbic acid (Dua et al. 2010). Among these, the hydrazine is considered for the best for making thin graphene sheets. Stankovich et al. reported the reduction of GO using hydrazine dispersed in water (Stankovich et al. 2007).

Recently, the various eco-friendly reducing agent and stabilizer of GO (or rGO) such as biomolecules are employed as substitutes of chemical agents to improve the biocompatibility and avoid hazardous effects (Agharkar et al. 2014; Thakur and Karak 2015; Kindalkar et al. 2021). For instance, it has been reported that bovine serum albumin (BSA) can act as surfactant while simultaneously reducing GO to produce hydrophilic rGO that can be used as an imaging agent (Sheng et al. 2013). Moreover, surface functionalization of them using biocompatible polymers and biomacromolecules like polyethylene glycol (PEG), gelatin, chitosan, DNA, and proteins has been studied for better solubility, selectivity, and biocompatibility of graphene-based nanomaterials (Zhang et al. 2013). Shi et al. reported that well-dispersible rGO in the physiological solution can be obtained by utilizing gelatin as reducing materials, and the produced gelatin-rGO can be successfully used as in vitro imaging agents (Liu et al. 2011b).

3.3 Graphene Quantum Dot and Nano-Graphene Oxide

GQDs are a small fragment of thermo-oxidatively cut GOs or other carbon precursors that have smaller size in the lower nanometer range and quantum properties. Unlike natural graphene, which is a zero band-gap semiconductor with a little overlapping valence and conductance bands, GQDs are highly heterogeneous and show band gap mediated and size tunable optical properties, so differences exist between them in terms of optoelectronic properties (Bacon et al. 2014; Geim and MacDonald 2007). Because the physical and chemical properties of GQDs are affected by the differentiation of size, edges, defects, extent of oxidation, and so on, the presence of functional groups like oxygen generated by rough, oxidative syntheses process can sometimes result in GOQDs that show different optical properties compared with GQDs.

Due to the size, edges dependent photoluminescence properties, and high quantum confinements, the synthesis of GQDs and its derivative have attracted the interest from researchers and various synthesis methods have been used to control the size and optical properties (Lerf et al. 1998). Generally, GQDs are prepared by oxidative processing of graphene oxide, an extension of the Hummers method, or by exfoliating the graphite nanoparticles (Shen et al. 2012a; Ha et al. 2015; Liu et al. 2013b). In other ways, GQDs also can be prepared by partial pyrolysis of organic materials like sugar, citric acid, etc. (Dong et al. 2011; Simpson et al. 2002; Wu et al. 2004; Tang et al. 2012; Liu et al. 2011a; Schroeder et al. 2016; Gu et al. 2020) (Fig. 3). Bacon et al. reported summary of several different synthesis pathways of GQDs (Chen et al. 2014). The GQD, synthesized through hydrothermal cutting, have an excess of oxygenated groups that affect their dispersibility and surface functionalization. Pan et al. assumed that epoxy and carboxyl groups in the oxidized graphene are easily reacted and targeted under hydrothermal treatments, observed the pH dependency of photoluminescence properties, and explained the synthesis mechanism of GQDs from graphene sheets under hydrothermal treatment (Bacon et al. 2014). Zhu et al. reported the synthesis of GQDs through solvothermal fabrication methods for biomedical imaging and further applications (Zhu et al. 2011). The resulting GQDs with an average diameter of 5.3 nm exhibit well-dispersion in polar solvents including cell culture medium. Moreover, it is confirmed that cellular uptake of GQDs was occurred without significant cytotoxicity by incubating MC3T3 cell with GQDs.

The surface functionalization of GQD can be used for enhanced optical properties. Wu et al. demonstrated fabrication of hydrophilic nitrogen-doped GQDs by bottom-up synthesis methods using L-glutamic acid as a precursor. In this work, the quantum yield of GQDs were reaching 54.5% and successfully imaging both in vitro and in vivo analysis. Besides, various GQD fabrication methods including ultrasonication and plasma assisted synthesis, nanolithography, and electrochemical exfoliation are also used (Feng et al. 2014; Luk et al. 2014; Ponomarenko et al. 2008).

4 Various Biomedical Imaging Modality Using Graphene-Based Nanomaterials

4.1 Fluorescence Bioimaging

Fluorescence imaging is a non-invasive technique using the photons emission of fluorescent probes (Baker and Baker 2010; Janib et al. 2010; Zhu et al. 2013) that should satisfy certain conditions (Zhang et al. 2002) and should be excitable with enough high quantum yield. In addition, the probe should not exhibit cytotoxicity and be appropriately resistant for preserving the original properties in biological system without photobleaching. In this aspect, graphene-based nanomaterials can be potential candidate for fluorescence imaging. Some studies reported that the quantum yield of the fluorescence imaging based on the intrinsic photoluminescence of graphene-based nanomaterials is not enough. Therefore, many studies utilized other fluorescent dyes and inorganic quantum dots to graphene-based nanomaterials for in vivo and in vitro bioimaging. Gollavelli et al. presented an eco-friendly strategy to prepare multi-functional graphene (MFG) using microwave-heated and sonication-assisted process as a fluorescent marker (Gollavelli and Ling 2012). In cellular imaging test in Hela cells, the MFG were especially localized in the intracellular region without any surface agonist, and the stability of MFG in the lysosomal region was confirmed via Lysotracker staining. And in whole-animal imaging test in zebrafish, the MFG were evenly distributed throughout the body. Yan and coworkers successfully prepared sinoporphyrin sodium (DVDMS) loaded PEGylated GO (PEG-GO/DVDMS), as photo-theranostic agents, which has improved fluorescence property. In addition, they confirmed that PEG-GO/DVDMS was enough to be an imaging marker in photodynamic therapy (PDT) (Yan et al. 2015) (Fig. 4) Through fluorescence analysis, it was confirmed that the emission peaks of PEG-GO/DVDMS were shifted to 644 and 670 nm, and the FL intensity increased 3–8-folds depending on the weight ratio between PEG-GO and DVDMS. Moreover, the PEG-GO/DVDMS showed stronger fluorescence signal in the tumor area compared to the free DVDMS, which shows a signal in the skin.

Furthermore, photoluminescent GQDs have been widely investigated for various biomedical application including in situ bioimaging due to their properties such as photostability, safety, and easy chemical modification. Nahain et al. reported rGOs and GQDs based anticancer theranostic application (Nahain et al. 2013a, b). Both of rGOs and GQDs were conjugated with hyaluronic acid for targeting CD44-overexpressed cancer cells. While HA-rGO was used as a fluorescence marker by attaching other fluorescent agents, HA-GQD with intrinsic fluorescence property could be used for theranostic application by attaching the anticancer drugs on the GQDs. In addition, Ge et al. developed several types of GQDs that absorb wide range of the UV region and the entire visible region and have strong deep-red emission peak at 680 nm (Ge et al. 2014).

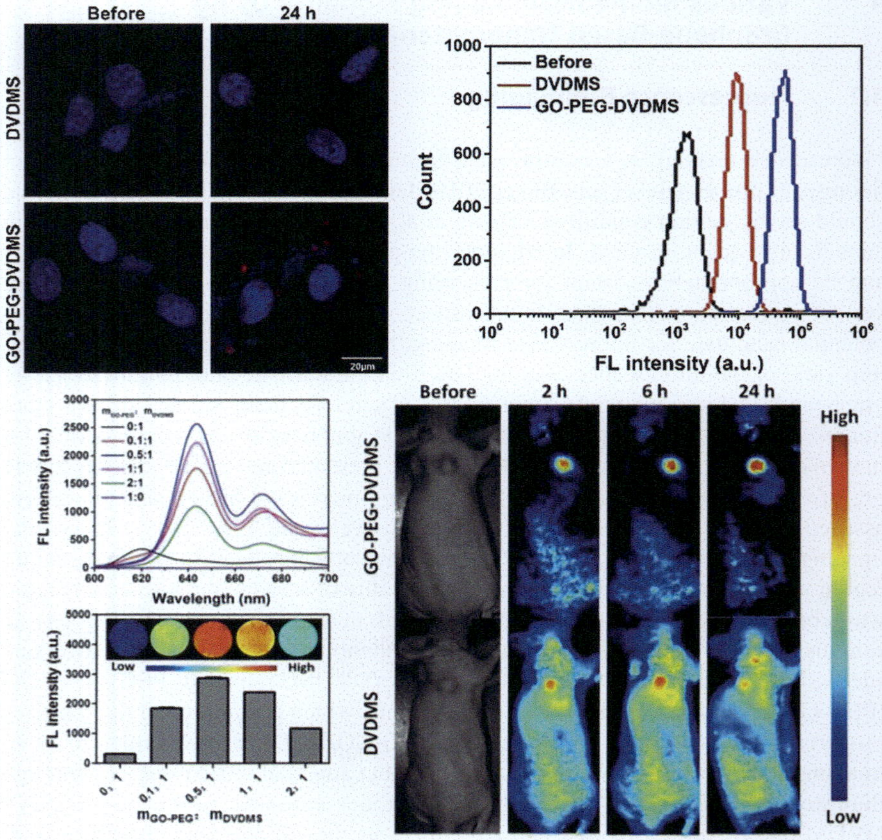

Fig. 4 Enhanced fluorescence imaging using sinoporphyrin sodium loaded graphene oxide (Reproduced from Yan et al. 2015)

4.2 Multiphoton Fluorescence Imaging

Current bioimaging mostly utilize fluorescent molecules with UV-vis emission (generally 400–600 nm). However, for non-invasive, the imaging studies for longer wavelengths are preferred, because they provide less damage analysis as well as enable deep tissue imaging. For these reasons, there is increasing interest in NIR-emitted fluorescent probes and there are attempts to synthesize GQDs that emit NIR fluorescence. However, these approaches often are challenging for many reasons, so multiphoton imaging is considered to be a great alternative.

Indeed, multiphoton FL imaging exhibits a small autofluorescence background and larger image depth due to low Rayleigh scattering and NIR tissue absorption, compared to single photon imaging, which has attracted much attention in promising applications in basic and biomedical research (Yoo et al. 2015). Compared to single photon excitation using simple continuous-wave laser, multiphoton nonlinear excitation generally requires the high reflux of excitation photons by femtosecond laser and generates relatively high levels of spatial resolution. Moreover, the wavelength for multiphoton excitation is known to be in the range of 700–1350 nm and is quite suitable for imaging of deep organs and tissues.

Qian et al. reported in vitro and in vivo cell imaging induced by two-photon and three-photon excitation using PEG–NGO (Qian et al. 2012). In that study, the 3D distribution of PEG–NGO fluorescence was clearly visualized in deep tissue imaging. In addition, Wang and Gu et al. reported for three-dimensional multiphoton FL imaging and laser-based cancer microsurgery (Li et al. 2012). After then, ultra-small sized nitrogen-doped GQDs (N-GQDs) as efficient two-photon fluorescent probes are reported by Gong group for the application of deep tissue cellular imaging (Liu et al. 2013a) (Fig. 5). Remarkably, the N-GQDs exhibited two-photon absorption and the penetration imaging depth of N-GQDs was considerable as deep as 1800 mm. As demonstrated in previous studies on multiphoton cell imaging with GQDs and other graphene derivatives, graphene-based materials can be promising candidates for non-invasive bioimaging probes with exceptional photostability and non-toxicity to be designed in the near future.

4.3 Raman Imaging

Fluorescence imaging is a common bioimaging technique, but it has some limitations such as high excitation energy, photobleaching, and broad excitation/emission peak widths. On the other hand, Raman imaging uses light scattering that occurs in the state of molecular vibrational excitation, so photon energy is not a prerequisite for the electronic excitation energy, and a low energy laser that does not cause biological damage can be available. In Raman imaging, photobleaching is reduced, and excitation/emission peak width is narrow, enabling more stable Raman bioimaging (Keren et al. 2008; Qian et al. 2008; Liu et al. 2008b). Although Raman imaging is not a common technology because of its low efficiency, advanced Raman imaging including surface-enhanced Raman spectroscopy (SERS) has been developed and used in bioimaging applications (Zavaleta et al. 2008, 2009). Unlike small organic molecules, GOs exhibit intrinsically strong Raman peaks without any enhancements, which can be further improved by incorporating metal NPs (Liu et al. 2012, 2013c; Kim et al. 2012). Liu et al. showed markedly enhanced Raman peaks under 632.8 nm laser illumination using a directly grown Au–graphene oxide composite.

Recently, Zhang et al. presented the complexes with GO, Au nanoparticles (AuNPs), Thioglycolic acid (TGA), and folic acid (FA) as Raman imaging markers which can target FA receptor overexpressed HeLa cells (Zhang et al. 2016). HeLa cells were incubated with each GO, GO/Pt, GO/AuNPs, GO/AuNPs/TGA, and GO/AuNPs/FA complexes for 12 h, and then SERS imaging was carried out. When the cells were incubated with GO/AuNPs, GO/AuNPs/TGA, and GO/AuNPs/FA composites, a continuous and distinguishable Raman imaged were

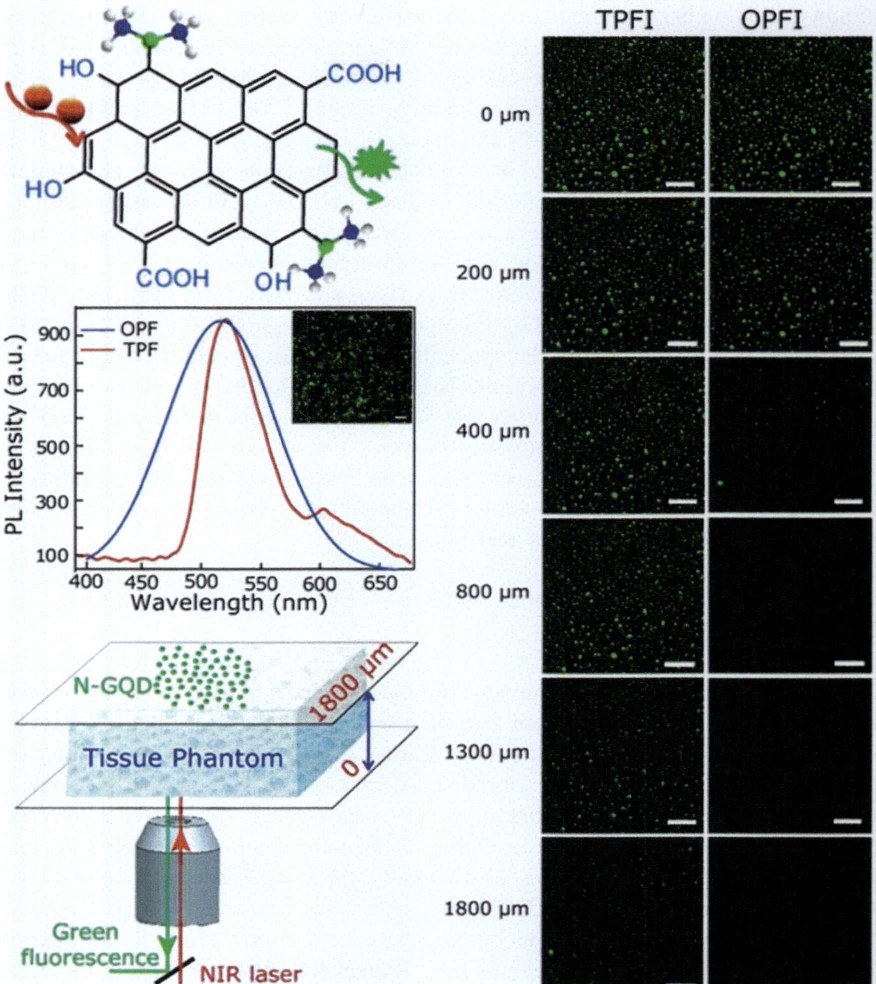

Fig. 5 Nitrogen-doped graphene quantum dots (N-GQDs) as efficient two-photon fluorescent probes for cellular and deep-tissue imaging (Reproduced from Liu et al. 2013a)

obtained compared to other groups. These results revealed that AuNP can be used to improve Raman imaging of cells by increasing the Raman intensity of GO (Fig. 6).

4.4 MRI

MRI, a powerful imaging technique in biomedical fields, has been widely used to image the anatomy as well as the function of tissues due to its excellent spatial

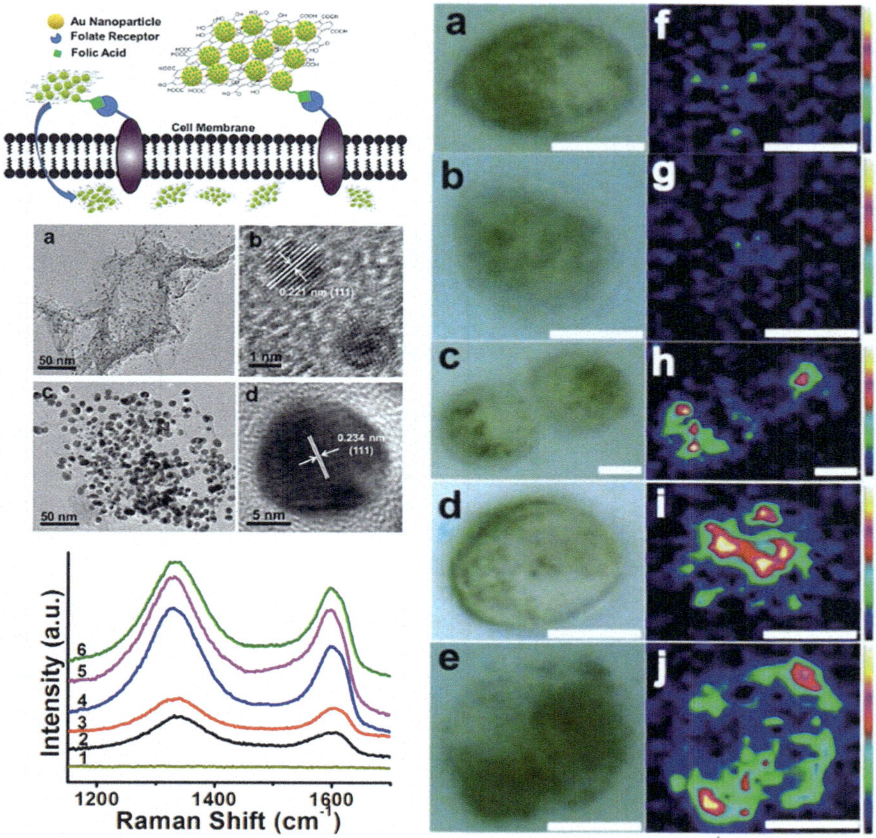

Fig. 6 FA receptor overexpressed HeLa cells targeting Raman imaging using graphene oxide-based hybrids (Reproduced from Zhang et al. 2016)

resolution (Mansfield 2004). This imaging technique indexes the difference between lattice-spin (longitudinal) relaxation time T1 and spin-spin (horizontal) relaxation time T2 in normal and lesion tissues and generates an image (Strijkers et al. 2007). Therefore, contrast agents have been developed and used in clinical practice to improve MRI visibility. The most commonly used MRI contrast agents are paramagnetic metal complexes and paramagnetic nanoparticles. However, pure graphenes cannot be used as MRI contrast agent because it does not have intrinsic paramagnetism. Graphene oxides (GO) can be paramagnetic when they are incorporated with other contrast agents (Borrás et al. 2020). In addition, their readily functionalized edges enable the targeting ability simultaneously.

Gizzatov and coworkers developed Gd^{3+}coordinated carboxy-phenylated graphene nanoribbon nanocomposites (CP-GNR/Gd^{3+} NCs) to enhance MRI relaxivity. CP-GNR/Gd^{3+} NCs showed better MRI contrasts in both of T1 and

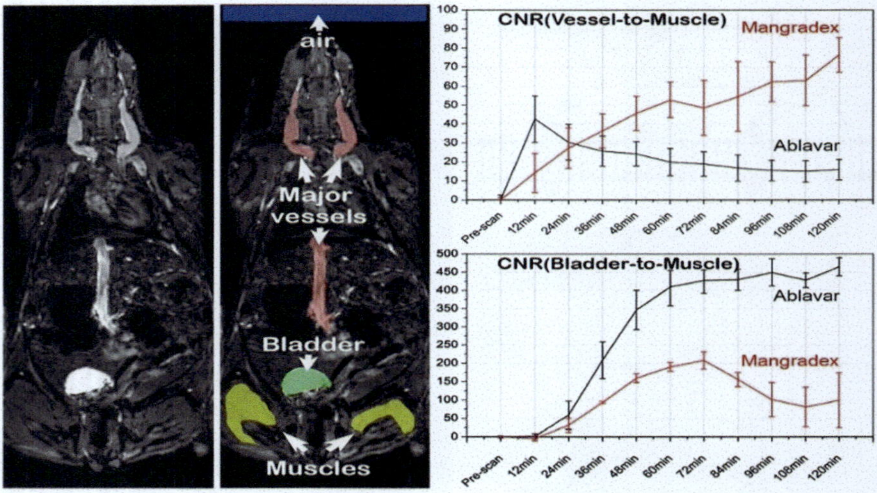

Fig. 7 Advanced graphene-based magnetic resonance imaging contrast agent studies in small animals (Reproduced from Kanakia et al. 2015)

T2-weighed images compared to Gd^{3+} based contrast agents which is clinically used (Gizzatov et al. 2014). In addition, Kanakia et al. developed manganese (Mn^{2+}) embedded graphene nanoparticles functionalized with dextran (hereafter, Mangradex) for more efficacious high field T_1 MRI contrasts (Kanakia et al. 2015). The nanocomposites neither induce an inflammatory response in any tissue, nor cause any noticeable adverse effects in hematological parameters and exhibit contrast enhancing capabilities, so it could be used as clinical MRI contrasts agents (Fig. 7).

Superparamagnetic iron oxide nanoparticles (SPION) which are widely used as T2 contrast agents can be directly grown on or easily combined with graphene-based materials to produce graphene-based T2 contrast agents (Na et al. 2009). In 2011, Chen et al. developed dextran coated Fe_3O_4 NPs loaded graphene oxide nanocomposites (DEX-Fe_3O_4/GO NC) for MRI contrast agents. The cellular uptake of DEX-Fe_3O_4/GO NC into the HeLa cells was confirmed depending on the concentration of the NCs. In addition, with significantly higher T2 relaxation rate, the DEX-Fe_3O_4/GO NC exhibited more improved T2 weighted MRI contrast compared to Fe_3O_4 NPs. Similarly, other kinds of paramagnetic NPs such as manganese ferrite ($MnFe_3O_4$) can also sufficiently combine with GO and exhibit satisfactory effect (Chen et al. 2013). Taken together, above-described approaches have shown the potential that GO-based paramagnetic nanoparticles can be utilized as MRI contrast agents (Chen et al. 2011).

4.5 Photoacoustic Imaging

Optical imaging in range of visible light provides a low penetration depth because the speed of light scattering in the tissues is highly fast, which results in limitation of measurement performed on the tissue surface (Ntziachristos 2010). The low-energy electromagnetic waves exhibit deeper penetration than the short wavelengths, and ultrasound or radio frequency waves show much lower light scattering in the tissue; therefore, these waves could be suitable for use in deep tissue imaging (Xu and Wang 2006). Photoacoustic imaging (PAI) is biomedical imaging technique based on the photoacoustic (PA) effect, in which the captured short-pulsed electromagnetic energy generated from the materials that are thermally expanded by laser pulses is converted into heat, leading to acoustic emission (Huang et al. 2014; Kim et al. 2010; Yapici et al. 2012; Park et al. 2016, 2021). However, generally, it is difficult to distinguish the PA signals of pathological tissues from that of normal tissues (De La Zerda et al. 2008). Thus, the contrast agents in PAI that could enhance the light absorption in the NIR region are required for appropriate PA contrasts. Some organic contrast materials and gold nanoclusters were investigated as candidates of contrast agents, but there are several limitations precluding their use.

The intrinsic optical properties of graphene-based nanomaterials provide excellent PA contrast (Yang et al. 2012b). Because GOs with disconnected small sp^2 domains have higher transition energy between HOMO and LUMO than the larger sp^2 domains, which resulted in the low absorbance in the NIR region. Although rGOs that have larger sp^2 domains than GOs can absorb NIR light more efficiently, there is a limit to being used as an imaging agent due to decreased hydrophilicity of rGOs after the reduction resulting in poor solubility (Lalwani et al. 2013). To resolve this, various approaches have been studied. Liu et al. developed solvothermally reduced GOs decorated with magnetic iron oxide nanoparticle (IONP) and PEGylated for solubility improvement (Yang et al. 2012b). The resulting RGO–IONP–PEG exhibited increased light absorption in the NIR range and high passive tumor accumulation in in vivo tumor imaging. Patel et al. demonstrated that the less oxygenated nanosize graphene sheets can well-disperse in water and exhibit enhanced PA effect due to strong NIR light absorption (Patel et al. 2013). Moreover, the one-step reduction, stabilization and functionalization of nano-rGO by BSA was reported by Sheng et al. (2013). The color of BSA functionalized nano-rGO solution was changed from yellow to black, which indicated an increased light absorption. And it was confirmed that the tumor was successfully imaged by using BSA functionalized nano-rGO as a photoacoustic contrast agent (Fig. 8).

5 Conclusions and Perspectives

Since their biocompatibility, surface functionalization, and unique electrical, chemical, mechanical, and physical properties, graphene-based nanomaterials have been studied in many biological fields. Among these, the use of graphene-based nanomaterials as bioimaging applications has highly attracted interest of researchers

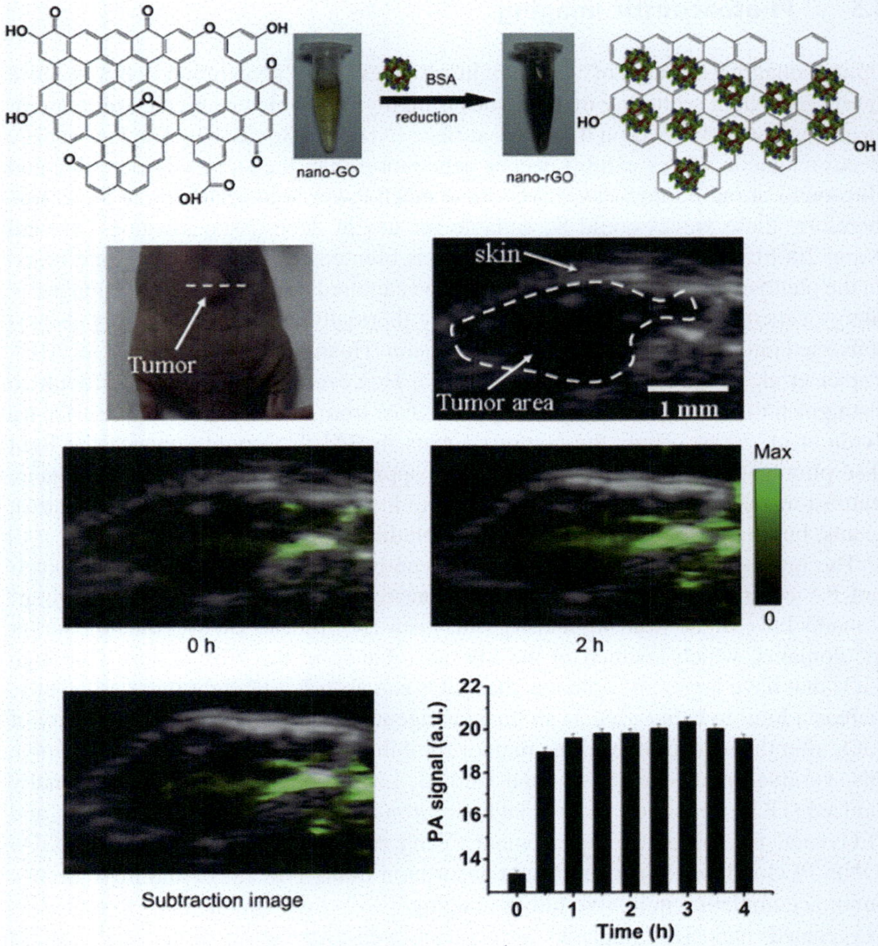

Fig. 8 BSA functionalized nano-reduced graphene oxide for combined in vivo photoacoustic imaging and photothermal therapy (Reproduced from Sheng et al. 2013, Copyright (2013))

and has been rapidly progressed. As mentioned before, the graphene-based nanomaterials have various advantages for bioimaging applications: Raman scattering, NIR absorbance, and the intrinsic photoluminescent properties. In comparison with Raman scattering of CNT which influenced by its chirality and diameter, consequently causes the nonuniform properties, the graphene is not chirality dependent and have small band gap in electronic structure that permit a broad range of photons (visible~IR) to be used in Raman imaging. The NIR absorbance of graphene-based nanomaterials is good for photoacoustic imaging. Among them, enhanced NIR absorbance of rGO, compared to GO and GQDs, is advantageous for photoacoustic imaging. The photoluminescent properties of graphene-based nanomaterials can be increased by functionalization or cutting graphene sheets into smaller sheets for fluorescence imaging.

Despite these advantages, however, there are various challenges such as the broad emission band, weak fluorescence intensity, quantum yield, and batch-to-batch variation which can be caused by the top-down based method currently used for synthesizing the graphene-based materials resulting in nonuniform size distribution. Therefore, for the more advanced performance of graphene-based nanomaterials in the field of bioimaging, minimizing batch-to-batch variation by controlling the size distribution, preventing photoluminescent quenching, improving the quantum yield, and reducing toxicity such as surface modification by PEGylation (Nurunnabi et al. 2014) are the future goals. Additionally, as far as the toxicity of graphene-based materials is concerned, it is one of the critical issues that the optimization of the graphene-based nanomaterials dose and precise in vivo distribution where we want. Because, until recently, the potential of graphene-based nanomaterials has not been entirely investigated and there are several challenges as mentioned before; combination biomedical approach could be novel future direction of graphene-based nanomaterials research for various applications.

Acknowledgments This research was supported by National Research Foundation of Korea (NRF) funded by the Ministry of Science (2019R1A4A1024116 and 2021M3H4A4079509) and Korea Environment Industry & Technology Institute (KEITI) through project to develop eco-friendly new materials and processing technology derived from wildlife, funded by Korea Ministry of Environment (MOE) (2021003270006).

References

Agarwal S, Zhou X, Ye F, He Q, Chen GCK, Soo J, Boey F, Zhang H, Chen P (2010) Interfacing live cells with nanocarbon substrates. Langmuir 26(4):2244–2247

Agharkar M, Kochrekar S, Hidouri S, Azeez MA (2014) Trends in green reduction of graphene oxides, issues and challenges: a review. Mater Res Bull 59:323–328

Akhavan O, Ghaderi E, Rahighi R (2012) Toward single-DNA electrochemical biosensing by graphene nanowalls. ACS Nano 6(4):2904–2916

Athanasios B, Bourlinos DG, Dimitrios P, Tamas S, Anna S, Imre D (2003) Graphite oxide: chemical reduction to graphite and surface modification with aliphatic amines and amino acids. Langmuir 19(15):6050–6055

Bacon M, Bradley SJ, Nann T (2014) Graphene quantum dots. Part Part Syst Charact 31(4): 415–428

Baker SN, Baker GA (2010) Luminescent carbon nanodots: emergent nanolights. Angew Chem Int Ed 49(38):6726–6744

Balaji M, Jegatheeswaran S, Nithya P, Boomi P, Selvam S, Sundrarajan M (2018) Photoluminescent reduced graphene oxide quantum dots from latex of Calotropis gigantea for metal sensing, radical scavenging, cytotoxicity, and bioimaging in Artemia salina: a greener route. J Photochem Photobiol 178:371–379

Balandin AA, Ghosh S, Bao W, Calizo I, Teweldebrhan D, Miao F, Lau CN (2008) Superior thermal conductivity of single-layer graphene. Nano Lett 8(3):902–907

Banerjee AN (2018) Graphene and its derivatives as biomedical materials: future prospects and challenges. Interface Focus 8(3):20170056

Becerril HA, Man J, Liu Z, Stoltenberg RM, Bao Z, Chen Y (2008) Evaluation of solution-processed reduced graphene oxide films as transparent conductors. ACS Nano 2(3):463–470
Bhaviripudi S, Jia X, Dresselhaus MS, Kong J (2010) Role of kinetic factors in chemical vapor deposition synthesis of uniform large area graphene using copper catalyst. Nano Lett 10(10): 4128–4133
Borisov SM, Wolfbeis OS (2008) Optical biosensors. Chem Rev 108(2):423–461
Borrás A, Fraile J, Rosado A, Marbán G, Tobias G, López-Periago AM, Domingo C (2020) Green and solvent-free supercritical CO2-assisted production of superparamagnetic graphene oxide aerogels: application as a superior contrast agent in MRI. ACS Sustain Chem Eng 8(12): 4877–4888
Cançado L, Pimenta M, Neves B, Dantas M, Jorio A (2004) Influence of the atomic structure on the Raman spectra of graphite edges. Phys Rev Lett 93(24):247401
Cao H, Yu Q, Colby R, Pandey D, Park C, Lian J, Zemlyanov D, Childres I, Drachev V, Stach EA (2010) Large-scale graphitic thin films synthesized on Ni and transferred to insulators: structural and electronic properties. J Appl Phys 107(4):044310
Chae SJ, Güneş F, Kim KK, Kim ES, Han GH, Kim SM, Shin HJ, Yoon SM, Choi JY, Park MH (2009) Synthesis of largearea graphene layers on poly-nickel substrate by chemical vapor deposition: wrinkle formation. Adv Mater 21(22):2328–2333
Chen W, Yi P, Zhang Y, Zhang L, Deng Z, Zhang Z (2011) Composites of aminodextran-coated Fe3O4 nanoparticles and graphene oxide for cellular magnetic resonance imaging. ACS Appl Mater Interfaces 3(10):4085–4091
Chen Y, Vedala H, Kotchey GP, Audfray A, Cecioni S, Imberty A, Vidal S, Star A (2012) Electronic detection of lectins using carbohydrate-functionalized nanostructures: graphene versus carbon nanotubes. ACS Nano 6(1):760–770
Chen ML, Shen LM, Chen S, Wang H, Chen XW, Wang JH (2013) In situ growth of β-FeOOH nanorods on graphene oxide with ultra-high relaxivity for in vivo magnetic resonance imaging and cancer therapy. J Mater Chem B 1(20):2582–2589
Chen D, Chen W, Ma L, Ji G, Chang K, Lee JY (2014) Graphene-like layered metal dichalcogenide/graphene composites: synthesis and applications in energy storage and conversion. Mater Today 17(4):184–193
Cheung F (2020) Engineering of new graphene-based materials as potential materials to assist near-infrared photothermal therapy cancer treatment. Heliyon 6(6):e04131
Chien CT, Li SS, Lai WJ, Yeh YC, Chen HA, Chen IS, Chen LC, Chen KH, Nemoto T, Isoda S (2012) Tunable photoluminescence from graphene oxide. Angew Chem Int Ed 51(27): 6662–6666
Close DM, Xu T, Sayler GS, Ripp S (2011) In vivo bioluminescent imaging (BLI): noninvasive visualization and interrogation of biological processes in living animals. Sensors 11(1):180–206
De La Zerda A, Zavaleta C, Keren S, Vaithilingam S, Bodapati S, Liu Z, Levi J, Smith BR, Ma TJ, Oralkan O (2008) Carbon nanotubes as photoacoustic molecular imaging agents in living mice. Nat Nanotechnol 3(9):557
Domaille DW, Que EL, Chang CJ (2008) Synthetic fluorescent sensors for studying the cell biology of metals. Nat Chem Biol 4(3):168–175
Dong A, Ye X, Chen J, Kang Y, Gordon T, Kikkawa JM, Murray CB (2011) A generalized ligand-exchange strategy enabling sequential surface functionalization of colloidal nanocrystals. J Am Chem Soc 133(4):998–1006
Dua V, Surwade SP, Ammu S, Agnihotra SR, Jain S, Roberts KE, Park S, Ruoff RS, Manohar SK (2010) Allorganic vapor sensor using inkjet-printed reduced graphene oxide. Chem Int Ed 49(12):2154–2157
Eda G, Fanchini G, Chhowalla M (2008) Large-area ultrathin films of reduced graphene oxide as a transparent and flexible electronic material. Nat Nanotechnol 3(5):270–274
Eda G, Lin YY, Mattevi C, Yamaguchi H, Chen HA, Chen IS, Chen CW, Chhowalla M (2010) Blue photoluminescence from chemically derived graphene oxide. Adv Mater 22(4):505–509

Feng Y, Zhao J, Yan X, Tang F, Xue Q (2014) Enhancement in the fluorescence of graphene quantum dots by hydrazine hydrate reduction. Carbon 66:334–339

Ferrari AC, Basko DM (2013) Raman spectroscopy as a versatile tool for studying the properties of graphene. Nat Nanotechnol 8(4):235

Galiotis C, Frank O, Koukaras EN, Sfyris D (2015) Graphene mechanics: current status and perspectives. Annu Rev Chem Biomol Eng 6:121–140

Ge J, Lan M, Zhou B, Liu W, Guo L, Wang H, Jia Q, Niu G, Huang X, Zhou H (2014) A graphene quantum dot photodynamic therapy agent with high singlet oxygen generation. Nat Commun 5(1):1–8

Geim AK (2009) Graphene: status and prospects. Science 324(5934):1530–1534

Geim AK, Kim P (2008) Carbon wonderland. Sci Am 298(4):90–97

Geim AK, MacDonald AH (2007) Graphene: exploring carbon flatland. Phys Today 60(8):35–41

Georgakilas V, Otyepka M, Bourlinos AB, Chandra V, Kim N, Kemp KC, Hobza P, Zboril R, Kim KS (2012) Functionalization of graphene: covalent and noncovalent approaches, derivatives and applications. Chem Rev 112(11):6156–6214

Georgakilas V, Tiwari JN, Kemp KC, Perman JA, Bourlinos AB, Kim KS, Zboril R (2016) Noncovalent functionalization of graphene and graphene oxide for energy materials, biosensing, catalytic, and biomedical applications. Chem Rev 116(9):5464–5519

Ghuge AD, Shirode AR, Kadam VJ (2017) Graphene: a comprehensive review. Curr Drug Targets 18(6):724–733

Gizzatov A, Keshishian V, Guven A, Dimiev AM, Qu F, Muthupillai R, Decuzzi P, Bryant RG, Tour JM, Wilson LJ (2014) Enhanced MRI relaxivity of aquated Gd^{3+} ions by carboxyphenylated water-dispersed graphene nanoribbons. Nanoscale 6(6):3059–3063

Goenka S, Sant V, Sant S (2014) Graphene-based nanomaterials for drug delivery and tissue engineering. J Control Release 173:75–88

Gokus T, Nair R, Bonetti A, Bohmler M, Lombardo A, Novoselov K, Geim A, Ferrari A, Hartschuh A (2009) Making graphene luminescent by oxygen plasma treatment. ACS Nano 3(12): 3963–3968

Gollavelli G, Ling YC (2012) Multi-functional graphene as an in vitro and in vivo imaging probe. Biomaterials 33(8):2532–2545

Gomez-Navarro C, Weitz RT, Bittner AM, Scolari M, Mews A, Burghard M, Kern K (2007) Electronic transport properties of individual chemically reduced graphene oxide sheets. Nano Lett 7(11):3499–3503

Gu S, Hsieh CT, Yuan CY, Gandomi YA, Chang JK, Fu CC et al (2020) Fluorescence of functionalized graphene quantum dots prepared from infrared-assisted pyrolysis of citric acid and urea. J Lumin 217:116774

Ha HD, Jang MH, Liu F, Cho YH, Seo TS (2015) Upconversion photoluminescent metal ion sensors via two photon absorption in graphene oxide quantum dots. Carbon 81:367–375

Heo C, Yoo J, Lee S, Jo A, Jung S, Yoo H, Lee YH, Suh M (2011) The control of neural cell-to-cell interactions through non-contact electrical field stimulation using graphene electrodes. Biomaterials 32(1):19–27

Hirata M, Gotou T, Horiuchi S, Fujiwara M, Ohba M (2004) Thin-film particles of graphite oxide 1: high-yield synthesis and flexibility of the particles. Carbon 42(14):2929–2937

Hong Y, Lam JWY, Tang BZ (2011) Aggregation-induced emission. Chem Soc Rev 40(11): 5361–5388

Hu W, Peng C, Luo W, Lv M, Li X, Li D, Huang Q, Fan C (2010) Graphene-based antibacterial paper. ACS Nano 4(7):4317–4323

Huang P, Xu C, Lin J, Wang C, Wang X, Zhang C, Zhou X, Guo S, Cui D (2011) Folic acid-conjugated graphene oxide loaded with photosensitizers for targeting photodynamic therapy. Theranostics 1:240–250

Huang P, Rong P, Lin J, Li W, Yan X, Zhang MG, Nie L, Niu G, Lu J, Wang W (2014) Triphase interface synthesis of plasmonic gold bellflowers as nearinfrared light mediated acoustic and thermal theranostics. J Am Chem Soc 136(23):8307–8313

Hummers WS Jr, Offeman RE (1958) Preparation of graphitic oxide. J Am Chem Soc 80(6):1339–1339

Janib SM, Moses AS, MacKay JA (2010) Imaging and drug delivery using theranostic nanoparticles. Adv Drug Deliv Rev 62(11):1052–1063

Jeong S, Pinals RL, Dharmadhikari B, Song H, Kalluri A, Debnath D et al (2020) Graphene quantum dot oxidation governs noncovalent biopolymer adsorption. Sci Rep 10(1):1–14

Jin SH, Kim DH, Jun GH, Hong SH, Jeon S (2013) Tuning the photoluminescence of graphene quantum dots through the charge transfer effect of functional groups. ACS Nano 7(2):1239–1245

Jo G, Choe M, Lee S, Park W, Kahng YH, Lee T (2012) The application of graphene as electrodes in electrical and optical devices. Nanotechnology 23(11):112001

Kanakia S, Toussaint J, Hoang DM, Chowdhury SM, Lee S, Shroyer KR, Moore W, Wadghiri YZ, Sitharaman B (2015) Towards an advanced graphene-based magnetic resonance imaging contrast agent: sub-acute toxicity and efficacy studies in small animals. Sci Rep 5:17182

Kang L, Chu J, Zhao H, Xu P, Sun M (2015) Recent progress in the applications of graphene in surface-enhanced Raman scattering and plasmon-induced catalytic reactions. J Mater Chem C 3(35):9024–9037

Kavitha MK, Jaiswal M (2016) Graphene: a review of optical properties and photonic applications. Asian J Phys 7:809–831

Keren S, Zavaleta C, Cheng Z, de La Zerda A, Gheysens O, Gambhir S (2008) Noninvasive molecular imaging of small living subjects using Raman spectroscopy. Proc Natl Acad Sci 105(15):5844–5849

Kim KS, Zhao Y, Jang H, Lee SY, Kim JM, Kim KS, Ahn J-H, Kim P, Choi J-Y, Hong BH (2009) Large-scale pattern growth of graphene films for stretchable transparent electrodes. Nature 457(7230):706–710

Kim C, Favazza C, Wang LV (2010) In vivo photoacoustic tomography of chemicals: high-resolution functional and molecular optical imaging at new depths. Chem Rev 110(5):2756–2782

Kim YK, Han SW, Min DH (2012) Graphene oxide sheath on Ag nanoparticle/graphene hybrid films as an antioxidative coating and enhancer of surface-enhanced Raman scattering. ACS Appl Mater Interfaces 4(12):6545–6551

Kindalkar VS, Kumara K, Bhat S, Dharmaprakash SM (2021) An eco-friendly approach for the reduction of graphene oxide using Syzygium samarangense fruit extract. Mater Chem Phys 261:124224

Kobayashi H, Ogawa M, Alford R, Choyke PL, Urano Y (2010) New strategies for fluorescent probe design in medical diagnostic imaging. Chem Rev 110(5):2620–2640

Kuila T, Bose S, Mishra AK, Khanra P, Kim NH, Lee JH (2012) Chemical functionalization of graphene and its applications. Prog Mater Sci 57(7):1061–1105

Lalwani G, Cai X, Nie L, Wang LV, Sitharaman B (2013) Graphene-based contrast agents for photoacoustic and thermoacoustic tomography. Photo-Dermatology 1(3–4):62–67

Larson N, Ghandehari H (2012) Polymeric conjugates for drug delivery. Chem Mater 24(5):840–853

Latil S, Henrard L (2006) Charge carriers in few-layer graphene gilms. Phys Rev Lett 97(3):036803–036806

Lee S, Lee K, Zhong Z (2010a) Wafer scale homogeneous bilayer graphene films by chemical vapor deposition. Nano Lett 10(11):4702–4707

Lee Y, Bae S, Jang H, Jang S, Zhu S-E, Sim SH, Song YI, Hong BH, Ahn J-H (2010b) Wafer-scale synthesis and transfer of graphene films. Nano Lett 10(2):490–493

Lerf A, He H, Forster M, Klinowski J (1998) Structure of graphite oxide revisited. J Phys Chem B 102(23):4477–4482

Li X, Cai W, An J, Kim S, Nah J, Yang D, Piner R, Velamakanni A, Jung I, Tutuc E (2009) Large-area synthesis of highquality and uniform graphene films on copper foils. Science 324(5932):1312–1314

Li JL, Bao HC, Hou XL, Sun L, Wang XG, Gu M (2012) Graphene oxide nanoparticles as a nonbleaching optical probe for two-photon luminescence imaging and cell therapy. Angew Chem 51(8):11830–11834

Lin J, Chen X, Huang P (2016) Graphene-based nanomaterials for bioimaging. Adv Drug Deliv Rev 105:242–254

Liu Z, Robinson JT, Sun X, Dai H (2008a) PEGylated nanographene oxide for delivery of water-insoluble cancer drugs. J Am Chem Soc 130(33):10876–10877

Liu Z, Li X, Tabakman SM, Jiang K, Fan S, Dai H (2008b) Multiplexed multicolor Raman imaging of live cells with isotopically modified single walled carbon nanotubes. J Am Chem Soc 130(41):13540–13541

Liu R, Wu D, Feng X, Müllen K (2011a) Bottom-up fabrication of photoluminescent graphene quantum dots with uniform morphology. J Am Chem Soc 133(39):15221–15223

Liu K, Zhang J-J, Cheng F-F, Zheng T-T, Wang C, Zhu J-J (2011b) Green and facile synthesis of highly biocompatible graphene nanosheets and its application for cellular imaging and drug delivery. J Mater Chem 21(32):12034–12040

Liu Q, Wei L, Wang J, Peng F, Luo D, Cui R, Niu Y, Qin X, Liu Y, Sun H (2012) Cell imaging by graphene oxide based on surface enhanced Raman scattering. Nanoscale 4(22):7084–7089

Liu Q, Guo B, Rao Z, Zhang B, Gong JR (2013a) Strong two-photon-induced fluorescence from photostable, biocompatible nitrogen-doped graphene quantum dots for cellular and deep-tissue imaging. Nano Lett 13(6):2436–2441

Liu F, Jang MH, Ha HD, Kim JH, Cho YH, Seo TS (2013b) Facile synthetic method for pristine graphene quantum dots and graphene oxide quantum dots: origin of blue and green luminescence. Adv Mater 25(27):3657–3662

Liu Z, Guo Z, Zhong H, Qin X, Wan M, Yang B (2013c) Graphene oxide based surface-enhanced Raman scattering probes for cancer cell imaging. Phys Chem Chem Phys 15(8):2961–2966

Luk C, Chen B, Teng K, Tang L, Lau S (2014) Optically and electrically tunable graphene quantum dot–polyaniline composite films. J Mater Chem C 2(23):4526–4532

Ma J, Zhang J, Xiong Z, Yong Y, Zhao XS (2011) Preparation, characterization and antibacterial properties of silver-modified graphene oxide. J Mater Chem 21(10):3350–3352

Malard L, Pimenta M, Dresselhaus G, Dresselhaus M (2009) Raman spectroscopy in graphene. Phys Rep 473(5–6):51–87

Mansfield P (2004) Snapshot magnetic resonance imaging (Nobel lecture). Angew Chem Int Ed 43(41):5456–5464

Mohanty N, Berry V (2008) Graphene-based single-bacterium resolution biodevice and DNA transistor: interfacing graphene derivatives with nanoscale and microscale biocomponents. Nano Lett 8(12):4469–4476

Na HB, Song IC, Hyeon T (2009) Inorganic nanoparticles for MRI contrast agents. Adv Mater 21(21):2133–2148

Nahain A-A, Lee J-E, In I, Lee H, Lee KD, Jeong JH, Park SY (2013a) Target delivery and cell imaging using hyaluronic acid-functionalized graphene quantum dots. Mol Pharm 10(10): 3736–3744

Nahain A-A, Lee J-E, Jeong JH, Park SY (2013b) Photoresponsive fluorescent reduced graphene oxide by spiropyran conjugated hyaluronic acid for in vivo imaging and target delivery. Biomacromolecules 14(11):4082–4090

Novoselov KS, Geim AK, Morozov SV, Jiang D, Zhang Y, Dubonos SV, Grigorieva IV, Firsov AA (2004) Electric field in atomically thin carbon films. Science 306(5696):666–669

Novoselov KS, Fal V, Colombo L, Gellert P, Schwab M, Kim K (2012) A roadmap for graphene. Nature 490(7419):192–200

Ntziachristos V (2010) Going deeper than microscopy: the optical imaging frontier in biology. Nat Methods 7(8):603

Nurunnabi M, Khatun Z, Reeck GR, Lee DY, Lee YK (2014) Photoluminescent graphene nanoparticles for cancer phototherapy and imaging. ACS Appl Mater Interfaces 6(15): 12413–12421

Olorunkosebi AA, Eleruja MA, Adedeji AV, Olofinjana B, Fasakin O, Omotoso E et al (2021) Optimization of graphene oxide through various Hummers' methods and comparative reduction using green approach. Diam Relat Mater 117:108456
Palmieri V, Spirito MD, Papi M (2020) Graphene-based scaffolds for tissue engineering and photothermal therapy. Nanomedicine 15(14):1411–1417
Pan D, Zhang J, Li Z, Wu M (2010) Hydrothermal route for cutting graphene sheets into blue-luminescent graphene quantum dots. Adv Mater 22(6):734–738
Park S, Mohanty N, Suk JW, Nagaraja A, An J, Piner RD, Cai W, Dreyer DR, Berry V, Ruoff RS (2010) Biocompatible, robust freestanding paper composed of a TWEEN/graphenecomposite. Adv Mater 22(15):1736–1740
Park J, Jeon S, Meng J, Song L, Lee JS, Kim C (2016) Delay-multiplyand-sum-based synthetic aperture focusing in photoacoustic microscopy. J Biomed Opt 21(3):036010–036010
Park B, Bang CH, Lee C, Han JH, Choi W, Kim J et al (2021) 3D wide-field multispectral photoacoustic imaging of human melanomas in vivo: a pilot study. J Eur Acad Dermatol Venereol 35(3):669–676
Patel MA, Yang H, Chiu PL, Mastrogiovanni DD, Flach CR, Savaram K, Gomez L, Hemnarine A, Mendelsohn R, Garfunkel E (2013) Direct production of graphene nanosheets for near infrared photoacoustic imaging. ACS Nano 7(9):8147–8157
Ponomarenko LA, Schedin F, Katsnelson MI, Yang R, Hill EW, Novoselov KS, Geim AK (2008) Chaotic Dirac billiard in graphene quantum dots. Science 320(5874):356–358
Qian X, Peng X-H, Ansari DO, Yin-Goen Q, Chen GZ, Shin DM, Yang L, Young AN, Wang MD, Nie S (2008) In vivo tumor targeting and spectroscopic detection with surface-enhanced Raman nanoparticle tags. Nat Biotechnol 26(1):83–90
Qian J, Wang D, Cai FH, Xi W, Peng L, Zhu ZF, He H, Hu ML, He S (2012) Observation of multiphoton-induced fluorescence from graphene oxide nanoparticles and applications in in vivo functional bioimaging. Angew Chem Int Ed 124(42):10570–10575
Rao C, Biswas K, Subrahmanyam K, Govindaraj A (2009a) Graphene, the new nanocarbon. J Mater Chem 19(17):2457–2469
Rao CNR, Sood AK, Subrahmanyam KS, Govindaraj A (2009b) Graphene: the new two-dimensional nanomaterial. Angew Chem Int Ed 48(42):7752–7777
Robinson JT, Tabakman SM, Liang Y, Wang H, Sanchez Casalongue H, Vinh D, Dai H (2011) Ultrasmall reduced graphene oxide with high near-infrared absorbance for photothermal therapy. J Am Chem Soc 133(17):6825–6831
Schedin F, Lidorikis E, Lombardo A, Kravets VG, Geim AK, Grigorenko AN, Novoselov KS, Ferrari AC (2010) Surface-enhanced Raman spectroscopy of graphene. ACS Nano 4(10):5617–5626
Schniepp HC, Li J-L, McAllister MJ, Sai H, Herrera-Alonso M, Adamson DH, Prud'homme RK, Car R, Saville DA, Aksay IA (2006) Functionalized single graphene sheets derived from splitting graphite oxide. J Phys Chem B 110(17):8535–8539
Schroeder KL, Goreham RV, Nann T (2016) Graphene quantum dots for theranostics and bioimaging. Pharm Res 33(10):2337–2357
Seo TH, Lee S, Cho H, Chandramohan S, Suh E-K, Lee HS, Bae SK, Kim SM, Park M, Lee JK (2016) Tailored CVD graphene coating as a transparent and flexible gas barrier. Sci Rep 6(1):1–8
Shao Y, Wang J, Wu H, Liu J, Aksay IA, Lin Y (2010) Graphene based electrochemical sensors and biosensors: a review. Electroanalysis 22(10):10271036
Shen J, Zhu Y, Yang X, Li C (2012a) Graphene quantum dots: emergent nanolights for bioimaging, sensors, catalysis and photovoltaic devices. Chem Commun 48(31):3686–3699
Shen J, Zhu Y, Yang X, Zong J, Zhang J, Li C (2012b) One-pot hydrothermal synthesis of graphene quantum dots surface-passivated by polyethylene glycol and their photoelectric conversion under near-infrared light. New J Chem 36(1):97–101

Sheng Z, Song L, Zheng J, Hu D, He M, Zheng M, Gao G, Gong P, Zhang P, Ma Y (2013) Protein-assisted fabrication of nano-reduced graphene oxide for combined in vivo photoacoustic imaging and photothermal therapy. Biomaterials 34(21):5236–5243

Shin HJ, Kim KK, Benayad A, Yoon SM, Park HK, Jung IS, Jin MH, Jeong HK, Kim JM, Choi JY (2009) Efficient reduction of graphite oxide by sodium borohydride and its effect on electrical conductance. Adv Funct Mater 19(12):1987–1992

Simpson CD, Brand JD, Berresheim AJ, Przybilla L, Räder HJ, Müllen K (2002) Synthesis of a giant 222 carbon graphite sheet. Chem Eur J 8(6):1424–1429

Somani PR, Somani SP, Umeno M (2006) Planer nano-graphenes from camphor by CVD. Chem Phys Lett 430(1–3):56–59

Stankovich S, Dikin DA, Dommett GH, Kohlhaas KM, Zimney EJ, Stach EA, Piner RD, Nguyen ST, Ruoff RS (2006) Graphene-based composite materials. Nature 442(7100):282–286

Stankovich S, Dikin DA, Piner RD, Kohlhaas KA, Kleinhammes A, Jia Y, Wu Y, Nguyen ST, Ruoff RS (2007) Synthesis of graphene-based nanosheets via chemical reduction of exfoliated graphite oxide. Carbon 45(7):1558–1565

Strijkers GJ, Mulder M, Willem J, Van Tilborg F, Geralda A, Nicolay K (2007) MRI contrast agents: current status and future perspectives. Anti Cancer Agents Med Chem 7(3):291–305

Sun X, Liu Z, Welsher K, Robinson JT, Goodwin A, Zaric S, Dai H (2008) Nano-graphene oxide for cellular imaging and drug delivery. Nano Res 1(3):203–212

Tang L, Ji R, Cao X, Lin J, Jiang H, Li X, Teng KS, Luk CM, Zeng S, Hao J (2012) Deep ultraviolet photoluminescence of water-soluble self-passivated graphene quantum dots. ACS Nano 6(6): 5102–5110

Thakur S, Karak N (2015) Alternative methods and nature-based reagents for the reduction of graphene oxide: a review. Carbon 94:224–242

Tian B, Wang C, Zhang S, Feng L, Liu Z (2011) Photothermally enhanced photodynamic therapy delivered by nano-graphene oxide. ACS Nano 5(9):7000–7009

VanEngelenburg SB, Palmer AE (2008) Fluorescent biosensors of protein function. Curr Opin Chem Biol 12(1):60–65

Wang JJ, Zhu MY, Outlaw RA, Zhao X, Manos DM, Holoway BC (2004a) Free-standing subnanometer graphite sheets. Appl Phys Lett 85(7):1265–1267

Wang JJ, Zhu MY, Outlaw RA, Zhao X, Manos DM, Holoway BC (2004b) Synthesis of carbon nanosheets by inductively coupled radio-frequency plasma enhanced chemical vapor deposition. Carbon 42(14):2867–2872

Wang Y, Shyy JY-J, Chien S (2008) Fluorescence proteins, live-cell imaging, and mechanobiology: seeing is believing. Annu Rev Biomed Eng 10:1–38

Wang Y, Li Z, Wang J, Li J, Lin Y (2011) Graphene and graphene oxide: biofunctionalization and applications in biotechnology. Trends Biotechnol 29(5):205–212

Wang X, Huang P, Feng L, He M, Guo S, Shen G, Cui D (2012) Green controllable synthesis of silver nanomaterials on graphene oxide sheets via spontaneous reduction. RSC Adv 2(9): 3816–3822

Weissleder R, Pittet MJ (2008) Imaging in the area of molecular oncology. Nature 452(7187): 580–589

Whitener KE Jr, Sheehan PE (2014) Graphene synthesis. Diam Relat Mater 46:25–34

Wu J, Tomović Ž, Enkelmann V, Müllen K (2004) From branched hydrocarbon propellers to C3-symmetric graphite disks. J Org Chem 69(16):5179–5186

Xie J, Liu G, Eden HS, Ai H, Chen X (2011) Surface-engineered magnetic nanoparticle platforms for cancer imaging and therapy. Acc Chem Res 44(10):883–892

Xin G, Meng Y, Ma Y, Ho D, Kim N, Cho SM, Chae H (2012) Tunable photoluminescence of graphene oxide from near-ultraviolet to blue. Mater Lett 74:71–73

Xu M, Wang LV (2006) Photoacoustic imaging in biomedicine. Rev Sci Instrum 77(4):041101

Yan X, Niu G, Lin J, Jin AJ, Hu H, Tang Y, Zhang Y, Wu A, Lu J, Zhang S (2015) Enhanced fluorescence imaging guided photodynamic therapy of sinoporphyrin sodium loaded graphene oxide. Biomaterials 42:94–102

Yang X, Zhang X, Ma Y, Huang Y, Wang Y, Chen Y (2009) Superparamagnetic graphene oxide-Fe_3O_4 nanoparticles hybrid for controlled targeted drug carriers. J Mater Chem 19(18): 2710–2714

Yang K, Wan J, Zhang S, Tian B, Zhang Y, Liu Z (2012a) The influence of surface chemistry and size of nanoscale graphene oxide on photothermal therapy of cancer using ultra-low laser power. Biomaterials 33(7):2206–2214

Yang K, Hu L, Ma X, Ye S, Cheng L, Shi X, Li C, Li Y, Liu Z (2012b) Multimodal imaging guided photothermal therapy using functionalized graphene nanosheets anchored with magnetic nanoparticles. Adv Mater 24(14):1868–1872

Yang K, Feng L, Shi X, Liu Z (2013) Nano-graphene in biomedicine: theranostic applications. Chem Soc Rev 42(2):530–547

Yapici MK, Kim C, Chang C-C, Jeon M, Guo Z, Cai X, Zou J, Wang LV (2012) Parallel acoustic delay lines for photoacoustic tomography. J Biomed Opt 17(11):116019–116019

Yoo JM, Kang JH, Hong BH (2015) Graphene-based nanomaterials for versatile imaging studies. Chem Soc Rev 44(14):4835–4852

Yoon KY, Dong G (2020) Liquid-phase bottom-up synthesis of graphene nanoribbons. Mater Chem Front 4(1):29–45

Zavaleta C, De La Zerda A, Liu Z, Keren S, Cheng Z, Schipper M, Chen X, Dai H, Gambhir S (2008) Noninvasive Raman spectroscopy in living mice for evaluation of tumor targeting with carbon nanotubes. Nano Lett 8(9):2800–2805

Zavaleta CL, Smith BR, Walton I, Doering W, Davis G, Shojaei B, Natan MJ, Gambhir SS (2009) Multiplexed imaging of surface enhanced Raman scattering nanotags in living mice using noninvasive Raman spectroscopy. Proc Nal Acad Sci 106(32):13511–13516

Zhang J, Campbell RE, Ting AY, Tsien RY (2002) Creating new fluorescent probes for cell biology. Nat Rev Mol Cell Biol 3(12):906–918

Zhang J, Yang H, Shen G, Cheng P, Zhang J, Guo S (2010a) Reduction of graphene oxide via L-ascorbic acid. Chem Commun (Camb) 46(7):1112–1114

Zhang J, Zhang F, Yang H, Huang X, Liu H, Zhang J, Guo S (2010b) Graphene oxide as a matrix for enzyme immobilization. Langmuir 26(9):6083–6085

Zhang K, Zhang LL, Zhao X, Wu J (2010c) Graphene-polyaniline nanofiber composites as supercapacitor electrodes. Chem Mater 22(4):1392–1401

Zhang W, Guo Z, Huang D, Liu Z, Guo X, Zhong H (2011) Synergistic effect of chemo-photothermal therapy using PEGylated graphene oxide. Biomaterials 32(33):8555–8561

Zhang H, Grüner G, Zhao Y (2013) Recent advancements of graphene in biomedicine. J Mater Chem B 1(20):2542–2567

Zhang Z, Wang M, Gao D, Luo D, Liu Q, Yang J, Li Y (2016) Targeted Raman imaging of cells using graphene oxide-based hybrids. Langmuir 32(40):10253–10258

Zhou L, Wei S, Ge X, Zhou J, Jiang H, Li F, Shen J (2014) Combination of chemotherapy and photodynamic therapy using graphene oxide as drug delivery system. J Photochem Photobiol B 135:7–16

Zhu S, Zhang J, Qiao C, Tang S, Li Y, Yuan W, Li B, Tian L, Liu F, Hu R (2011) Strongly green-photoluminescent graphene quantum dots for bioimaging applications. Chem Commun 47(24): 6858–6860

Zhu S, Meng Q, Wang L, Zhang J, Song Y, Jin H, Zhang K, Sun H, Wang H, Yag B (2013) Highly photoluminescent carbon dots for multicolor patterning, sensors, and bioimaging. Angew Chem Int Ed 52(14):3953–3957

Zhu J, Yang D, Yin Z, Yan Q, Zhang H (2014) Graphene and graphene based materials for energy storage applications. Small 10(17):3480–3498

Zhuo S, Shao M, Lee S-T (2012) Upconversion and downconversion fluorescent graphene quantum dots: ultrasonic preparation and photocatalysis. ACS Nano 6(2):1059–1064

Graphene: A Promising Theranostic Agent

S. M. Shatil Shahriar, Md Nafiujjaman, Jeong Man An, Vishnu Revuri, Md. Nurunnabi, Dong-Wook Han, and Yong-kyu Lee

S. M. S. Shahriar
Department of Cancer Research, College of Medicine, University of Nebraska Medical Center, Omaha, NE, USA

M. Nafiujjaman
Department of Biomedical Engineering, Institute for Quantitative Health Science and Engineering, Michigan State University, East Lansing, MI, USA

J. M. An
Department of Bioengineering, College of Engineering, Hanyang University, Seoul, Republic of Korea

V. Revuri
Department of Green Bioengineering, Korea National University of Transportation, Chungju, South Korea

Department of Pharmaceutical Sciences, Temple University School of Pharmacy, Philadelphia, PA, USA
e-mail: vishnu.revuri@temple.edu

M. Nurunnabi
Department of Pharmaceutical Sciences, School of Pharmacy, University of Texas at El Paso, El Paso, TX, USA
e-mail: mnurunnabi@utep.edu

D.-W. Han
Department of Cogno-Mechatronics Engineering, College of Nanoscience and Nanotechnology, Pusan National University, Busan, South Korea
e-mail: nanohan@pusan.ac.kr

Y.-k. Lee (✉)
Department of Green Bioengineering, Korea National University of Transportation, Chungju, South Korea

Department of Chemical and Biological Engineering, Korea National University of Transportation, Chungju, South Korea
e-mail: leeyk@ut.ac.kr

D.-W. Han, S. W. Hong (eds.), *Multifaceted Biomedical Applications of Graphene*,
Advances in Experimental Medicine and Biology 1351,
https://doi.org/10.1007/978-981-16-4923-3_8

Abstract

Graphene has drawn tremendous interest in the field of nanoscience as a superior theranostic agent owing to its high photostability, aqueous solubility, and low toxicity. This monoatomic thick building block of a carbon allotrope exhibits zero to two-dimensional characteristics with a unique size range within the nanoscale. Their high biocompatibility, quantum yield, and photoluminescent properties make them more demandable in biomedical research. Its application in biomedical sciences has been limited due to its small-scale production. Large-scale production with an easy synthesis process is urgently required to overcome the problem associated with its translational application. Despite all possible drawbacks, the graphene-based drug/gene delivery system is gaining popularity day by day. To date, various studies suggested its application as a theranostic agent for target-specific delivery of chemotherapeutics or antibiotics against various diseases like cancer, Alzheimer's diseases, multidrug resistance diseases, and more. Also, studying the toxicological profile of graphene derivatives is very important before starting its practical use in clinical applications. This chapter has tried to abbreviate several methods and their possible incoming perspective as claimed by researchers for mass production and amplifying graphene-based treatment approaches.

Keywords

Graphene · Synthesis of graphene · Biomedical applications of graphene · Toxicity of graphene · Next-generation graphene derivatives

1 Introduction

To begin with, the excellent research article of Novoselvo KS and coworkers who discovered this novel theranostic agent in 2004, (Novoselov et al. 2004) graphene and its derivatives are still a promising material in the thirsty fields of nanomaterials as well as biomedical research because of their identical properties like attractive quantum confinement, potential edge yielding, biocompatibility, and photostability (Hasan et al. 2018). However, some of its derivatives are also highly water-soluble, usually comprised of H_2 and O_2 with umpteen molecular tiers that can easily tune bandgap and absorb optic light (Chen et al. 2020). Its usages in immunological assays such as electrochemical, amperometric, optical, magnetic, or thermometric and in piezoelectric immunosensors, and in multiple nucleic acid assays increased its acceptability as an iconic nanomaterial (Bhatnagar et al. 2016; Malekzad et al. 2017; Keegan et al. 2011; Chen et al. 2017; Razalli et al. 2017).

Graphene and its derivatives are being widely used not only as in vivo and in vitro imaging agent but also as a noninvasive drug nanocarrier in targeted therapeutic delivery systems. As an effective anticancer drug carrier, graphene and its derivatives can modulate pH-sensitive control release of chemotherapeutics in targeted cell lines by facilitating defending mechanism to prevent the biological

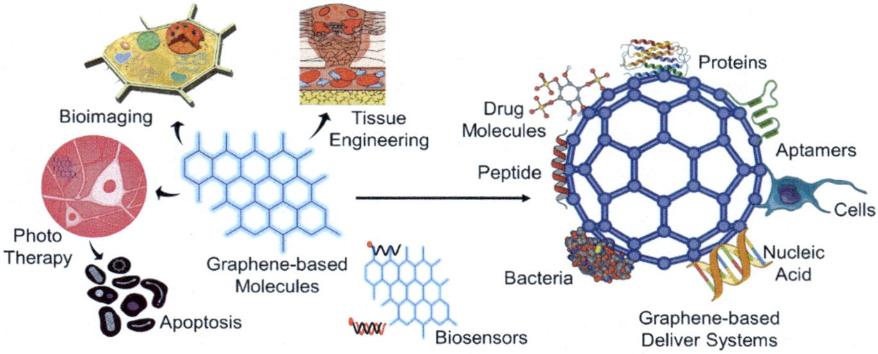

Fig. 1 Illustration depicts the utilization of graphene-based nanomaterials in various biomedical applications

degradation of anticancer drugs before arriving in the targeted cells or tissue, and thus modulate an enhanced cellular uptake profile (Liu et al. 2015). An example would be the introduction of GQDs and carbon quantum dots (CQDs) that can not only load and release the high drug payloads but also superintended cellular uptake mechanisms of delivered active components in the absence of external dyes (Wang et al. 2017; Singh et al. 2017). Thus, graphene and its derivatives can enhance the treatment proficiency of various disorders by using them as effective drug delivery platforms. However, large-scale synthesis of pure graphene derivatives such as fluorinated graphene (fluorographene), oxidized graphene (graphene oxide), graphyne, and graphdiyne and understanding their toxicological study are still unclear. Recently, researchers are focusing more on the synthesis of bio-friendly, cost-effective, and large-scale production of graphene derivatives. However, these novel nanomaterials showed toxicity with targeted cell, tissue or organ, and even cleavage DNA structure though that is very rare (Wang et al. 2015a). From the starting period, researchers have progressed on graphene research, and every year, many projects based on graphene and its derivatives have been running. This review summarizes the synthesis process, characterization, their possible usages as novel drug nanocarriers, and their toxicological studies of graphene-based nanomaterials to understand their cellular mechanisms better. We believe that further development of graphene and its derivatives will ensure more safety in biomedical sciences (Fig. 1).

2 Synthesis of Graphene Derivatives

Despite its excellent biocompatibility and unique properties, it has limitations in synthesizing cost-effective graphene derivatives. Earlier, graphene-based materials were mainly synthesized from highly-cost raw materials, including graphene (Pan et al. 2010) and photonic crystals (Guo et al. 2012). Even the synthesis process of its derivatives was expensive, such as laser ablation, electron beam lithography, and

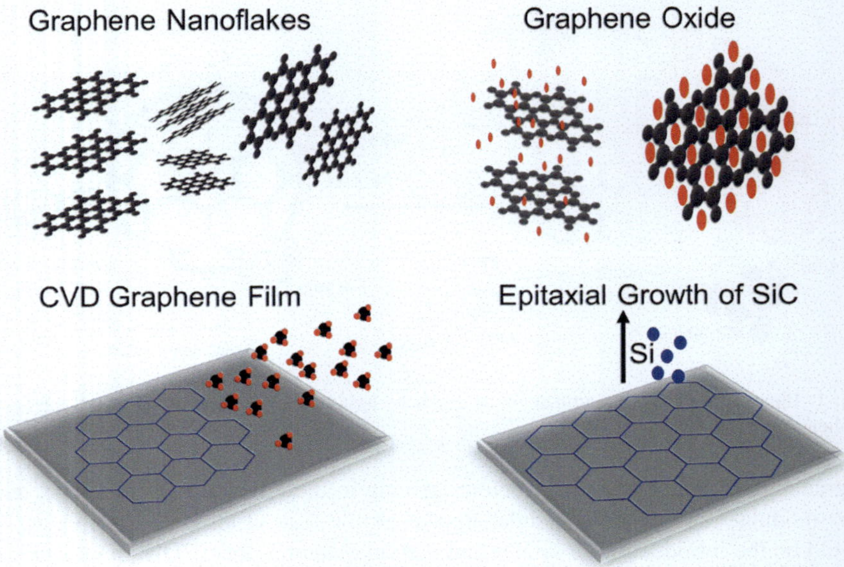

Fig. 2 Commercially available graphene derivatives such as graphene nanoflakes, graphene oxide, CVD graphene film, and epitaxial growth of silicon carbide (SiC)

electrochemical synthesis (Li et al. 2013; Sun et al. 2006; Li et al. 2011). Recently, researchers have identified variations of inexpensive sources, for example, citric acid/urea, from where graphene-based materials are produced and used on a large scale (Li et al. 2015). Another example of the least expensive materials for the synthesis of graphene-based nanomaterials would be coal that facilitates the synthesis of water-soluble and fluorescence graphene derivatives as described in Ye et al. (Ye et al. 2013) Various commercial graphene-based therapeutic agents are available in the clinics, which are being used for various applications **(Fig. 2)**.

However, the synthesis approaches of graphene and its derivatives can be folded into two categories. The first one is top-down methods where large graphene sheets, carbon nanotubes, carbon fibers, and graphite are cut down into small pieces of graphene sheets using various treatments to prepare various graphene derivatives. The primary advantages of this method are the synthesized graphene materials which influence solubility and oxygen-containing functional groups of graphene. On the contrary, the second method is the bottom-up method in which small molecules are needed to act as starting material for graphene-based material synthesis. This method is more appropriate to prepare graphene-based nanodepot using organic reactions and purifications. In comparing these two approaches, the top-down method is more cost-effective and has multiple synthesis steps than bottom-up approaches (Fig. 3).

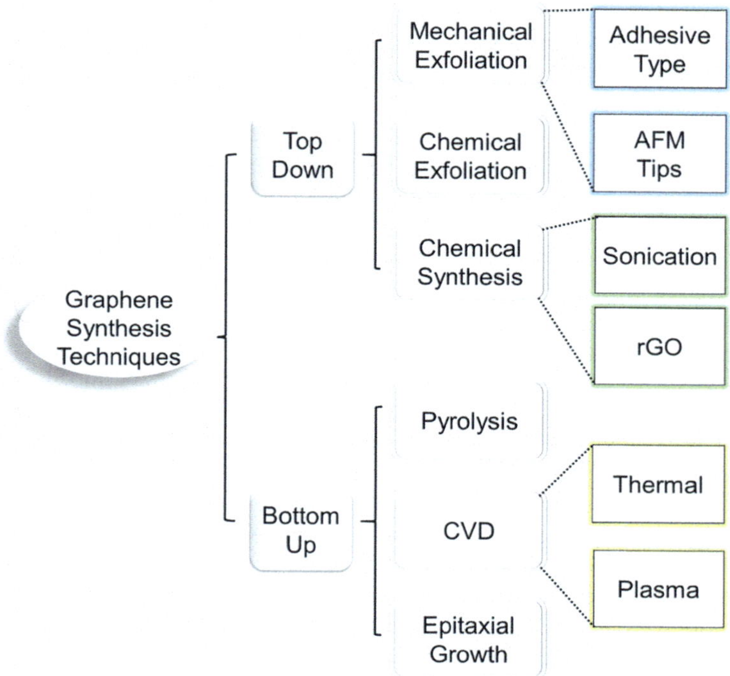

Fig. 3 Overview on the synthesis of graphene and its derivatives till date

2.1 Top-Down Method

The top-down process is also known as defect-mediated fragmentation process (Tong et al. 2016; Markovic et al. 2012; Joshi et al. 2016) and exemplified as nanolithography technique, acidic oxidation, hydrothermal or solvothermal microwave-assisted, sonication-assisted, electrochemical, Photo-Fenton reaction, selective plasma oxidation, and chemical exfoliation methods (Chen et al. 2017). Pan et al. described the hydrothermal procedure using an aqueous solvent as a green solvent to synthesize water-dispersible GQDs (Pan et al. 2010). This group also synthesized water-soluble and well-crystallized 3.0 nm GQDs with high temperatures (Ye et al. 2013). In another study, amino-functionalized graphene derivatives synthesized by using oxidized graphene sheets and ammonia in which the luminescence of graphene derivatives vary from violet emission spectra to yellow emission spectra mainly depended on the variation of concentration of ammonia (Tetsuka et al. 2012). On the other hand, Feng et al. described a method to synthesize fluorinated-graphene by using xenon difluoride at high temperatures from this hydrothermal synthesis process (Feng et al. 2013). Hu et al. developed a new methodology to obtain robust blue emission nitrogen-doped graphene derivatives from graphene oxide at 180 °C in the presence of ammonia by the hydrothermal treatment procedure for bioimaging of cancer cells (Hu et al. 2013). Lawsone (2-hydroxy-1,4-naphthoquinone), a widely used reducing agent, was adopted by Nigam and coworkers to synthesize 3–6 nm green fluorescence-graphene derivatives from graphene oxide in hydrothermal reaction (Nigam et al. 2014). Khodadadei et al. used citric acid and urea as a carbon and nitrogen source to synthesize blue fluorescent nitrogen-doped graphene derivatives where Fourier Transform Infrared Spectroscopy (FTIR) characterization confirmed the formation of doped nitrogen in the newly synthesized graphene materials (Khodadadei et al. 2017). In the solvothermal method, dimethyl sulfoxide (DMSO), dimethylformamide (DMF), and benzene are used as the solvent. Zhu et al. used DMF as a solvent to obtain strongly fluorescent graphene materials through solvothermal treatment (Zhu et al. 2011). They also synthesized graphene-based derivatives with tunable surface chemistry by increasing reaction time up to 8 h and investigating PL mechanisms based on surface chemistry and PL behaviors (Zhu et al. 2012). To obtain zero-dimensional graphene derivatives, Shin et al. proposed a novel acid-free and oxone oxidant-assisted solvothermal process using graphite, multiwall carbon nanotubes, carbon fibers, and charcoal. These synthesized graphene quantum dots exhibited robust blue photoluminescence properties (Shin et al. 2015).

Luk et al. described a "one-pot" microwave-assisted hydrothermal synthesis of nitrogen-doped graphene materials and studied how their optical properties change (Khan 2017). A low-cost approach was proposed to obtain nitrogen and sulfur co-doped graphene materials (N, S-GD) as a metal-free electrocatalyst for the oxygen reduction reaction with high fluorescent characteristics through microwave-assisted solvothermal method (Luo et al. 2014). Glucose-derived water-soluble self-passivated graphene quantum dots are synthesized by using a facile microwave-assisted hydrothermal method. The synthesized graphene derivative showed the shortest deep ultraviolet emission and was able to convert from blue emission to white emission, especially when graphene materials were coated with blue light emission diode (Pan et al. 2012). Recently, top-down method is also used for preparing Hofmann metal organic framework graphene nanosheet (Córdova Wong et al. 2019). The stacking interaction between the monolayers of the nanosheet have been shown in Fig. 4.

Li et al. used microwave irradiation to prepare greenish-yellow luminescent graphene materials (g-GD), where the derivatives were reduced with sodium borohydride to synthesize robust blue luminescent graphene-based nanomaterials (b-GD). The quantum yields of these two derivatives (g-GD and b-GD) were 11.7% and 22.9%, respectively (Tang et al. 2012; Li et al. 2012a). Zhu and coworkers who used only graphene oxide and potassium permanganate proposed a one-step ultrasonic synthesis of graphene derivatives, and those graphene derivatives exhibited the characteristics properties of 3 nm in diameter, with high quantum yield (Zhu et al. 2015). A research study shows that the electrochemical method was used to cleavage graphene film and obtained 3–5 nm graphene materials. Besides, oxygen plasma treatment was conducted to increase hydrophilicity, whereas another research claimed that large-scale production of aqueous graphene derivative was

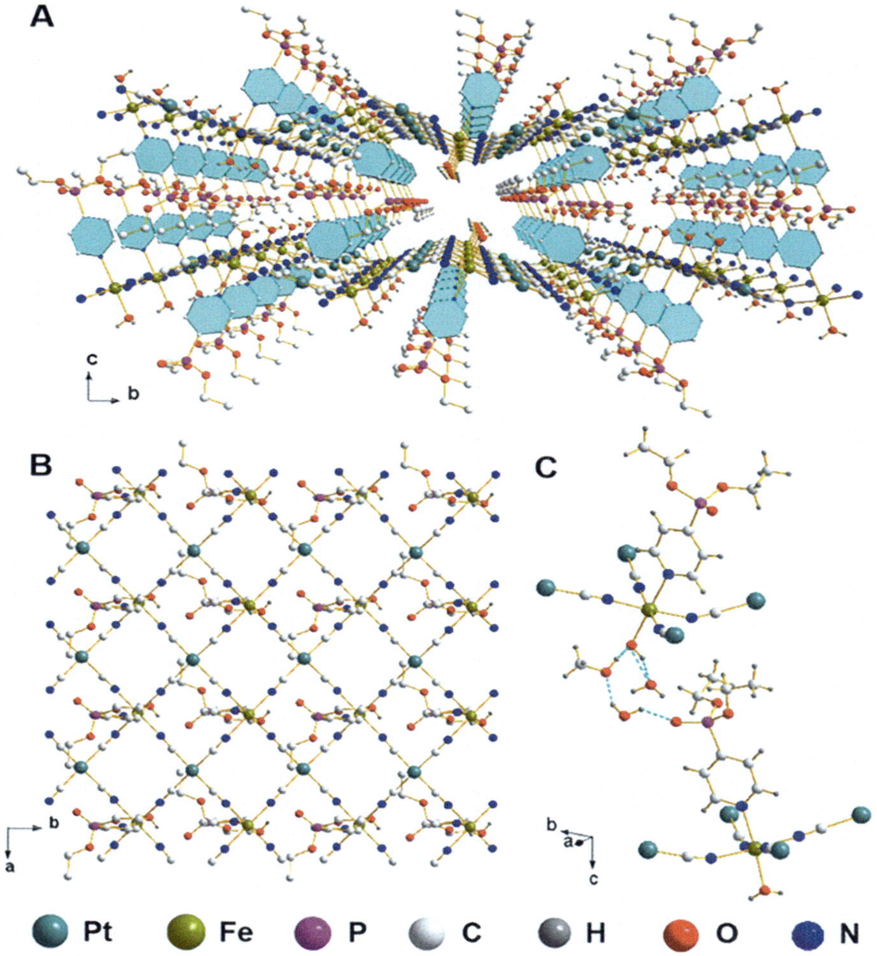

Fig. 4 (**a**) Crystal structure of hMOF-1 and (**b**) its layer viewed along the *c*-axis. (**c**) Stacking interaction of two adjacent layers through H bonds in hMOF-1. (Reproduced from ref. (Córdova Wong et al. 2019) with permission from ACS Publications 2012)

possible even without polymeric or surfactant stabilizers (Li et al. 2011). For the first time, a facile electrochemical method was used to obtain such nanoscale graphene derivatives with high yellow emission spectra at 14% quantum yield (Zhang et al. 2012a). Shinde et al. described an electrochemical process to prepare green luminescent graphene derivatives from multiwalled carbon nanotubes (MWCNT) (Shinde and Pillai 2012). The toxicity effects of carbon nanotubes were one of the main barriers to limit its application. Shinde et al. further described a two-step electrochemical method for size-selective nitrogen-doped graphene materials (N-GDs) synthesis from MWCNT (Shinde et al. 2015). These metal-free electrocatalysts

(N-GDs) were identified for their optoelectronic properties. Recently, Tan et al. reported electrochemical exfoliation of graphene derivatives in potassium persulphate solution to synthesize water-soluble graphene derivatives with a particle size of 3 nm and red emission spectra (Tan et al. 2015). This red fluorescent graphene has in vivo applications as biological labels. An ideal example of a nanolithography method to obtain graphene-based materials from graphene sheets was developed by Ponomarenko et al., where quantum dot devices could transport electron (Ponomarenko et al. 2008). Alternatively, Lee et al. synthesized uniform graphene materials patterned from self-assembled block copolymer (BCP) as an etch mask on graphene films using a chemical vapor deposition method (Lee et al. 2012).

2.2 Bottom-Up Routes

In this process, graphene materials are usually synthesized from small organic molecules like benzene derivatives, carbonation, fullerenes, and unsubstituted Hexa-peri-hexabenzocoronene as starting materials. Zhu et al. introduced a novel approach to synthesize simple graphene molecules and heteroatom-doped graphene materials through the free-radical polymerization of oxygen-containing aromatic compounds as starting materials under ultraviolet irradiation. The synthesized graphene molecules exhibited diverse optical and biological applications (Zhu et al. 2017). However, it is also possible to prepare glucose-derived water-soluble graphene materials through the carbonization process in the microwave-assisted hydrothermal process. In another study, it has been shown that fullerenes and carbon nanotubes could be used to synthesize nanomolecular graphene as starting materials (Teradal and Jelinek 2017). Graphene, along with its derivatives and their possible energy applications, showed that hydrogen peroxide and graphene sheets could be used as starting material by using the one-step solvothermal process for mass production of hybrid graphene materials, which are expected to be suitable for various biomedical applications (Bak et al. 2016). However, many impurities are one of the most significant concerns among scientists who are used to synthesizing graphene-based biomaterials by considering various chemical methods. To overcome this problem, Tian et al., for the first time, introduced a one-step solvothermal approach to produce pure GQDs with the application of hydrogen peroxide in N, N-Dimethylformamide. They claim that there is no need to introduce concentrated sulfuric acid or nitric acid to react with raw materials, even, they did not find any impurities in the whole process. As-produced GQDs exhibit a strong blue emission at 15% quantum yield (Tian et al. 2016). Nevertheless, Liu et al. used unsubstituted Hexa-peri-hexabenzocoronene as a starting material to achieve multicolor photoluminescent graphene derivatives. This novel strategy is good enough to fabricate monodispersed hybrid graphene-based biomaterials (Fig. 5) (Liu et al. 2011).

In another study, a one-step pyrolysis method was described to obtain glutathione (GSH)-characterized GDs (graphene dots) (GDs@GSH) as selective fluorescent materials with increased biocompatibility and a high quantum yield of about

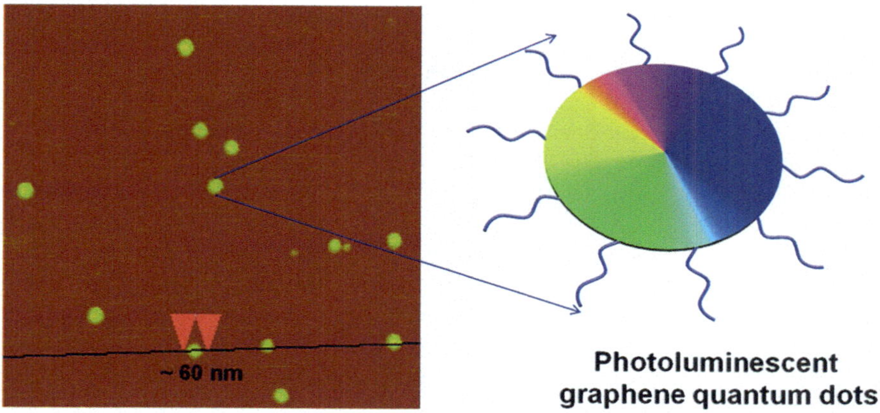

Fig. 5 Multicolor PL GQDs were prepared by using hexa-peri-hexabenzocoronene (Reproduced from ref. (Liu et al. 2011) with permission from ACS Publications 2011)

33.6% (Liu et al. 2013). Wang et al. proposed a facile molecular fusion route under mild and green hydrothermal conditions to synthesize single-crystalline graphene derivatives. The synthesis process of the high-quality graphene molecules involved nitration of pyrene in an aqueous alkaline solution where alkaline was responsible for tuning size, functionalization, and high optical properties (Wang et al. 2014a). Moreover, highly luminescent nitrogen-doped graphene derivatives with high quantum yield (almost 59.2%) were synthesized using a facile bottom-up method. Interestingly, the produced graphene derivatives emitted a robust blue light under 365 nm UV light and were used to analyze 2,4,6-trinitrophenol in natural water samples (Lin et al. 2015). Nevertheless, a bottom-up approach could be defined by using fullerene. For example, C60 molecules are used as a starting material to prepare geometrically well-defined GQDs on a ruthenium surface (Lu et al. 2011).

3 Characteristics of Graphene-Based Theranostic Agents

The derivatives of graphene are carbonaceous material and usually crystalline semiconductor in nature which is synthesized from graphene and very small in diameter between 1 and 20 nm that can cause excitation confinement (Paulo et al. 2016; Cayuela et al. 2016; Nakano et al. 2018; Zhang et al. 2012b) and have opened a new window in the field of nanotechnology which has been highly concerned by researchers due to its unique characteristics. For graphene and its hybrid derivatives, various characterization techniques are widely used, such as transmission electron microscopy (TEM), scanning electron microscope, atomic force microscope (AFM), Raman spectrum, thermal gravimetric analysis (TGA), UV-vis spectrum, and X-ray powder diffraction (XRD) which are illustrated here (Fig. 6).

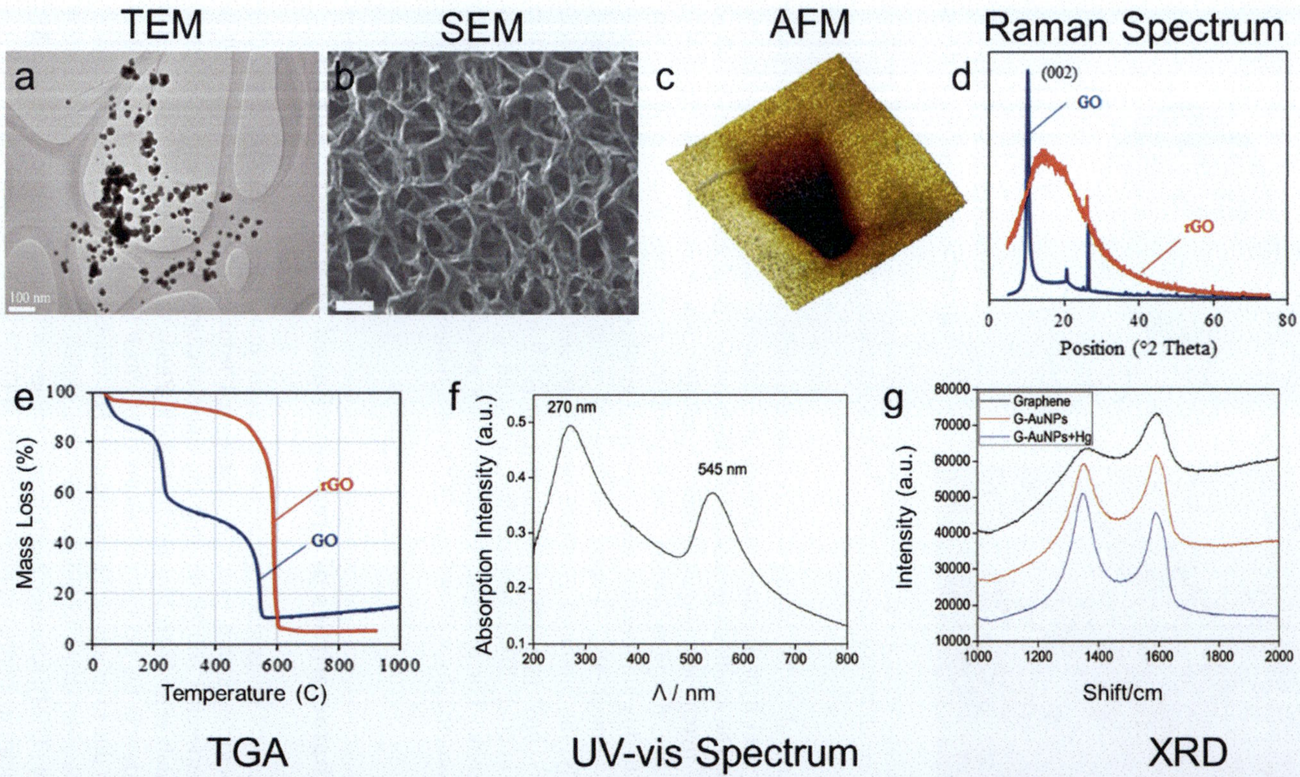

Fig. 6 The characterization techniques of graphene and its derivatives. TEM (**a**), Raman spectrum (**d**), UV-vis spectrum of multifunctional graphene-gold nanocomposite for environment-friendly enriching, separating, and detecting Hg^{2+} simultaneously (reproduced from ref. (Yan et al. 2014) with permission from ACS publications 2014), SEM analysis of graphene foam (reproduced from ref. (Goh et al. 2014) with permission from ACS publications 2014) (**b**); Graphene

Nanopore Support System for Simultaneous High-Resolution AFM Imaging and Conductance Measurements (reproduced from ref. (Connelly et al. 2014) with permission from ACS publications 2014) (**c**); TGA (**e**) and XRD (**g**) analysis of graphene oxide and reduced graphene oxide (reproduced from ref. (Gholampour et al. 2017) with permission from ACS publications 2017)

Peng et al. claimed that few derivatives are zero-dimensional graphene segments in the range of 1–4 nm that can quickly point out quantum confinement and their particle size effects. These graphene derivatives also exhibited electrooptical properties, two-dimensional morphological structure, luminescence stability, nanosecond lifetime, biocompatibility, and low to toxicity, (Peng et al. 2012) which is why various graphene derivatives are being used as nanocarriers to ameliorate the delivery of operable chemotherapeutics to cancer cells. Most importantly, in graphene oxides, the bandgap is tunable by modifying the size, which is strong enough to show high photoluminescence for bioimaging, light-emitting diodes, and biosensors compared with zero-band gap graphene sheets (Wu et al. 2013). Even like conventional graphene, its hybrid derivatives express a single atom layered structure (Chhabra et al. 2018; Zhao et al. 2018; Zhou et al. 2012). Their superior biocompatibility and multimodal conjugation characteristics usually deal with resistance mechanism, systemic toxicity, and undesirable side effects of anticancer drug (Iannazzo et al. 2017). These nano-dimensional materials are nontoxic with having a molecular-like structure. The chemical inertness of graphene can minimize potential toxicity associated with traditional theranostic agents induced by intrinsic heavy metals (Derfus et al. 2004). Miao et al. elucidated various fascinating characteristics of graphene-based materials, for example, photoluminescence emission, chemiluminescence, electrochemical luminescence, and peroxidase-like activity (Miao et al. 2015). As functional groups like carboxylic acid groups are present at the border of the surface-modified graphene, it allows graphene-based hybrid theranostic agents to be highly water-soluble and favorable for further modification with ions, biomolecules, genetic molecules, proteins, enzymes, and even whole cells (Singh and Nalwa 2011). Graphene as planar nanomaterials indicates surface defects and edge structure (Ponomarenko et al. 2008). Interestingly, various graphene derivatives absorption spectra show absorption peaks between 230 and 370 nm due to different types of transitions (Chen et al. 2006). Besides, the previous study pointed out that graphene has exceptional electrothermal conductivity compared to traditional semiconductor theranostic agents due to the presence of π–π bonds at both levels of the atomic plane (Chen et al. 2017).

4 Theranostic Applications of Graphene

Widely graphene-based biomaterials have taken great attention from researchers due to their high cellular imaging and labeling properties. Graphene derivatives are being used in biomedical applications because of their highly water-soluble properties and target-specific treatment approach (Zhu et al. 2014; Wo et al. 2016; Guo et al. 2018). As an essential part of the theranostic application, the roles of graphene and its derivatives are undeniable. To unravel the essential roles of graphene as theranostic agents, Lei et al. claimed that organic fluorophores are responsible for visualizing drug delivery systems. At the same time, cellular uptake is now so far so good due to the promising biogenic properties of graphene and its derivatives (Li et al. 2011). In another application, Xiao et al. developed a new combination where graphene

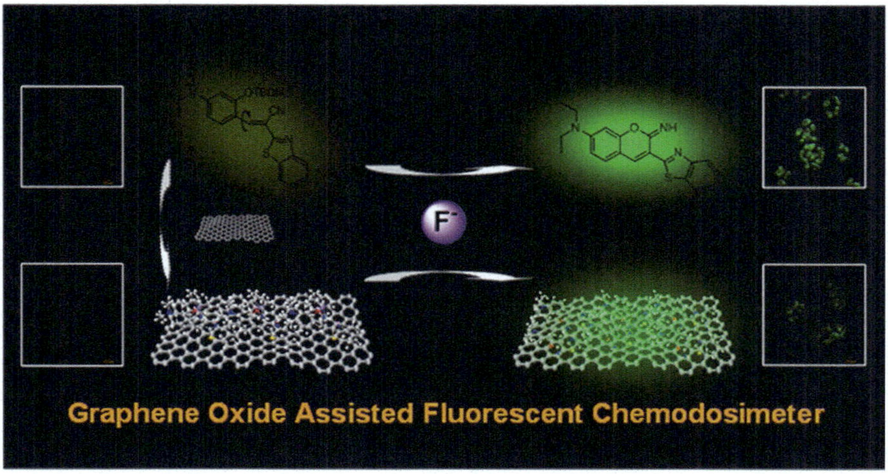

Fig. 7 Graphene oxide assisted fluorescent chemodosimeter for high-performance sensing and bioimaging of fluoride ions (Reproduced from ref. (Wang et al. 2014b) with permission from ACS publications 2014)

materials were labeled with neuroprotective peptide glycine-proline-glutamate to act against Alzheimer's disease. In this study, the APP/PS1 transgenic mice model was used. The authors concluded that hybrid graphene materials could be effective in nanomedicine as a novel theranostic agent. The novel combination of graphene and therapeutics could enhance learning capacity, amount of dendritic spine, and anti-inflammatory cytokines and decrease the aggregation of Aβ1–42 fibrils and various cytokines (Xiao et al. 2016). Also, a potential application of graphene-based materials in the future could be the identification or sensing of various biological markers (Wang et al. 2014b). For example, Fig. 7 illustrates the application of GO in bioimaging of fluoride ions. Previous studies have shown the biomedical application of natural products in diabetes, cancer, and many complex diseases through traditional drug delivery systems (Shahriar et al. 2018; Shipa et al. 2018; Shahriar and Rahmatullah 2015a; Shahriar and Rahmatullah 2015b; Amirunnesa et al. 2018; Zahan et al. 2013; Hasan and Shahriar 2018). However, current advances in science believe that the next generation of microfluidic devices can be used more effectively against diabetes. Such devices will facilitate real-time monitoring of blood glucose levels through a graphene-based sensor, and the other arm will be used to deliver drugs derived from natural products to the body (Hasan and Shahriar 2018; Ghosh et al. 2018; Khan et al. 2018).

In recent days, researchers tried to use graphene-based semiconductor materials along with various chemotherapeutics. Recently, Sui et al. demonstrated that cytotoxicity, cell cycle arrest, and DNA fragmentation of cisplatin could be increased by graphene. The combination of CDDP-graphene increased the cellular uptake via increasing cellular permeability even in the CDDP resistance cells. Moreover, graphene-based therapeutic agents are also reported to enhance the interaction between CDDP and DNA. Thus, this combination of graphene and therapeutics could be used to enhance anticancer activity (Sui et al. 2016). Gemcitabine is another anticancer agent primarily used against pancreatic cancer. Nigam et al. studied that encapsulated Gemcitabine in graphene-based biomaterials could increase this anticancer drug's bioavailability and sustain release properties into pancreatic cancer cells (Nigam et al. 2014). However, Doxorubicin is one of the most promising chemotherapy medications against different forms of cancer, either alone or in combination with other therapeutics. For the time being, researchers have used this drug in conjugation with various graphene derivatives. Figure 1 illustrated that blue-color emitting graphene derivatives were developed with Herceptin (HER) as a target stimuli of HER2-overexpressed breast cancer cell lines and β-cyclodextrin (β-CD) as drug loading site of Doxorubicin, to propose this nanoconjugate as a novel theranostic probe against breast cancer. Control release characteristics of Doxorubicin as a hydrophobic anticancer drug had taken place within the intracellular acidic conditions of breast cancer cells and inhibited uncontrolled cell division. Nafiujjaman et al. reported graphene-based nano complex could have high potentiality as a promising theranostic agent against spontaneous malignancy (Ko et al. 2017). They developed a blue-color emitting graphene derivative from organic molecules that exhibited highly biocompatible and quantum yield properties (almost 24%), which is far better in comparison with traditional graphene derivatives. In addition, epithelial carcinoma cell lines could easily absorb their developed graphene derivative without showing any toxicity or cell membrane damage properties in human erythrocytes. This study again suggested that graphene-based hybrid nanomaterials could be used as safe and potential theranostic agents.

In another study, graphene quantum dots were covalently linked with tumor-targeting module biotin (BTN) to deliver DNA intercalating drug doxorubicin (DOX). The synthesized nanoconjugation could easily trace biotin receptors that over-expressed on A549 lung cancer cells. The acid-sensitive release of Doxorubicin from its depot caused cytotoxic effects to cancer cells rather than normal tissues; hence, this system was free from non-desirable (Iannazzo et al. 2017). Graphene and its derivatives have attracted wide attention for phototherapy against cancer. Recently a promising imaging tool was studied for the monitoring and early prognosis of phototherapy. The study concluded that an iron deposited and PEGylated graphene oxide nanoparticle could accurately evaluate phototherapy in a cancer-induced experimental animal model (Fu et al. 2016). Since bile acid derivatives such as taurocholic acid (TCA) could target the liver via the enterohepatic circulation pathway (Shahriar et al. 2017; Nurunnabi and Shahriar 2021; Nafiujjaman et al. 2021; Shahriar et al. 2021b; Hasan et al. 2021), TCA-functionalized graphene derivative could be a promising option to develop a successful therapeutic agent targeting the liver metastasis (An et al. 2021a; Shahriar et al. 2021a). Furthermore, the promising results of FcRn receptor-targeted therapeutics, (Hasan et al. 2020) electrospinning nanofibers, (Shahriar et al. 2019) and hydrogel-based therapeutics (An et al. 2021b) open a new window to utilize the possibility of GQDs in drug delivery purposes (Fig. 8).

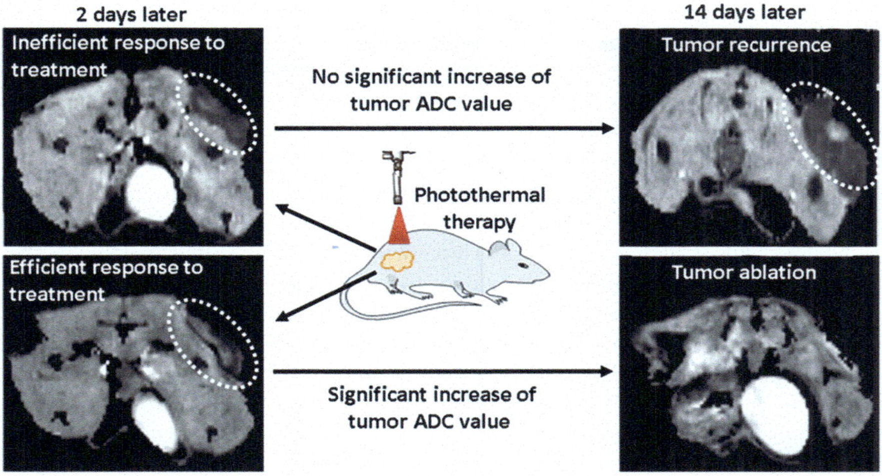

Fig. 8 Diffusion-weighted magnetic resonance imaging for therapy response monitoring and early treatment prediction of photothermal therapy (Reproduced from ref. (Fu et al. 2016) with permission from ACS publications 2016)

To enhance cancer therapy, carboxyl-terminated graphene derivative (CG) was linked with iron oxide conjugated silicon dioxide (Fe_3O_4@SiO_2)to form Fe_3O_4@SiO_2@CG). Finally, the anticancer therapeutic doxorubicin was loaded with cancer-targeting moiety folic acid-modified Fe_3O_4@SiO_2@G-FA conjugation. This Fe_3O_4@SiO_2@G-FA/DOX nanoprobe proved itself as a potential theranostic agent to accurately diagnose and treat cancer disease on HeLa cells effectively (Su et al. 2017). In such a way, graphene and its derivatives are being used for a long back as a theranostic agent either alone or in combination with chemotherapeutics. Abdullah Al Nahain and coworkers used hyaluronic acid (HA) as an organ-specific homing ligand to achieve efficiency and target-specific delivery of graphene-based materials. Here, hybrid graphene material was used to load and deliver hydrophobic drug doxorubicin in specific targets and release in mildly acidic conditions. The in vivo and in vitro studies showed that GQDs could be used as non-toxic and truly biocompatible theranostic agent (Al-Nahain et al. 2013).

Recently, folic acid conjugated graphene derivatives (FA-G) were synthesized to act as dual padlock theranostic agents with the anticancer drug doxorubicin (DOX). This DOX-G-FA nano-assembly could discriminate between cancerous cells and normal cells. Hence, target-specific delivery of anticancer drug doxorubicin is achieved by FA-G. The DOX-G-FA conjugation could easily target and enter into cells through receptor-mediated endocytosis of HeLa cells. This study also suggested that real-time monitoring of intracellular intake of this nanoprobe and constant release of DOX could occur due to the strong fluorescence of graphene-based materials. The synthesized nanoprobe exhibited an enhanced accumulation profile in targeted tissues with sustained-release properties (Wang et al. 2014c). The authors reported a vast difference in cellular uptake and toxicological profile among naked DOX and graphene-inspired DOX. In a similar study, Wang et al. developed DOX-G nanoparticles to enhance nuclear uptake, cytotoxicity, and DNA cleavage activity of doxorubicin in chemotherapeutic resistance cancerous cells. Graphene-based biomaterials boosted the anticancer activity of therapeutics due to their super unique properties as a theranostic agent (Wang et al. 2013). Moreover, the graphene-based derivative is also reported as a light-responsive nano construct for in vivo imaging in cancer therapy. In combination with load gadolinium texaphyrin and lutetium texaphyrin, it could show a synergistic effect against solid tumors. Its ability to load and target-specific delivery, adsorb high drug payload and magnetic resonance imaging, produce free radicals, and activate near-infrared-fluorescence make it more acceptable as a theranostic agent (Yang et al. 2017). In previous scientific validation, graphene-based nanoprobe shows a synergistic effect against spontaneous malignancy rather than naked therapeutics (Khodadadei et al. 2017).

Drug resistance is a big issue in cancer nanotechnology, (Shi et al. 2017; Casals et al. 2017) and various multidrug resistance (MDR) genes can cause the failure of multiple chemotherapeutics (Shahriar et al. 2019; Alfarouk et al. 2015). To overcome these limitations, researchers are finding a novel single reagent to target ATP-binding cassette (ABC) transporters for inhibiting multidrug genes. In this perspective, Luo et al. showed that a hybrid graphene derivative could interact with C-rich promoters to inhibit multidrug resistance protein-like multidrug-resistant protein-1 (MRP1), P-glycoprotein, breast cancer resistance genes and alter the functions of doxorubicin resistance cells (Luo et al. 2017). For the significant delivery process of anticancer molecule berberine hydrochloride (BHC), cow milk-derived highly water-soluble graphene derivatives were conjugated with Cysteamine hydrochloride (Cys) to attach BHC and form G@Cys-BHC. The G@Cys-BHC Nanoprobe studied various cancerous cell line models and found that G@Cys-BHC nanoprobe has highly cytotoxic effects on both cervical and human breast cancer cells (Fig. 6) (Thakur et al. 2016). Nowadays, bacterial resistance to antibiotics is a severe issue due to its irregular and long-term uses. Sometimes, patients cannot take antibiotics because of improper health conditions, age, or high dosages. To overcome this problem, Pan et al. suggested a new approach, graphene-based antibiotic delivery to deep-seated tissues/organs. In this study, graphene materials were conjugated with antibacterial agent vancomycin (Van) and finally bound with Protoporphyrin IX (PpIX), a photo/radiation sensitizer, to form Van-Gs/PpIX construct. The Van-GQDs/PpIX usually target *E. coli* and exhibited bactericidal actions via using a low-dose x-ray. Interestingly, *E. coli* cannot form a resistance against this new complex at least in the first week seven repetitive dose (Pan et al. 2017). To date, various multimodal drug delivery systems have been studied to overcome all the barriers related to multidrug resistance. A hyaluronic acid-inspired and rhodamine B conjugated graphene biomaterial was synthesized for an imaging-guided targeted delivery system against drug resistance lung cancer tissues. As a model anticancer drug, Doxorubicin was loaded with this

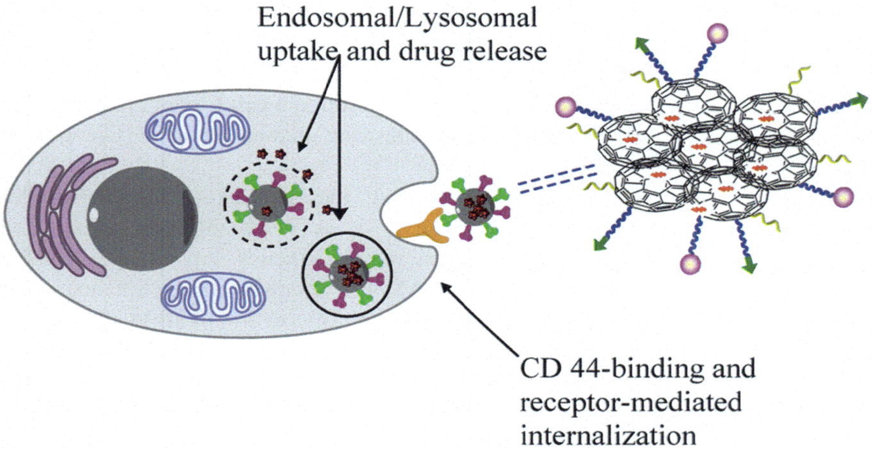

Fig. 9 Hyaluronic acid-modified multifunctional Q-graphene for targeted killing of drug-resistant lung cancer cells (Reproduced from ref. (Luo et al. 2016) with permission from ACS publications 2016)

dual padlock, multifunctional graphene-based nanocarrier via Π-Π stacking (Luo et al. 2016) (Fig. 9).

In a traceable drug delivery system, graphene and its derivatives are used as drug carriers because of targeted and pH-sensitive drug delivery (Yao et al. 2017; Cherian et al. 2014). For instance, Lv and coworkers synthesized graphene-labeled chitosan (CS) hybrid xerogels as a drug carrier with a porous 3D network. In vivo study suggested three distinct strong luminescence properties of G-CS such as blue, green, and red luminescence at 405, 430, and 435 nm wavelength, respectively, and also confirmed 43% drug payload in the xerogel. G-CS hybrid xerogel was used to target specific drug delivery due to its pH-dependent drug release characteristics (Lv et al. 2016). On the other hand, Zhou et al. synthesized periphery carboxylic groups containing graphene derivative by Photo-Fenton chemistry of graphene oxide under UV irradiation. Hence, these graphene derivatives could play an essential role as drug or gene porter and a theranostic agent in nanomedicine (Zhou et al. 2012). The reactive oxygen species (ROS), on the other hand, are significant concerns among scientists due to the immunosuppression of acute myeloid leukemia. To take this concern in hand, Cherukula et al. prepared a stable graphene-based theranostic agent that facilitates ROS inhibiting properties along with histamine dihydrochloride (HDC) to prevent immunosuppression in the leukemic cells. The authors further coated the synthesized form HDC-G nanoprobe with hyaluronic acid (HA) to form HDC-G_HA so that this proposed conjugation could easily target cancer cells, K-562 cells. The high drug loading graphene derivative showed sustained drug release properties and cytotoxic effects at K-562 cells (Cherukula et al. 2018). In another study, Oh, et al. examined ROS regulation in macrophages with thiolated-graphene derivative (SH-GDs) as a theranostic nanocarrier. This study concluded that the expression level of ATP-binding cassette transporters (ABCA1) in macrophages was dramatically increased when cells treated with SH-GDs and macrophage scavenger receptors (MSR) were on the surface of the macrophage was down-regulated. Hence, to prevent foam cell formation and plaque development, SH-GDs could be used as a promising theranostic agent (Oh and Lee 2016).

Moreover, Zheng et al. capped fluorescent mesoporous silica nanoparticles (FMSNs) with a hybrid graphene derivative via ATP aptamer for cell-targeted drug delivery and real-time monitoring of release kinetics. Aptamer-Gs-FMSNs could easily target cancer cells due to the AS1411 aptamer, and then nanocarriers released the drug resulting in severe cytotoxicity. Due to low ATP level in extracellular portions, that complex never releases the drug, but in intracellular portions, enhanced ATP in the cytoplasm could cause drug release from its carrier. Thus, AS1411-Gs work as good fluorescent resonance energy transfer-based nanocarrier in cancer therapy (Zheng et al. 2015). Various studies like Zhao et al. described graphene-based drug/gene delivery and their possible toxicity (Zhao et al. 2017). Hence, scientists agreed in the statement that GQDs are one of the most promising theranostic agents.

5 Toxicological Study

Despite all potential biomedical applications of graphene derivatives such as drug delivery systems, drug carriers, and bioimaging, toxicity evaluation is a significant concern among researchers. In nanomedicine, toxicological study of raw, chemically doped and chemically functionalized graphene is one of the key requirements before starting its use in this highly prospective area. Both in vitro and in vivo studies suggest that the toxicity associated with graphene and its derivatives mainly depends on the concentration of graphene derivatives, their synthesis procedures, diameter, surface biochemistry, and chemical doping (Singh and Nalwa 2007; Wang et al. 2016).

To start, previously reported Nafiujjaman et al. found that graphene-based hybrid materials which were prepared from pyrolyzing L-glutamic acid exhibited non-toxicity on cells at a concentration of 1000 μg per milliliter. Nevertheless, no damage and changes in the number of erythrocytes were found up to 750 μg per milliliter (Nafiujjaman et al. 2018). Hereafter, Luo et al. showed the graphene-based theranostic complexes as toxic free molecules with the ability to enter and accumulate within the cells targeting human breast cancer cells, but they do not react with cell division (Luo et al. 2017). Another study validated the non-toxic properties of graphene derivatives. For example, a high concentration of graphene derivative exhibited no toxicity in K-562 cell lines (Cherukula et al. 2018). In another study, Nahain et al. developed a novel drug carrier where graphene derivatives were conjugated with HA for target-specific delivery. The in vitro and in vivo studies suggested that this new conjugation could be used as a highly biocompatible, non-toxic theranostic agent in the biological application of graphene (Al-Nahain

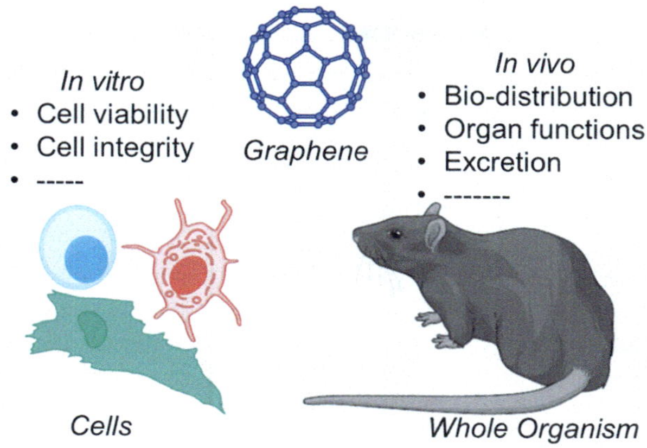

Fig. 10 The toxicity of graphene and its derivatives in various cell lines and laboratory animals

et al. 2013). Also, the interaction between graphene oxide NPs and gastric cancer cells or skin cells showed no toxicity (Li et al. 2012b) (Fig. 10).

In contrast, ROS such as singlet oxygen could be raised by photo-excited graphene derivative resulting in apoptosis and autophagy programmed cell death in U251 human glioma cells (Markovic et al. 2012). In biological systems, graphene and its derivatives can sometimes interfere with DNA, resulting in DNA damage. A few years back, Wang et al. investigated such genotoxicity of graphene materials in NIH-3 T3 cells. Graphene and its derivatives could enhance the expression of tumor antigen p53, eukaryotic gene Rad 51 and (8-oxoguanine glycosylase) OGG1 proteins in cells by increasing the generation of ROS; hence, some of the derivatives of graphene such as GQDs or CQDs could cause more intracellular DNA damage compared with untreated cells (Wang et al. 2015a). The chemical modification of graphene derivatives with various functional groups, such as primary amine (NH_2) group, carboxylic acid (COOH) group, and CO-N $(CH_3)_2$ group, showed low toxicity in lung carcinoma epithelial A549 cells and spindle-like neural glioma C6 cell lines. Interestingly, G-[CO-N $(CH_3)_2$] functional group exhibits less toxicity in comparison with G-NH_2 and G-COOH functional groups in both cell types. On the other hand, G-NH_2 was less toxic than G-COOH in human A549 lung carcinoma cells. These modified graphene derivatives with three functional groups were suitably biocompatible even with a high concentration (200 μg per milliliter) (Yuan et al. 2014). An in vivo study suggests that the toxicity of graphene-based biomaterials maintains dose-dependent enhancement properties. Wang et al. investigated the toxicity in embryonic zebrafish by incubating a graphene derivative (0, 12.5, 25, 50, 100, and 200 μg per milliliter) for 4–96 h post-fertilization. At lower concentrations, graphene derivative was toxic-free in zebrafish embryo (Wang et al. 2015b). Another study suggests that the nanostructure and high oxygen content of graphene and its derivatives are responsible for its toxicity. Besides, the in vivo study indicates the same statistics whether they form no accumulation in the major organs of the experimental mice model due to its unique ultrasmall size. In comparison with traditional graphene oxide, GQDs and CQDs do not generate aggregation inside mice (Chong et al. 2014). Due to their nanostructure and excellent targeting properties can target differentially and efficiently and leave biological systems carefully. This unique mechanism showed how GQDs and CQDs reduce cytotoxicity to non-target cells (Wang et al. 2014c).

6 To-Be Perspective

Due to the excellent physicochemical and biological properties, graphene and its derivatives are widely used as a potential theranostic agent by conjugating with other nanomolecules, but to translate the main aim of graphene-based nanomaterials in broad term clinical applications, some issues should be addressed, including synthesis protocols, manufacturing cost, biological behavior, toxicity, body clearance, and so one (Shahriar 2018).

To date, there are few reports that claimed the large-scale production of highly water-soluble graphene derivatives (Zhou et al. 2012). Despite these studies, it is now a great concern to overcome the problems related to small-scale productions. Moreover, an easy synthesis process to prepare high-quality graphene-based derivatives is now raising questions to expand the theranostic applications of this new nanoconstruct (Chen et al. 2017). Still, the proposed quantum yield of graphene and its derivatives ranged from 10% to 55%, which needs to be improved. Easy purification approaches are needed to synthesize a high quantum yield of graphene-based theranostic agents (Joshi et al. 2016). Sometimes, graphene and its derivatives increase ROS activities or interfere with DNA that may cause toxicity and other possible side effects. Dose-dependent toxicity of graphene-based theranostic agent is one of the critical difficulties that researchers are facing nowadays. Hence, studying the toxicology of graphene and its derivatives in a biological environment is more important. Though some of its derivatives show low toxicity rather than nanomaterials, more understanding of biodistribution, accumulation, and genotoxicity is necessary.

Last but not least, graphene and its derivatives could be used to treat neurodegenerative disorders. We can also think about graphene-based gene therapy regarding brain application. For this purpose, a scientist has to develop new GQDs to cross the blood-brain barrier and demonstrate their activities easily (Xu et al. 2013; Zhou et al. 2018).

Acknowledgments This research was supported by a National Research Foundation (NRF) grant funded by the Korean government (no. 2019R1A4A1024116, 2020R1I1A1A01058267, and 2021R1A6A1A03046418).

References

Alfarouk KO, Stock C-M, Taylor S, Walsh M, Muddathir AK, Verduzco D, Bashir AHH, Mohammed OY, Elhassan GO, Harguindey S, Reshkin SJ, Ibrahim ME, Rauch C (2015) Resistance to cancer chemotherapy: failure in drug response from ADME to P-gp. Cancer Cell Int 15:71. https://doi.org/10.1186/s12935-015-0221-1

Al-Nahain A, Lee JE, In I, Lee H, Lee KD, Jeong JH, Park SY (2013) Target delivery and cell imaging using hyaluronic acid-functionalized graphene quantum dots. Mol Pharm 10 (10):3736–3744. https://doi.org/10.1021/mp400219u

Amirunnesa SK, Akter K, Shipa SMS, Rahmatullah M (2018) Folk medicinal practices in Keraniganj Upazila, Dhaka district, Bangladesh. J Med Plants Stud 6:89–91. https://www.plantsjournal.com/archives/2018/vol6issue1/PartB/5-6-38-816.pdf

An JM, Shahriar SMS, Hwang YH, Hwang SR, Lee DY, Cho S, Lee Y (2021a) Oral delivery of parathyroid hormone using a triple-padlock nanocarrier for osteoporosis via an enterohepatic circulation pathway. ACS Appl Mater Interfaces. https://doi.org/10.1021/acsami.0c22170

An JM, Shahriar SMS, Hasan MN, Cho S, Lee Y (2021b) Carboxymethyl cellulose, pluronic, and pullulan-based compositions efficiently enhance antiadhesion and tissue regeneration properties without using any drug molecules. ACS Appl Mater Interfaces 13:15992–16006. https://doi.org/10.1021/acsami.0c21938

Bak S, Kim D, Lee H (2016) Graphene quantum dots and their possible energy applications: a review. Curr Appl Phys 16:1192–1201. https://doi.org/10.1016/j.cap.2016.03.026

Bhatnagar D, Kumar V, Kumar A, Kaur I (2016) Graphene quantum dots FRET based sensor for early detection of heart attack in human. Biosens Bioelectron 79:495–499. https://doi.org/10.1016/j.bios.2015.12.083

Casals E, Gusta MF, Cobaleda-Siles M, Garcia-Sanz A, Puntes VF (2017) Cancer resistance to treatment and antiresistance tools offered by multimodal multifunctional nanoparticles. Cancer Nanotechnol 8:7. https://doi.org/10.1186/s12645-017-0030-4

Cayuela A, Soriano ML, Carrillo-Carrión C, Valcárcel M (2016) Semiconductor and carbon-based fluorescent nanodots: the need for consistency. Chem Commun 52:1311–1326. https://doi.org/10.1039/C5CC07754K

Chen K-Y, Hsieh C-C, Cheng Y-M, Lai C-H, Chou P-T, Chow TJ (2006) Tuning excited-state electron transfer from an adiabatic to nonadiabatic type in donor−bridge−acceptor systems and the associated energy-transfer process. J Phys Chem A 110:12136–12144. https://doi.org/10.1021/jp063038s

Chen F, Gao W, Qiu X, Zhang H, Liu L, Liao P, Fu W, Luo Y (2017) Graphene quantum dots in biomedical applications: recent advances and future challenges. Front Lab Med 1:192–199. https://doi.org/10.1016/j.flm.2017.12.006

Chen F, Jin X, Jia D, Cao Y, Duan H, Long M (2020) Efficient treament of organic pollutants over CdS/graphene composites photocatalysts. Appl Surf Sci 504:144422. https://doi.org/10.1016/j.apsusc.2019.144422

Cherian R, Sreejith R, Syama S, Sruthi S, Gayathri TM, Sakthikumar D (2014) Evaluation of toxicity of Maura reduced graphene oxide using in vitro systems. J Nanomed Nanotechnol 05. https://doi.org/10.4172/2157-7439.1000200

Cherukula K, Nurunnabi M, Jeong YY, Lee YK, Park IK (2018) A targeted graphene nanoplatform carrying histamine dihydrochloride for effective inhibition of leukemia-induced immunosuppression. J Biomater Sci Polym Ed. https://doi.org/10.1080/09205063.2017.1390382

Chhabra VA, Kaur R, Kumar N, Deep A, Rajesh C, Kim K-H (2018) Synthesis and spectroscopic studies of functionalized graphene quantum dots with diverse fluorescence characteristics. RSC Adv 8:11446–11454. https://doi.org/10.1039/C8RA01148F

Chong Y, Ma Y, Shen H, Tu X, Zhou X, Xu J, Dai J, Fan S, Zhang Z (2014) The in vitro and in vivo toxicity of graphene quantum dots. Biomaterials 35(19):5041–5048. https://doi.org/10.1016/j.biomaterials.2014.03.021

Connelly LS, Meckes B, Larkin J, Gillman AL, Wanunu M, Lal R (2014) Graphene nanopore support system for simultaneous high-resolution AFM imaging and conductance measurements. ACS Appl Mater Interfaces 6:5290–5296. https://doi.org/10.1021/am500639q

Córdova Wong BJ, Xu D, Bao S-S, Zheng L-M, Lei J (2019) Hofmann metal–organic framework monolayer nanosheets as an axial coordination platform for biosensing. ACS Appl Mater Interfaces 11:12986–12992. https://doi.org/10.1021/acsami.9b00693

Derfus AM, Chan WCW, Bhatia SN (2004) Probing the cytotoxicity of semiconductor quantum dots. Nano Lett 4:11–18. https://doi.org/10.1021/nl0347334

Feng Q, Cao Q, Li M, Liu F, Tang N, Du Y (2013) Synthesis and photoluminescence of fluorinated graphene quantum dots. Appl Phys Lett 102:013111. https://doi.org/10.1063/1.4774264

Fu G, Zhu L, Yang K, Zhuang R, Xie J, Zhang F (2016) Diffusion-weighted magnetic resonance imaging for therapy response monitoring and early treatment prediction of photothermal therapy. ACS Appl Mater Interfaces 8:5137–5147. https://doi.org/10.1021/acsami.5b11936

Gholampour A, Valizadeh Kiamahalleh M, Tran DNH, Ozbakkaloglu T, Losic D (2017) From graphene oxide to reduced graphene oxide: impact on the physiochemical and mechanical properties of graphene–cement composites. ACS Appl Mater Interfaces 9:43275–43286. https://doi.org/10.1021/acsami.7b16736

Ghosh NK, Tanzin F, Manjur F, Rizwan F, Khanam M, Shahriar SMS (2018) Antipyretic effect of oral versus rectal acetaminophen in children - a randomized comparative study. World J Pharm Res 7:1722–1734. https://doi.org/10.20959/wjpr20187-11685

Goh B-M, Wang Y, Reddy MV, Ding YL, Lu L, Bunker C, Loh KP (2014) Filling the voids of graphene foam with graphene "eggshell" for improved lithium-ion storage. ACS Appl Mater Interfaces 6:9835–9841. https://doi.org/10.1021/am5022655

Guo X, Wang C-F, Yu Z-Y, Chen L, Chen S (2012) Facile access to versatile fluorescent carbon dots toward light-emitting diodes. Chem Commun 48:2692. https://doi.org/10.1039/c2cc17769b

Guo J, Mei T, Li Y, Hafezi M, Lu H, Li J, Dong G (2018) One-pot synthesis and lubricity of fluorescent carbon dots applied on PCL-PEG-PCL hydrogel. J Biomater Sci Polym Ed 29: 1549–1565. https://doi.org/10.1080/09205063.2018.1470736

Hasan MN, Shahriar SMS (2018) Oral vaccines-types, delivery strategies, current and future perspectives. Biomed J Sci Tech Res 11:8398–8404. https://doi.org/10.26717/BJSTR.2018.11.002075

Hasan MN, Shahriar SMS, Rabbi MF, Sammy LA, Jahan D, Hossen ME, Shah M-R, Islam M-R (2018) Nanoparticles and their utilization in cancer detection and treatment. Sch J Appl Sci Res 1:65–75. Direct Link: https://www.innovationinfo.org/Scholar-Journal-of-Applied-Sciences-and-Research/article/Nanoparticles-and-Their-Utilization-in-Cancer-Detection-and-Treatment#fulltext

Hasan MN, Hwang YH, An JM, Shahriar SMS, Cho S, Lee Y (2020) Oral GLP1 gene delivery by an antibody-guided nanomaterial to treat type 2 diabetes mellitus. ACS Appl Mater Interfaces. https://doi.org/10.1021/acsami.0c09814

Hasan MN, Shahriar SMS, Mondal J, Nurunnabi M, Lee Y (2021) Bioinspired and biomimetic materials for oral drug delivery. In: Nurunnabi M (ed) Bioinspired and biomimetic materials for drug delivery. Elsevier, pp 89–104. https://doi.org/10.1016/B978-0-12-821352-0.00001-0

Hu C, Liu Y, Yang Y, Cui J, Huang Z, Wang Y, Yang L, Wang H, Xiao Y, Rong J (2013) One-step preparation of nitrogen-doped graphenequantum dots from oxidized debris of graphene oxide. J Mater Chem B 1:39–42. https://doi.org/10.1039/C2TB00189F

Iannazzo D, Pistone A, Salamò M, Galvagno S, Romeo R, Giofré SV, Branca C, Visalli G, Di Pietro A (2017) Graphene quantum dots for cancer targeted drug delivery. Int J Pharm 518:185–192. https://doi.org/10.1016/j.ijpharm.2016.12.060

Joshi PN, Kundu S, Sanghi SK, Sarkar D (2016) Graphene quantum dots - from emergence to nanotheranostic applications. Smart Drug Deliv Syst 7:159–195. https://doi.org/10.5772/61932

Keegan J, O'Kennedy R, Crooks S, Elliott C, Brandon D, Danaher M (2011) Detection of benzimidazole carbamates and amino metabolites in liver by surface plasmon resonance-biosensor. Anal Chim Acta 700:41–48. https://doi.org/10.1016/j.aca.2010.12.041

Khan ZH (ed) (2017) Recent trends in nanomaterials. Springer Singapore, Singapore. https://doi.org/10.1007/978-981-10-3842-6

Khan AI, Kawser CA, Chakraborty B, Shahriar SMS (2018) Assessment of risk factors for obesity in selected school children aged 4-9 YEARS OLD. World J Pharm Res 7:39–51. https://doi.org/10.20959/wjpr20186-10624

Khodadadei F, Safarian S, Ghanbari N (2017) Methotrexate-loaded nitrogen-doped graphene quantum dots nanocarriers as an efficient anticancer drug delivery system. Mater Sci Eng C 79:280–285. https://doi.org/10.1016/j.msec.2017.05.049

Ko NR, Nafiujjaman M, Lee JS, Lim H-N, Lee Y-K, Kwon IK (2017) Graphene quantum dot-based theranostic agents for active targeting of breast cancer. RSC Adv 7:11420–11427. https://doi.org/10.1039/C6RA25949A

Lee J, Kim K, Park WI, Kim B-H, Park JH, Kim T-H, Bong S, Kim C-H, Chae G, Jun M, Hwang Y, Jung YS, Jeon S (2012) Uniform graphene quantum dots patterned from self-assembled silica nanodots. Nano Lett 12:6078–6083. https://doi.org/10.1021/nl302520m

Li Y, Hu Y, Zhao Y, Shi G, Deng L, Hou Y, Qu L (2011) An electrochemical avenue to green-luminescent graphene quantum dots as potential electron-acceptors for photovoltaics. Adv Mater 23:776–780. https://doi.org/10.1002/adma.201003819

Li L-L, Ji J, Fei R, Wang C-Z, Lu Q, Zhang J-R, Jiang L-P, Zhu J-J (2012a) A facile microwave avenue to electrochemiluminescent two-color graphene quantum dots. Adv Funct Mater 22: 2971–2979. https://doi.org/10.1002/adfm.201200166

Li JL, Bao HC, Hou XL, Sun L, Wang XG, Gu M (2012b) Graphene oxide nanoparticles as a nonbleaching optical probe for two-photon luminescence imaging and cell therapy. Angew Chemie Int Ed. 51(8):1830–1834. https://doi.org/10.1002/anie.201106102

Li L, Wu G, Yang G, Peng J, Zhao J, Zhu J-J (2013) Focusing on luminescent graphene quantum dots: current status and future perspectives. Nanoscale 5:4015. https://doi.org/10.1039/c3nr33849e

Li X, Zhang S, Kulinich SA, Liu Y, Zeng H (2015) Engineering surface states of carbon dots to achieve controllable luminescence for solid-luminescent composites and sensitive Be2+ detection. Sci Rep 4:4976. https://doi.org/10.1038/srep04976

Lin L, Rong M, Lu S, Song X, Zhong Y, Yan J, Wang Y, Chen X (2015) A facile synthesis of highly luminescent nitrogen-doped graphene quantum dots for the detection of 2,4,6-trinitrophenol in aqueous solution. Nanoscale 7:1872–1878. https://doi.org/10.1039/C4NR06365A

Liu R, Wu D, Feng X, Müllen K (2011) Bottom-up fabrication of photoluminescent graphene quantum dots with uniform morphology. J Am Chem Soc 133:15221–15223. https://doi.org/10.1021/ja204953k

Liu J-J, Zhang X-L, Cong Z-X, Chen Z-T, Yang H-H, Chen G-N (2013) Glutathione-functionalized graphene quantum dots as selective fluorescent probes for phosphate-containing metabolites. Nanoscale 5:1810. https://doi.org/10.1039/c3nr33794d

Liu H, Qiu J, Zhang R, Li J, Sang Y, Tang W, Rivera Gil P (2015) Fluorescent graphene quantum dots as traceable, pH-sensitive drug delivery systems. Int J Nanomedicine 6709. https://doi.org/10.2147/IJN.S91864

Lu J, Yeo PSE, Gan CK, Wu P, Loh KP (2011) Transforming C60 molecules into graphene quantum dots. Nat Nanotechnol 6:247–252. https://doi.org/10.1038/nnano.2011.30

Luo Z, Yang D, Qi G, Shang J, Yang H, Wang Y, Yuwen L, Yu T, Huang W, Wang L (2014) Microwave-assisted solvothermal preparation of nitrogen and sulfur co-doped reduced graphene oxide and graphene quantum dots hybrids for highly efficient oxygen reduction. J Mater Chem A 2:20605–20611. https://doi.org/10.1039/C4TA05096G

Luo Y, Cai X, Li H, Lin Y, Du D (2016) Hyaluronic acid-modified multifunctional Q-graphene for targeted killing of drug-resistant lung cancer cells. ACS Appl Mater Interfaces 8:4048–4055. https://doi.org/10.1021/acsami.5b11471

Luo C, Li Y, Guo L, Zhang F, Liu H, Zhang J, Zheng J, Zhang J, Guo S (2017) Graphene quantum dots downregulate multiple multidrug-resistant genes via interacting with their c-rich promoters. Adv Healthc Mater 6(21):1700328. https://doi.org/10.1002/adhm.201700328

Lv O, Tao Y, Qin Y, Chen C, Pan Y, Deng L, Liu L, Kong Y (2016) Highly fluorescent and morphology-controllable graphene quantum dots-chitosan hybrid xerogels for in vivo imaging and pH-sensitive drug carrier. Mater Sci Eng C. https://doi.org/10.1016/j.msec.2016.05.031

Malekzad H, Hasanzadeh M, Shadjou N, Jouyban A (2017) Highly sensitive immunosensing of prostate specific antigen using poly cysteine caped by graphene quantum dots and gold nanoparticle: a novel signal amplification strategy. Int J Biol Macromol 105:522–532. https://doi.org/10.1016/j.ijbiomac.2017.07.069

Markovic ZM, Ristic BZ, Arsikin KM, Klisic DG, Harhaji-Trajkovic LM, Todorovic-Markovic BM, Kepic DP, Kravic-Stevovic TK, Jovanovic SP, Milenkovic MM, Milivojevic DD, Bumbasirevic VZ, Dramicanin MD, Trajkovic VS (2012) Graphene quantum dots as autophagy-inducing photodynamic agents. Biomaterials 33(29):7084–7092. https://doi.org/10.1016/j.biomaterials.2012.06.060

Miao P, Han K, Tang Y, Wang B, Lin T, Cheng W (2015) Recent advances in carbon nanodots: synthesis, properties and biomedical applications. Nanoscale 7:1586–1595. https://doi.org/10.1039/C4NR05712K

Nafiujjaman M, Kim J, Park HK, Lee YK (2018) Preparation of blue-color-emitting graphene quantum dots and their in vitro and in vivo toxicity evaluation. J Ind Eng Chem. https://doi.org/10.1016/j.jiec.2017.08.019

Nafiujjaman M, Shahriar SMS, Hasan MN (2021) Bile acid-inspired oral small molecules drug delivery. In: Nurunnabi M (ed) Bioinspired and biomimetic materials for drug delivery. Elsevier, pp 171–186. https://doi.org/10.1016/B978-0-12-821352-0.00017-4

Nakano H, Tetsuka H, Spencer MJS, Morishita T (2018) Chemical modification of group IV graphene analogs. Sci Technol Adv Mater 19:76–100. https://doi.org/10.1080/14686996.2017.1422224

Nigam P, Waghmode S, Louis M, Wangnoo S, Chavan P, Sarkar D (2014) Graphene quantum dots conjugated albumin nanoparticles for targeted drug delivery and imaging of pancreatic cancer. J Mater Chem B 2:3190–3195. https://doi.org/10.1039/C4TB00015C

Novoselov KS, Geim AK, Morozov SV, Jiang D, Zhang Y, Dubonos SV, Grigorieva IV, Firsov AA (2004) Electric field effect in atomically thin carbon films. Science (80-.) 306:666–669. https://doi.org/10.1126/science.1102896

Nurunnabi M, Shahriar SMS (2021) Bioinspired and biomimetic materials for drug delivery, 1st edn. Woodhead Publishing. https://www.elsevier.com/books/bioinspired-and-biomimetic-materials-for-drug-delivery/nurunnabi/978-0-12-821352-0

Oh B, Lee CH (2016) Development of thiolated-graphene quantum dots for regulation of ROS in macrophages. Pharm Res. https://doi.org/10.1007/s11095-016-2000-7

Pan D, Zhang J, Li Z, Wu M (2010) Hydrothermal route for cutting graphene sheets into blue-luminescent graphene quantum dots. Adv Mater 22:734–738. https://doi.org/10.1002/adma.200902825

Pan D, Guo L, Zhang J, Xi C, Xue Q, Huang H, Li J, Zhang Z, Yu W, Chen Z, Li Z, Wu M (2012) Cutting sp2 clusters in graphene sheets into colloidal graphene quantum dots with strong green fluorescence. J Mater Chem 22:3314. https://doi.org/10.1039/c2jm16005f

Pan CL, Chen MH, Tung FI, Liu TY (2017) A nanovehicle developed for treating deep-seated bacteria using low-dose X-ray. Acta Biomater 47:159–169. https://doi.org/10.1016/j.actbio.2016.10.003

Paulo S, Palomares E, Martinez-Ferrero E (2016) Graphene and carbon quantum dot-based materials in photovoltaic devices: from synthesis to applications. Nano 6:157. https://doi.org/10.3390/nano6090157

Peng J, Gao W, Gupta BK, Liu Z, Romero-Aburto R, Ge L, Song L, Alemany LB, Zhan X, Gao G, Vithayathil SA, Kaipparettu BA, Marti AA, Hayashi T, Zhu J-J, Ajayan PM (2012) Graphene quantum dots derived from carbon fibers. Nano Lett 12:844–849. https://doi.org/10.1021/nl2038979

Ponomarenko LA, Schedin F, Katsnelson MI, Yang R, Hill EW, Novoselov KS, Geim AK (2008) Chaotic dirac billiard in graphene quantum dots. Science (80-.) 320:356–358. https://doi.org/10.1126/science.1154663

Razalli RL, Abdi MM, Tahir PM, Moradbak A, Sulaiman Y, Heng LY (2017) Polyaniline-modified nanocellulose prepared from semantan bamboo by chemical polymerization: preparation and characterization. RSC Adv 7:25191–25198. https://doi.org/10.1039/C7RA03379F

Shahriar SMS (2018) siRNA-based nanoparticles: a novel approach for gene silencing to enhance cancer therapy. World J Pharm Res 7:149–165. https://doi.org/10.20959/wjpr201815-11778

Shahriar SMS, Rahmatullah M (2015a) Evaluation of antihyperglycemic activity of a polyherbal formulation Diabeto-VLC. World J Pharm Pharm Sci 5:191–198. https://www.wjpps.com/Wjpps_controller/abstract_id/4356

Shahriar SMS, Rahmatullah M (2015b) Oral glucose tolerance tests in mice with a polyherbal formulation Diabeto-MLF. World J Pharm Pharm Sci 5:199–206. https://www.wjpps.com/Wjpps_controller/abstract_id/4357

Shahriar SMS, Shoma JF, Rahmatullah M (2017) Oral glucose tolerance test (OGTT) with whole plants of Alocasia fronicata Roxb. (Araceae). World J Pharm Res 6:48–53. https://doi.org/10.20959/wjpr201712-9719

Shahriar SMS, Sammy LA, Rabbi MF, Bhuiyan B, Rahman MA, Nafiujjaman M (2018) First time reported in vitro pharmacological activity of Heliconia rostrata (Heliconiaceae) revealed its potency in pain management over a standard analgesic drug. Biol Eng Med 3:1–6. https://doi.org/10.15761/BEM.1000159

Shahriar S, Mondal J, Hasan M, Revuri V, Lee D, Lee Y-K (2019) Electrospinning nanofibers for therapeutics delivery. Nano 9:532. https://doi.org/10.3390/nano9040532

Shahriar SMS, An JM, Hasan MN, Surwase SS, Kim Y-C, Lee DY, Cho S, Lee Y (2021a) Plasmid DNA nanoparticles for nonviral oral gene therapy. Nano Lett. https://doi.org/10.1021/acs.nanolett.1c00832

Shahriar SMS, Hasan MN, Nurunnabi M, Lee Y (2021b) Bile acid transporter as a bioinspired method for oral therapeutics delivery system. In: Nurunnabi M (ed) Bioinspired and biomimetic materials for drug delivery. Elsevier, pp 105–138. https://doi.org/10.1016/B978-0-12-821352-0.00008-3

Shi J, Kantoff PW, Wooster R, Farokhzad OC (2017) Cancer nanomedicine: progress, challenges and opportunities. Nat Rev Cancer 17:20–37. https://doi.org/10.1038/nrc.2016.108

Shin Y, Park J, Hyun D, Yang J, Lee J-H, Kim J-H, Lee H (2015) Acid-free and oxone oxidant-assisted solvothermal synthesis of graphene quantum dots using various natural carbon materials as resources. Nanoscale 7:5633–5637. https://doi.org/10.1039/C5NR00814J

Shinde DB, Pillai VK (2012) Electrochemical preparation of luminescent graphene quantum dots from multiwalled carbon nanotubes. Chem A Eur J 18:12522–12528. https://doi.org/10.1002/chem.201201043

Shinde DB, Vishal VM, Kurungot S, Pillai VK (2015) Electrochemical preparation of nitrogen-doped graphene quantum dots and their size-dependent electrocatalytic activity for oxygen reduction. Bull Mater Sci 38:435–442. https://doi.org/10.1007/s12034-014-0834-3

Shipa A, Koli S, Akter K, Shahriar SMS, Rahmatullah M (2018) Phytotherapeutic practices of a folk medicinal practitioner in Kishoreganj district, Bangladesh. J Med Plants Stud 6:240–242. https://www.plantsjournal.com/archives/?year=2018&vol=6&issue=1&part=D&ArticleId=772

Singh S, Nalwa HS (2007) Nanotechnology and health safety – toxicity and risk assessments of nanostructured materials on human health. J Nanosci Nanotechnol 7:3048–3070. https://doi.org/10.1166/jnn.2007.922

Singh R, Nalwa HS (2011) Medical applications of nanoparticles in biological imaging, cell labeling, antimicrobial agents, and anticancer nanodrugs. J Biomed Nanotechnol 7(4):489–503. https://doi.org/10.1166/jbn.2011.1324

Singh E, Meyyappan M, Nalwa HS (2017) Flexible graphene-based wearable gas and chemical sensors. ACS Appl Mater Interfaces 9:34544–34586. https://doi.org/10.1021/acsami.7b07063

Su X, Chan C, Shi J, Tsang M-K, Pan Y, Cheng C, Gerile O, Yang M (2017) A graphene quantum dot@Fe3O4 @SiO2 based nanoprobe for drug delivery sensing and dual-modal fluorescence and MRI imaging in cancer cells. Biosens Bioelectron 92:489–495. https://doi.org/10.1016/j.bios.2016.10.076

Sui X, Luo C, Wang C, Zhang F, Zhang J, Guo S (2016) Graphene quantum dots enhance anticancer activity of cisplatin via increasing its cellular and nuclear uptake. Nanomed Nanotechnol Biol Med 12:1997–2006. https://doi.org/10.1016/j.nano.2016.03.010

Sun Y-P, Zhou B, Lin Y, Wang W, Fernando KAS, Pathak P, Meziani MJ, Harruff BA, Wang X, Wang H, Luo PG, Yang H, Kose ME, Chen B, Veca LM, Xie S-Y (2006) Quantum-sized carbon dots for bright and colorful photoluminescence. J Am Chem Soc 128:7756–7757. https://doi.org/10.1021/ja062677d

Tan X, Li Y, Li X, Zhou S, Fan L, Yang S (2015) Electrochemical synthesis of small-sized red fluorescent graphene quantum dots as a bioimaging platform. Chem Commun 51:2544–2546. https://doi.org/10.1039/C4CC09332A

Tang L, Ji R, Cao X, Lin J, Jiang H, Li X, Teng KS, Luk CM, Zeng S, Hao J, Lau SP (2012) Deep ultraviolet photoluminescence of water-soluble self-passivated graphene quantum dots. ACS Nano 6:5102–5110. https://doi.org/10.1021/nn300760g

Teradal NL, Jelinek R (2017) Carbon nanomaterials in biological studies and biomedicine. Adv Healthc Mater 6:1700574. https://doi.org/10.1002/adhm.201700574

Tetsuka H, Asahi R, Nagoya A, Okamoto K, Tajima I, Ohta R, Okamoto A (2012) Optically tunable amino-functionalized graphene quantum dots. Adv Mater 24:5333–5338. https://doi.org/10.1002/adma.201201930

Thakur M, Mewada A, Pandey S, Bhori M, Singh K, Sharon M, Sharon M (2016) Milk-derived multi-fluorescent graphene quantum dot-based cancer theranostic system. Mater Sci Eng C 67: 468–477. https://doi.org/10.1016/j.msec.2016.05.007

Tian R, Zhong S, Wu J, Jiang W, Shen Y, Jiang W, Wang T (2016) Solvothermal method to prepare graphene quantum dots by hydrogen peroxide. Opt Mater (Amst) 60:204–208. https://doi.org/10.1016/j.optmat.2016.07.032

Tong X, Wei Q, Zhan X, Zhang G, Sun S (2016) The new graphene family materials: synthesis and applications in oxygen reduction reaction. Catalysts 7:1. https://doi.org/10.3390/catal7010001

Wang C, Wu C, Zhou X, Han T, Xin X, Wu J, Zhang J, Guo S (2013) Enhancing cell nucleus accumulation and DNA cleavage activity of anti-cancer drug via graphene quantum dots. Sci Rep 3:2852. https://doi.org/10.1038/srep02852

Wang L, Wang Y, Xu T, Liao H, Yao C, Liu Y, Li Z, Chen Z, Pan D, Sun L, Wu M (2014a) Gram-scale synthesis of single-crystalline graphene quantum dots with superior optical properties. Nat Commun 5:5357. https://doi.org/10.1038/ncomms6357

Wang C, Yang S, Yi M, Liu C, Wang Y, Li J, Li Y, Yang R (2014b) Graphene oxide assisted fluorescent chemodosimeter for high-performance sensing and bioimaging of fluoride ions. ACS Appl Mater Interfaces 6:9768–9775. https://doi.org/10.1021/am502142d

Wang X, Sun X, Lao J, He H, Cheng T, Wang M, Wang S, Huang F (2014c) Multifunctional graphene quantum dots for simultaneous targeted cellular imaging and drug delivery. Colloids Surfaces B Biointerfaces 122:638–644. https://doi.org/10.1016/j.colsurfb.2014.07.043

Wang D, Zhu L, Chen JF, Dai L (2015a) Can graphene quantum dots cause DNA damage in cells? Nanoscale 7(21):9894–9901. https://doi.org/10.1039/c5nr01734c

Wang ZG, Zhou R, Jiang D, Song JE, Xu Q, Si J, Chen YP, Zhou X, Gan L, Li JZ, Zhang H, Liu B (2015b) Toxicity of graphene quantum dots in zebrafish embryo. Biomed Environ Sci 28 (5):341–351. https://doi.org/10.3967/bes2015.048

Wang S, Cole IS, Li Q (2016) The toxicity of graphene quantum dots. RSC Adv 6:89867–89878. https://doi.org/10.1039/C6RA16516H

Wang Z, Colombi Ciacchi L, Wei G (2017) Recent advances in the synthesis of graphene-based nanomaterials for controlled drug delivery. Appl Sci 7:1175. https://doi.org/10.3390/app7111175

Wo F, Xu R, Shao Y, Zhang Z, Chu M, Shi D, Liu S (2016) A multimodal system with synergistic effects of magneto-mechanical, photothermal, photodynamic and chemo therapies of cancer in graphene-quantum dot-coated hollow magnetic nanospheres. Theranostics 6:485–500. https://doi.org/10.7150/thno.13411

Wu X, Tian F, Wang W, Chen J, Wu M, Zhao JX (2013) Fabrication of highly fluorescent graphene quantum dots using l-glutamic acid for in vitro/in vivo imaging and sensing. J Mater Chem C 1: 4676. https://doi.org/10.1039/c3tc30820k

Xiao S, Zhou D, Luan P, Gu B, Feng L, Fan S, Liao W, Fang W, Yang L, Tao E, Guo R, Liu J (2016) Graphene quantum dots conjugated neuroprotective peptide improve learning and memory capability. Biomaterials 106:98–110. https://doi.org/10.1016/j.biomaterials.2016.08.021

Xu G, Mahajan S, Roy I, Yong K-T (2013) Theranostic quantum dots for crossing blood–brain barrier in vitro and providing therapy of HIV-associated encephalopathy. Front Pharmacol 4: 140. https://doi.org/10.3389/fphar.2013.00140

Yan Z, Xue H, Berning K, Lam Y-W, Lee C-S (2014) Identification of multifunctional graphene–gold nanocomposite for environment-friendly enriching, separating, and detecting Hg 2+ simultaneously. ACS Appl Mater Interfaces 6:22761–22768. https://doi.org/10.1021/am506875t

Yang Y, Chen S, Liu L, Li S, Zeng Q, Zhao X, Li H, Zhang Z, Bouchard L-S, Liu M, Zhou X (2017) Increasing cancer therapy efficiency through targeting and localized light activation. ACS Appl Mater Interfaces 9:23400–23408. https://doi.org/10.1021/acsami.7b05463

Yao X, Tian Z, Liu J, Zhu Y, Hanagata N (2017) Mesoporous silica nanoparticles capped with graphene quantum dots for potential chemo–photothermal synergistic cancer therapy. Langmuir 33:591–599. https://doi.org/10.1021/acs.langmuir.6b04189

Ye R, Xiang C, Lin J, Peng Z, Huang K, Yan Z, Cook NP, Samuel ELG, Hwang C-C, Ruan G, Ceriotti G, Raji A-RO, Martí AA, Tour JM (2013) Coal as an abundant source of graphene quantum dots. Nat Commun 4:2943. https://doi.org/10.1038/ncomms3943

Yuan X, Liu Z, Guo Z, Ji Y, Jin M, Wang X (2014) Cellular distribution and cytotoxicity of graphene quantum dots with different functional groups. Nanoscale Res Lett. https://doi.org/10.1186/1556-276X-9-108

Zahan T, Ahmed I, Omi SI, Naher K, Islam S, Bin Mahmud ASMSH, Sultana S, Shahriar SMS, Khatun Z, Rahmatullah M (2013) Ethnobotanical uses of medicinal plants by the Tudu sub-clan of the Santal tribe in Joypurhat district of Bangladesh. Am J Sustain Agric 7:137–142. https://www.researchgate.net/publication/289748985_Ethnobotanical_uses_of_medicinal_plants_by_the_Tudu_subclan_of_the_Santal_tribe_in_Joypurhat_district_of_Bangladesh

Zhang M, Bai L, Shang W, Xie W, Ma H, Fu Y, Fang D, Sun H, Fan L, Han M, Liu C, Yang S (2012a) Facile synthesis of water-soluble, highly fluorescent graphene quantum dots as a robust biological label for stem cells. J Mater Chem 22:7461. https://doi.org/10.1039/c2jm16835a

Zhang Z, Zhang J, Chen N, Qu L (2012b) Graphene quantum dots: an emerging material for energy-related applications and beyond. Energy Environ Sci 5:8869. https://doi.org/10.1039/c2ee22982j

Zhao H, Ding R, Zhao X, Li Y, Qu L, Pei H, Yildirimer L, Wu Z, Zhang W (2017) Graphene-based nanomaterials for drug and/or gene delivery, bioimaging, and tissue engineering. Drug Discov Today 22:1302–1317. https://doi.org/10.1016/j.drudis.2017.04.002

Zhao S, Lavie J, Rondin L, Orcin-Chaix L, Diederichs C, Roussignol P, Chassagneux Y, Voisin C, Müllen K, Narita A, Campidelli S, Lauret J-S (2018) Single photon emission from graphene quantum dots at room temperature. Nat Commun 9:3470. https://doi.org/10.1038/s41467-018-05888-w

Zheng FF, Zhang PH, Xi Y, Chen JJ, Li LL, Zhu JJ (2015) Aptamer/graphene quantum dots nanocomposite capped fluorescent mesoporous silica nanoparticles for intracellular drug delivery and real-time monitoring of drug release. Anal Chem. https://doi.org/10.1021/acs.analchem.5b03131

Zhou X, Zhang Y, Wang C, Wu X, Yang Y, Zheng B, Wu H, Guo S, Zhang J (2012) Photo-Fenton reaction of graphene oxide: a new strategy to prepare graphene quantum dots for DNA cleavage. ACS Nano 6(8):6592–6599. https://doi.org/10.1021/nn301629v

Zhou Y, Peng Z, Seven ES, Leblanc RM (2018) Crossing the blood-brain barrier with nanoparticles. J Control Release 270:290–303. https://doi.org/10.1016/j.jconrel.2017.12.015

Zhu S, Zhang J, Qiao C, Tang S, Li Y, Yuan W, Li B, Tian L, Liu F, Hu R, Gao H, Wei H, Zhang H, Sun H, Yang B (2011) Strongly green-photoluminescent graphene quantum dots for bioimaging applications. Chem Commun 47:6858. https://doi.org/10.1039/c1cc11122a

Zhu S, Zhang J, Liu X, Li B, Wang X, Tang S, Meng Q, Li Y, Shi C, Hu R, Yang B (2012) Graphene quantum dots with controllable surface oxidation, tunable fluorescence and up-conversion emission. RSC Adv 2:2717. https://doi.org/10.1039/c2ra20182h

Zhu R-Y, He J, Wang X-C, Yu J-Q (2014) Ligand-promoted alkylation of C(sp 3)–H and C(sp 2)–H bonds. J Am Chem Soc 136:13194–13197. https://doi.org/10.1021/ja508165a

Zhu Y, Wang G, Jiang H, Chen L, Zhang X (2015) One-step ultrasonic synthesis of graphene quantum dots with high quantum yield and their application in sensing alkaline phosphatase. Chem Commun 51:948–951. https://doi.org/10.1039/C4CC07449A

Zhu J, Tang Y, Wang G, Mao J, Liu Z, Sun T, Wang M, Chen D, Yang Y, Li J, Deng Y, Yang S (2017) Green, rapid, and universal preparation approach of graphene quantum dots under ultraviolet irradiation. ACS Appl Mater Interfaces 9:14470–14477. https://doi.org/10.1021/acsami.6b11525

Graphene as Photothermal Therapeutic Agents

Vishnu Revuri, Jagannath Mondal, and Yong-kyu Lee

Abstract

Light-assisted hyperthermic therapy is a promising strategy to treat cancer. Graphene and their derivatives with unique physiochemical properties, intrinsic near infrared absorption, and ability to transduce the absorbed light energy into heat, have attracted researchers to use them for photothermal therapy (PTT). In addition, the presence of surface functional groups and large surface area that can facilitate interactions with hydrophobic molecules has favored the use of graphene allotropes for developing PTT-based combinatorial therapies. In this book chapter we have reviewed different graphene-based PTT-assisted photodynamic, gene, chemo, and immunotherapeutic strategies developed to improve the outcome of cancer treatment. We have also discussed how PTT from graphene derivatives can improve the therapeutic outcomes of gene, chemo, and immunotherapies. Finally, this book chapter provides promising insights to

V. Revuri
Department of Green Bioengineering, Korea National University of Transportation, Chungju, South Korea

Department of Pharmaceutical Sciences, Temple University School of Pharmacy, Philadelphia, PA, USA

J. Mondal
Department of Green Bioengineering, Korea National University of Transportation, Chungju, South Korea

Y.-k. Lee (✉)
Department of Green Bioengineering, Korea National University of Transportation, Chungju, South Korea

Department of Chemical and Biological Engineering, Korea National University of Transportation, South Korea
e-mail: leeyk@ut.ac.kr

D.-W. Han, S. W. Hong (eds.), *Multifaceted Biomedical Applications of Graphene*, Advances in Experimental Medicine and Biology 1351,
https://doi.org/10.1007/978-981-16-4923-3_9

develop novel graphene-based multifunctional PTT-assisted combinatorial therapeutics with both imaging and therapeutic regimens to treat cancer.

Keywords

Graphene photothermal therapy · Combined photothermal therapy · Theranostics · Bioimaging

1 Introduction

Advancements in medical technologies are essential for improving the quality of human life. Novel technologies and therapies with enhanced therapeutic efficiency and negligible systemic toxicity are required for treating complex diseases. Cancer is the second leading cause of death worldwide. Current treatment strategies like surgical resection, chemo and radiation therapies are considered ineffective and are resulting in life-threatening side effects. Furthermore, cancer cells have a tumor specific niche around them which not only enhances cancer cell proliferation, but also ingeniously develope resistance to the current therapeutic drugs. This suggests the need for developing novel therapeutic strategies and technologies to treat cancer.

Thermal ablation technologies have played a predominant role in cancer therapy where the generated hyperthermic temperatures can cause irrefutable damage to the cancer cells and result in cancer cell death (Huang et al. 2015). Photothermal therapy (PTT) has recently gained significance as one of the thermal ablation technologies to treat cancer. Here, a light sensitive photoactive agent absorbs light and transduce the absorbed energy into heat. The generated heat can cause irreversible damage to the cancer cells and ensue in cancer cell ablation (Datta et al. 2015; Jha et al. 2016; van Straten et al. 2017). Irradiated light plays a predominant role in enhancing the therapeutic efficiency of PTT. Light in the visible range (350–650 nm) have poor penetration depth due to their scattering and therefore cannot treat deep tumors. However, nanomaterials which can absorb the light in the near infrared (NIR) region and generate heat have gained significance in PTT. In addition, the heat generated during the PTT can also be used to develop technologies which can tune the release of encapsulated drugs or therapeutic agents, in response to the applied heat at the cancer site. This would minimize the off-targeting side effects and improve the therapeutic efficiency of the loaded therapeutics (Cheng et al. 2014; Zou et al. 2016). Several metallic nanomaterials (like gold, silver, platinum, tungsten, copper, etc.), organic dyes (IR 820, ICG), semiconducting quantum dots, upconversion nanomaterials, and transition metal oxide nanomaterials have been used as NIR photoactive agents for PTT.

Graphene-based nanomaterials have gained tremendous attention as efficient NIR photoactive agents for PTT (de Melo-Diogo et al. 2017). In particular, their physiochemical properties like larger surface area, availability of surface functional group and most importantly their ability to absorb light in NIR region and translate the absorbed energy into heat have attracted researchers to use graphene for PTT. Although PTT is considered as an efficient treatment strategy, the heterogeneity in the heat profiles around the tumors could not be sufficient to ablate all the tumor cells. Moreover, the tumor cells can activate cascade of signaling pathways that not only enhance the repair and tumor cell proliferation but also recruit immunosuppressive cell artillery that can prevent immune activation. Therefore, it is advisable to integrate PTT with other novel therapeutics to develop efficient hybrid strategies to treat cancer. In this book chapter, we will discuss the mechanism of PTT by graphene-based nanomaterials, graphene-based PTT, graphene-based hybrid technologies for synergistic PTT and will also discuss the limitations and prospects of using graphene nanomaterials for PTT in clinics.

2 Mechanism of Photothermal Therapy by Graphene

The fundamental rule of developing nanomaterials for PTT is the ability of the photoactive agents to absorb the irradiated light (mostly NIR) and release the absorbed energy through nonradioactive decay in the form of heat. The released kinetic energy, which is in the form of heat, can increase the local temperatures around the photoactive agents and assist in thermal ablation and cell death. Studies report that PTT can induce both intrinsic apoptosis and extrinsic necrosis for killing tumor cells (Martin et al. 2012). In addition, studies reported that low energy radiations generated during the PTT can activate the caspase-3 mediated apoptosis in the cancer cells, resulting in the tumor cell death. In addition, the higher energy radiation, i.e., higher heat generated during the PTT, can assist in the generation of tumor lysates, which can serve as tumor antigens and trigger the recruitment of pro-inflammatory immune cell artillery at the tumor site (Mocan et al. 2014).

Graphene is a 2D allotrope of carbon with sp^2-hybridized carbon atoms that form hexagonal single layer lattice structure. Owing to their excellent physicochemical properties such as light weight, excellent electrical conductivity, high thermal conductivity, ease of functionalization, high mechanical strength, and fluorescence properties, graphene nanomaterials are explored for wide range of applications. Graphene and its derivatives have been utilized extensively for diverse biomedical applications including drug delivery, bioimaging, phototherapy, antibacterial composites, and biosensors (Balapanuru et al. 2010; Cui et al. 2012; He et al. 2010; Lu et al. 2009).

The exceptional structural properties and strong interplay between graphene and infrared/terahertz photons can result in plasmonic photothermal conversion and generate heat from graphene-based nanomaterials (Koppens et al. 2011). Specifically, the excited surface plasmons on graphene upon NIR laser irradiations assist in the resonance of random dipoles which ultimately result in the creation of thermal photon energy output that is necessary for PTT (Koppens et al. 2011). Structure of graphene materials plays a significant role in achieving desired photothermal conversion efficiencies from graphene. For example, Robinson et al., demonstrated that the reduced graphene oxide (rGO), a graphene derivative with highly intact aromatic graphene ring structures, has a six-time higher photothermal conversion efficiency compared to graphene oxide (GO) (Robinson et al. 2011). This suggests that the graphene derivatives with resemblance of graphene structures have higher NIR absorbance and heat generation capacities compared to other carbon derivatives. This study indicates that apart from the NIR laser irradiation, the geometry of the nanoparticles plays a crucial role in enhancing the cellular uptake and therapeutic activity of the developed nano-vectors.

Currently, graphene and gold are the widely explored nanomaterials for the development of plasmonic photothermal therapeutics. Interestingly, graphene nanomaterials require lower power density (~2 W/cm^2) compared to gold nanomaterials (2–4 W/cm^2) for generating efficient PTT (Koppens et al. 2011). Higher power densities can cause damage and ablation of normal cells, resulting in undesired complications. Furthermore, graphene has larger surface area compared to gold nanomaterials, which can be used to load hydrophobic drugs or conjugate other functional groups for targeted therapeutic delivery. Finally, photothermal effects can cause a structural change of gold nanomaterials, which can result in irregular photothermal conversion efficiencies. These superlative properties of graphene favor their use for PTT.

3 Graphene Derivatives

Graphene is a 2D and thickness materials containing hexagonal carbon ring. Graphene derivatives were categorized based on chemical structure and shape such as graphene oxide (GO), reduced graphene oxide(rGO), graphene nanotubes (GNT), and graphene quantum dots (CQDs).

3.1 Graphene Oxide (GO)

Graphene oxide is considered as a backbone precursors of other derivatives, which has been extensively explored in the PTT applications (de Melo-Diogo et al. 2018). This nanomaterial is a highly oxygenated form of graphene, containing several types of oxygen functional groups such as hydroxyl, carboxyl, or epoxy groups embedded in their structure. GO is commonly produced by exfoliation and the chemical oxidation of graphite. H_2SO_4, H_3PO_4, and $KMnO_4$ have been widely used to produce GO from graphite(Hummers and Offeman 1958; Marcano et al. 2010).

3.2 Reduced Graphene Oxides (rGO)

rGO can be synthesized from GO through electrochemical reduction, chemical reduction, and thermal reduction(Ding et al. 2011; Stankovich et al. 2006; Wang et al. 2008a; Wang et al. 2008b; Tung et al. 2009). In this process, the graphitic aromatic lattice can be restored by removing oxygen functional groups by using reducing agents (e.g., hydrazine hydrate (Robinson et al. 2011), L-ascorbic acid (Lima-Sousa et al. 2018) glucose (Akhavan et al. 2012a)).

3.3 Graphene Nanoribbons (GNRs)

GNRs, having width less than 100 nm, can be produced from multiwall carbon nanotubes (using $KMnO_4$ and H_2SO_4) where unzipping of the nanotubes results in their synthesis (Kosynkin et al. 2009). Reduced graphene nanoribbons (rGNRs) can also be synthesized by using reducing agents (e.g., hydrazinehydrate) which displayed an enhanced NIR absorption (Akhavan et al. 2012b).

3.4 Graphene Quantum Dots (GQDs)

Graphene quantum dots are nanoscale structure of graphene containing quantum confinement effect and high density of sp2 sites which results in exceptional photoluminescence properties. Their size and surface chemistry control the energy bandgap of GQDs. The GQDs can be synthesized by two strategies such as top-down approach and bottom-up strategy. The top-down approach has been utilized extensively due to produce on large scale (Chen et al. 2018b). The GQDS are soluble in highly polar solvents such as water, ethanol, and *N,N*-dimethylformamide (DMF) due to having negatively charged and hydrophilic oxygen groups (Nakano et al. 2018). The GQDs can be synthesized in the range of 3-20 nm (Tetsuka et al. 2012).

4 Graphene-Based Combination Photothermal Therapy (PPT)

While heat generated by the photothermal sensitizers is sufficient to kill the cancer cells, limitations associated with reduced and nonspecific accumulation of photothermal sensitizers apart from the tumor site and heterogeneous heating profiles in the tumors, restrict their translation into clinical settings. In particular, the inability to completely ablate the large tumors due to heterogeneous heating profiles not only encourages the tumor growth but also creates immunosuppressive tumor niche around the tumors, which reduces the efficiency of conventional cancer therapeutics. In addition, the enhanced secretion of heat shock proteins post PTT can promote anti-apoptosis and thereby tumor growth of residual tumors. In addition, PTT induces the recruitment of myeloid derived suppressor cells (MDSCs) at the metastatic tumors, which promotes the growth of secondary metastatic tumors. Therefore, PTT alone is not sufficient for eliminating tumors. Combining PTT with other existing or novel therapies is a promising approach to overcome the abovementioned limitations (de Melo-Diogo et al. 2018). In this section we reviewed the advantages of integrating PTT with existing therapies. In particular, we highlighted the advantage of having graphene derivatives as photothermal sensitizers, and how inclusion of graphene derivatives in the combinatorial therapies improved the therapeutic efficiencies.

4.1 Enhanced Photothermal Therapy

Although graphene family demonstrate enhanced PTT, loading of drugs and functionalization can affect their heat generation profiles and limit their applications. To overcome this hurdle, researchers integrated graphene with other functional photothermal agents to not only accentuate the heating profiles but also achieve enhanced PTT with low laser power irradiations (Zhen et al. 2018). Ji et al. developed a nanogel to enhance the heating profiles of GO. They have initially signalized GO and later conjugated *N*-isopropylacrylamide (NIPAM) and loaded with DOX to demonstrate enhanced PTT and chemotherapy. Their results demonstrate that compared to GO, the GO loaded in nanogel can generate enhanced heating profiles with ΔT of 58 °C. In addition, these nanogels demonstrated pH, GSH, and temperature responsive release of DOX in the cancer cells (Ji et al. 2019).

Metallic and transition metal oxide nanoparticles and organic dyes have been coated to enhance the heating profiles of graphene allotropes. Wang et al. loaded surface of nano goxide (nGO) with copper sulfide quantum dots (CSQD) to enhance the heating profiles of the nGO (Wang and Yan 2019). Their results demonstrated that the CSQD-nGO demonstrated enhanced heating profiles, presented apoptosis morphologies, and augmented the levels of LDH, caspase 3, and ROS in the cancer cells. Zhang et al. coated the surface of gold nanorods (AuNR) with mesoporous silica and further coated with rGO to enhance the PTT activity of the nanocomposite (Zhang et al. 2018b). Prolonged exposure of AuNR to NIR laser irradiations demonstrated a decelerated heating profile, while coating of AuNR with rGO stabilized the heating profiles and enhanced the PTT activity in the cancer cells (Fig. 1) (Song et al. 2015). Guo et al. demonstrated pH responsive enhanced photothermal therapy by conjugating NIR fluorescent dye Cypate over GO via fluorescence resonance energy transfer (FRET) (Guo et al. 2015). The additional PTT is associated by the energy transfer between Cypate donor and GO acceptor under NIR laser irradiation. They have demonstrated that Cypate over the surface of GO can undergo conformational change with respect to the change in pH of the solution. This pH dependent conformational change resulted in enhanced FRET between Cypate and GO under acidic pH and enhanced the NIR laser mediated photothermal heat profiles under lysosomal pH conditions.

Ultrasound can also be used to enhance the PTT activity of GO. Liu et al. loaded the cores of liposomes with perfluoropropane to form a nanobubble, functionalized with avidin and coated with biotin functionalized rGO (Liu et al. 2018). Under ultrasound, the nanobubbles were destroyed, resulting in increasing the local concentrations of rGO and thereby enhancing the PTT effect of rGO under NIR laser irradiations. In addition, ultrasound treatment can result in transient opening of the cell membranes which can further improve the loading of the drugs in the cancer cells. Their results claim that only 2.2% cells were viable after ultrasound-assisted PTT in HepG2 cancer cells.

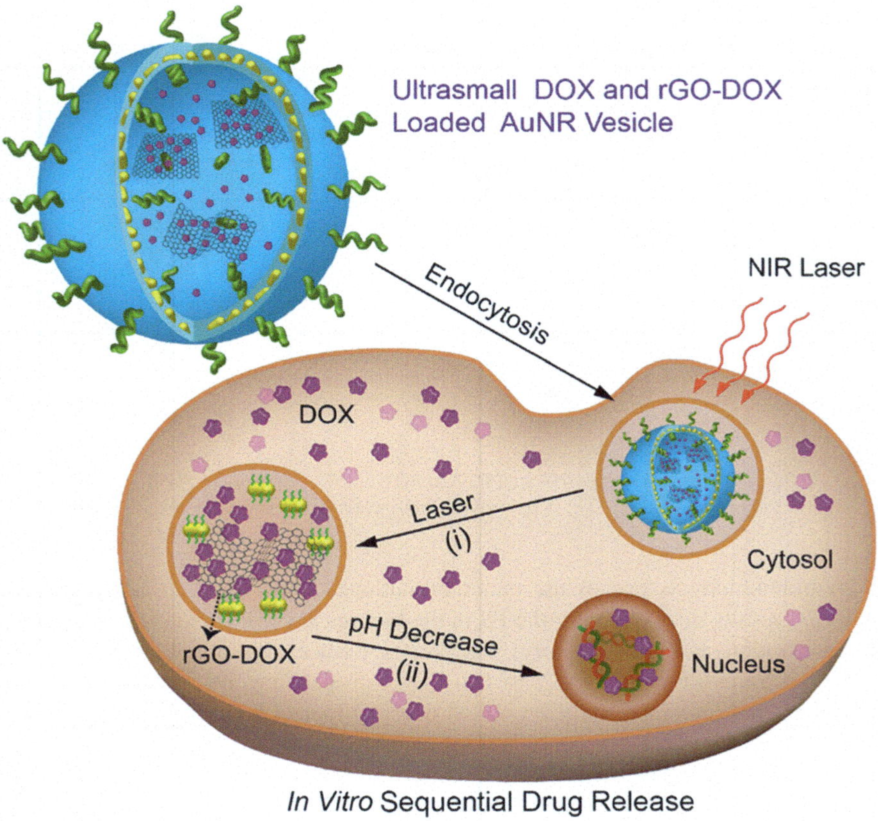

Fig. 1 Schematic illustration of Enhanced PPT therapy (Adapted with permission from (Song et al. 2015) Copyright 2015 American Chemical Society)

4.2 Photodynamic Photothermal Therapy

Photodynamic therapy (PDT) is a light mediated therapy like PTT, where a photosensitizer absorbs light and generates singlet oxygen species ($^1O_2$)/reactive oxygen species (ROS) in the cancer cells to trigger apoptosis and cancer cell death. Several photosensitizers like chlorin e6 (Ce6), (Nafiujjaman et al. 2016) Zinc phthalocyanine (ZNPC), and methylene blue (MB) (Liang et al. 2019) have been used as photosensitizers (PS) that can be used for PDT. Studies report that graphene quantum dots (GQDs), zero-dimensional fluorescent 2–3 layered graphene nanostructures, itself can function as a photosensitizer and can be used for PDT. The studies reported by Ge et al. demonstrate that GQDs can produce $^1O_2$ via multistate sensitization process with a quantum yield of ~1.3, which is highest compared to other reported PDT agents (Ge et al. 2014). However, this highest quantum yield is achieved only at wavelengths at visible range. The light in visible range suffers from deep tumor penetration, which prevents the access of the light to deep tumors and can reduce the efficiency of PDT.

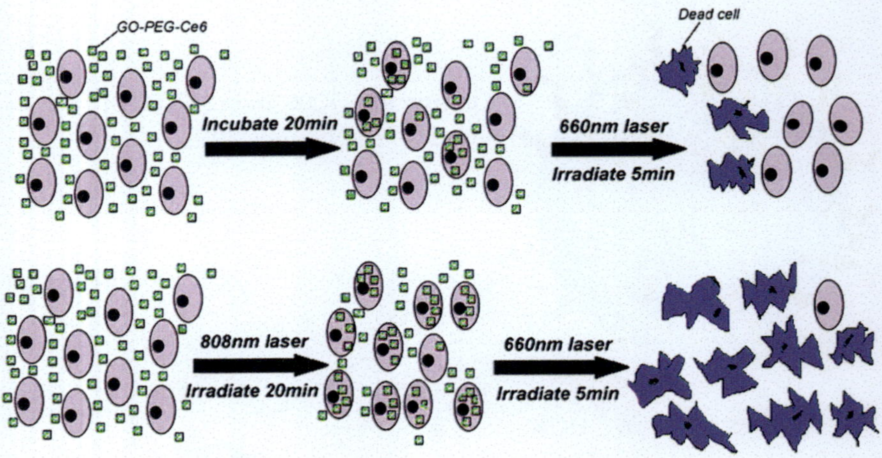

Fig. 2 Schematic illustration of combined PPT and PDT of GO-Ce6 nanocomposite (Adapted with permission from (Tian et al. 2011c) (Copyright 2011 American Chemical Society)

Hydrophobicity is one of the major limitations of the PS. The lack of water solubility results in aggregating the PS in the biological media, resulting in reduced $^1O_2$ generation and PDT activity. The larger surface area and the presence of aromatic groups in the PS can assist in the pi-pi stacking mediated binding and uptake of PS. Numerous studies reported that GO loaded PS increased their water solubility and enhanced $^1O_2$ generation from PS (F. Li et al. 2013; Xing et al. 2016; Zeng et al. 2015). Tian et al. initially demonstrated that mild PTT by GO can enhance the uptake of GO-Ce6 nanocomposite and facilitate in enhanced PDT activity (Tian et al. 2011a). Their results demonstrate that the cells treated with GO-Ce6 nanocomposite and irradiated with both 808 nm and 660 nm lasers resulted in complete tumor cell death compared to the 660 nm laser irradiated cancer cells (Fig. 2) (Tian et al. 2011c). Scientists have conjugated C_{60} to GO and used them as a hybrid nanomaterial that can generate both PDT and PTT (Hu et al. 2015, 2017). Interestingly, the developed hybrid nanomaterial can generate $^1O_2$ when irradiated with green, red, and NIR lasers (Li et al. 2017). In another study, gold nanospheres was coated with GO shell and loaded with ZnPC for dual PTT and PDT (Kim et al. 2015). With a reduced 808 nm laser power density of 0.6 W/cm^2 for PTT and 660 nm laser with power density of 0.2 W/cm^2 resulted in generation of heat and ROS production in the cancer cells respectively and facilitated 94% cell death in HeLa cells. Scientists have also used Ruthenium (II) complexes as PS and demonstrate dual PDT at 405 nm and PTT at 808 nm in tumor xenograft animal models (Zhang et al. 2017a). However, 660 nm laser show reduced penetration and can minimize the PDT activity in deep tumors. Therefore, scientists have developed new PS which can generate PDT when irradiated with a laser with a wavelength of 808 nm. Scientists have demonstrated that GQDs derived from withered leaves can itself generate heat as well as produce $^1O_2$ when irradiated with NIR laser 808 nm (Thakur et al. 2017). Interestingly, Kalluru et al. demonstrated that nano GO can generate both PDT and PTT when irradiated with NIR laser with an excitation wavelength of 980 nm (Kalluru et al. 2016). Their results demonstrate enhanced heat profiles at 808 nm but could generate higher $^1O_2$ when irradiated with 980 nm laser. Interestingly, the animals treated with nano GO-PEG and irradiated with 980 nm laser demonstrated negligible tumor growth, while the animals treated with 808 nm laser displayed tumor regrowth and scar due to PTT.

Scientists have further integrated GO with other NIR PS to develop synergistic PDT and PTT (Tian et al. 2011b). Luo et al. conjugated IR-808 dye to bPEI-PEG modified nano GO (nGO) and used them for dual PDT and PTT. Their results demonstrate enhanced tumor cell death with increased ROS generation and heating profiles ($\Delta T \sim 45$ °C) in both A549 and Lewis cancer cells (Luo et al. 2016). The developed nGO-808 demonstrated organic anion transporter peptide mediated cellular uptake and resulted in enhanced tumor accumulation and complete tumor reduction in both A549 and Lewis tumor xenograft animal models.

Upconversion nanomaterials, with their ability to absorb two or more lower energy photons and release one high energy photon, have attracted researchers as they can overcome the issues associated with penetration depth, autofluorescence, and photobleaching. Scientists have integrated NaYF4: Yb3þ, Er3þ, Tm3þ/NaYF core-shell upconversion nanomaterials, GO, loaded with ZnPC and used them as a NIR laser mediated dual PDT and PTT (Wang et al. 2013). The results demonstrate that synergistic PDT and PTT resulted in enhanced cell death (~85%) compared to singular therapy. Thapa et al. decorated GO with palladium nanoparticles (PdNp) and used them for dual PDT and PTT. The coating of PdNp over GO resulted in an eight-degree enhancement in the heating profiles after NIR laser irradiations in tumor bearing mice. In addition, enhanced ROS generation was observed from the GO-PdNp nanocomposites (Thapa et al. 2018).

4.3 Photothermal Gene Therapy

Gene therapy involves in the delivery of therapeutic genes (DNA/RNA) to prevent/stop desired activity in the target cell. Graphene nanomaterials function as an efficient carrier for the gene therapy as the large surface area assists in loading the genes on their surface and protects them from enzymatic degradations in the biological system. In addition, the heat generated during PTT can assist in transient loosening of the endo/lysosomal/cell membranes and assist in enhancing the transfection efficiency of the loaded gene. Furthermore, the gene therapy could further help in enhancing the activity of PTT by blocking the expression of heat shock proteins in the cells or mitigate the resistance offered by PTT induced heat damage. This cooperative synergism has attracted researcher to use graphene as a non-viral vector for gene therapy.

Studies report that PTT from graphene nanocomposites can control the release of the loaded gene in the cancer cells. Kim et al. fabricated bPEI modified rGO, functionalized with PEG and used it as a non-viral vector for gene delivery (Kim and Kim 2014). In this study, they have used bafilomycin as a proton pump inhibitor and proved that the local heat released during PTT by bPEI-rGO-PEG facilitated the endosomal escape, controlled the gene release, and facilitated 7 times higher gene transfection compared to the bPEI-rGO-PEG treated cells without laser irradiation. Similar results were reported by Feng et al., where they have demonstrated that transfection of Polo-like kinase 1 (Plk1) siRNA using PEI-rGO-PEG nanovector resulted in the reduced Plk-1 protein expression from the cells after NIR laser irradiations (Feng et al. 2013). Assali et al. fabricated Au core GO shell nanomaterials, functionalized with poly arginine and demonstrated the effect of shape on the transfection efficiency of miRNA-101 (Assali et al. 2018). Their results demonstrate that rod shaped nanomaterials showed enhanced cellular uptake and cytotoxicity compared to spherical nanoparticles. In addition, the delivery of miRNA-101 from the nanorod constructs resulted in enhanced apoptosis induction in the cell compared to nanosphere materials. Furthermore, the cells irradiated with 1.2 W/cm^2 laser power density demonstrated 835 times higher transfection compared to the control groups.

GO have also been to co-deliver multiple siRNAs to the cancer cells. Yin et al. functionalized GO with folic acid and further coated with poly-allylamine hydrochloride (PAH) to load both HDAC1 and K-Ras siRNAs and use them to treat pancreatic cancer (Yin et al. 2017). The co-delivery of HDAC1 and K-Ras siRNAs significantly reduced the expression of HDAC-1 and Kras protein expression, increased the expression of pro-apoptotic caspase 3 protein, and reduced the levels of anti-apoptotic BCL-2 proteins in the MiaPACA-2 cancer cells. The tumor xenograft In vivo animal models demonstrated 80% tumor reduction enhanced the mean survival rates of mice after intraperitoneal injection and PTT of GO-Kras-HDAC1 nanocomposite. Interestingly, one third of the treated mice showed complete tumor reduction after 60 days post PTT and gene therapy. However, intravenous injection of the developed nanocomposite demonstrated animal death indicating the facilitation of systemic toxicity of the developed nanocomposite.

4.4 Photothermal Chemotherapy

Chemotherapy is the conventional strategy used to treat cancers. Here, chemotherapeutic drugs like doxorubicin (DOX), paclitaxel (PTX), or methotrexate (MTX) neutralize the activity of the cancer cells and result in their death. Although chemotherapy is an efficient strategy, the systemic delivery of chemotherapeutics results in poor tumor accumulation and has resulted in adverse side effects. The larger surface area and pi-pi stacking mediated drug loading make graphene nanomaterials an optimal carrier for loading the chemotherapeutic drugs. In addition, the photothermal properties of graphene can facilitate dual treatment functionalities, which can boost the therapeutic efficiencies of these synergistic therapeutics.

Zhang et al. developed a hydrophilic rGO, by conjugating GO with hyperbranched amine terminated polymers (Zhang et al. 2018a). These hyperbranched polymers assisted in improving the hydrophilicity and water solubility of rGO. In addition, the loaded doxorubicin (DOX) in this hyperbranched polymer coated rGO resulted in both pH and temperature responsive release profiles. The protonation of amine groups of hyperbranched polymers resulted in repulsion

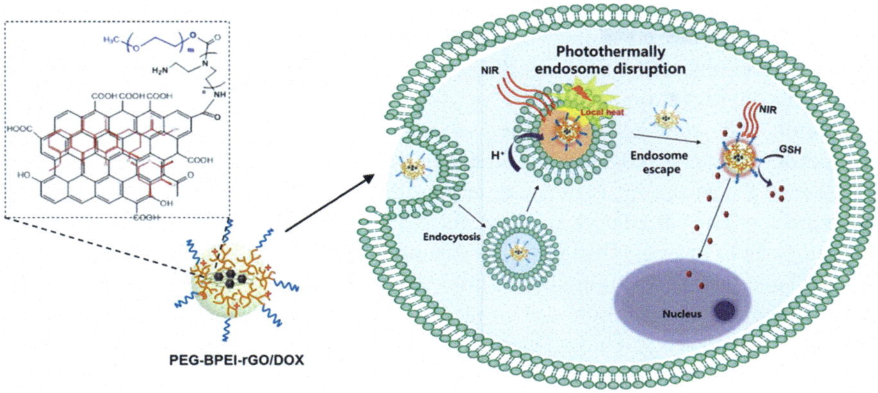

Fig. 3 Schematic illustration of DOX loaded, bPEI and PEG modified rGO enhancing PPT therapy (Adapted with permission from (Kim et al. 2013). Copyright 2013 American Chemical Society)

between the DOX and the NH_2 groups resulting in enhanced release of the drugs. In addition, the increased temperatures caused enhancement in the molecular motion, resulting in faster drug release from the developed nanoparticle system. However, studies report that positively charged polymers can attract the negatively charged opsonin proteins and result in immune activation and systemic toxicity. In another study, Roy et al. functionalized rGO with poly (allylamine hydrochloride) (PAH) and loaded with DOX to fabricate chemo-photothermal therapeutics (Roy et al. 2019). DOX release was significantly increased in acidic pH, generated reactive oxygen species (ROS), and resulted in apoptosis induced cancer cell death in MCF-7 breast cancer cells. Although PAH-rGO demonstrated ~60% cell death post PTT, DOX/PEG-rGO demonstrated 94% cell death post PTT at a concentration of 5 μg/mL. Studies have reported that PTT by GO can assist in endosomal disruption and favor the cytosolic release of the loaded drug in the cells (Kim et al. 2013). Kim et al. fabricated DOX loaded bPEI and PEG modified rGO and demonstrated both temperature and GSH responsive release of the loaded DOX from the nanocomplex in the cells. They have demonstrated that the DOX release was increased by 45% in the presence of both GSH and NIR laser irradiations. In addition, they have showed that PTT assisted in release of the nanocomplex from the endosomes and GSH assisted in the release of DOX into the cytosol. A 50% enhancement in the cell death was observed under NIR laser irradiations. Compared to passive delivery, targeted active delivery of drugs to the cancer site enhances the efficiency of the developed therapeutics with minimized systemic toxicity. Therefore, researchers conjugated cancer targeting ligands to the nanoparticles and employed them for the active delivery of biomolecules to treat cancer (Fig. 3) (Kim et al. 2013). Zhang et al. crosslinked GO with hyaluronic acid-MTX prodrug to develop pH responsive nanocarrier for targeted photothermal chemotherapy (Zhang et al. 2019). Hyaluronic acid (HA) favored the active delivery of the nanocomposites by targeting the CD44 receptors on the cancer cell surface. The presence of HA on the surface improved the accumulation and retention of the nanocomposites at the tumor site compared to the pristine GO nanoparticles. The developed nanocomposites demonstrated complete tumor reduction and no tumor recurrence after 21 days' post PTT. Even though studies report enhanced PTT from rGO compared to GO, the poor water solubility, irreversible aggregation and lack of functional group limit their biological applications. To address this issue, scientists coated the surface of rGO with mesoporous silica, a biocompatible, and non-toxic nanomaterial. In addition, the pore size of the msoporous silica can be modified and facilitate functionality to the nanoparticles. The varied pore sizes can further allow to load hydrophobic drugs into the nanoparticle system. Guo et al. used supramolecular interactions to graft SiO_2 over rGO and further functionalized with octadecanoic acid (C18)-grafted-mPEG. In addition, the pore adsorption and pi-pi stacking resulted in enhanced loading of DOX into the constructed nanocomposite. The synergistic photothermal chemotherapy resulted in almost 90% cell death post PTT.

Cell membranes collected from the native cells are currently regarded as an efficient carrier for the delivery of biomolecules to the targeted delivery site. Recently, researchers collected the red blood cell membranes, functionalized with folic acid and used them as a carrier to load GO, DOX, and indocyanine green (ICG) and use them as a functional nanocomposite for synergistic photothermal chemotherapy (Li et al. 2019). RBC membrane with intrinsic CD47 receptors expressed on their surface can help in enhancing the life time of the nanoparticle and can assist in immune evasion. Compared to GO, the ICG integrated nanocomposite demonstrated enhanced heat generation with a temperature of 48 °C, biocompatibility and accumulation of the DOX at the cancer site. Scientists have constructed a carboxymethyl chitosan functionalized rGO-aldehyde functionalized poly ethylene glycol-based hydrogel and loaded with DOX for pH responsive drug delivery and enhanced PTT (Liu et al. 2019). Interestingly, the hydrogel demonstrated a mean temperature rise (ΔT) of 50 °C and demonstrated photothermal conversion efficiency of 85%. In addition, the degradation of Schiff base linkages resulted in acidic pH responsive drug release from the hydrogels.

Metal nanoparticles or organic dyes can be coated over the surface of GO to enhance the PTT activity with reduced laser power intensity. Ma et al. coated the surface of rGO with gold (Au) clusters via electrostatic interactions and further functionalized with 3-(3-phenylureido) propanoic acid (PPA)-PEG (PPEG) to enhance their bioavailability (Ma et al. 2019). In addition, DOX was loaded over rGO via pi-pi interactions to facilitate enhanced photothermal chemotherapy. The dual chemo-photothermal nanocomposite demonstrated a maximum temperature of ~51 °C ($\Delta T \sim$ 15 °C) with a photothermal conversion efficiency of 19%. Furthermore, a pH dependent release of DOX from the nanocarrier was observed demonstrating an enhanced therapeutic efficiency from dual photothermal chemotherapy, compared to single therapies. Scientists have replaced Au with other nanomaterials, polymers, and dyes to accentuate the heat generation profiles from graphene allotropes. Polydopamine (pDOPA) is a versatile biogenic and biocompatible material that is abundant with functional groups and easy adherence to any substrate via catechol chemistry has attracted researchers to use them for diverse biomedical applications. Recently, it has been observed that pDOPA has the capability to absorb NIR laser irradiations and can release the absorbed energy in the form of heat. Therefore, scientists coated the surface of DOX loaded mesoporous silica coated rGO with pDOPA and used them as dual photothermal chemotherapy. The coating of pDOPA over the mesoporous silica coated rGO resulted in 4 °C higher temperatures compared to pristine materials. Further, scientists conjugated lactobionic acid to IR 820, an NIR responsive PTT dye, and further loaded over GO, along with DOX via pi-pi interactions, and used them for dual chemo-photothermal therapy (Huang et al. 2019). The presence of lactobionic acid over the surface of GO resulted in active delivery of the developed nanohybrids at the tumor site. The targeted nanohybrids demonstrated tumor temperatures of 51 °C, with 90% tumor reduction rate.

From the above discussed strategies, we can infer that graphene offers diverse advantages as an efficient carrier for loading the chemotherapeutic drugs and can improve their bioavailability. In addition, the presence of functional groups over graphene allotropes can facilitate easy functionalization that can deliver the developed hybrid nanocomposites at the tumor site and can help the chemotherapeutics to achieve fullest therapeutic efficiency with minimized systemic toxicity. The added PTT feature from graphene can result in dual therapeutic action and can favor complete tumor destruction.

4.5 Photothermal Immunotherapy

Immunotherapy is regarded as a best alternative strategy in cancer therapy, which has shifted the paradigm to treat cancer. Here, immunotherapeutic drugs, monoclonal antibodies, or specialized cells are injected into the body in order to sensitize and activate the immune cells in the lymph and anti-tumor immune cells around the tumor microenvironments to surveil, detect, and kill the cancer cells in the natural biological mechanisms (Zhang and Chen 2018). Compared to the conventional therapeutic strategies, immunotherapy offers reduced off-targeting side effects and are effective in treating later stages of cancer. Currently, Chimeric antigen receptor (CAR) T cell (T-cells obtained from patients are genetically reprogrammed and injected into the host to initiate immune activation and destroy cancer cells) and immune checkpoint blockade therapy (CBT: Monoclonal antibodies injected to enhance the activity of T-lymphocytes) have been approved by the FDA to treat different cancers in clinics (Riley et al. 2019).

Although immunotherapy is proved to be efficient, cancer cells developed a conductive tumor microenvironment (TME) that not only facilitates the cancer cell proliferation, tumor growth, and metastases, but also prevents the immunosurveillance by recruiting immunosuppressive cells like myeloid derived suppressor cells, M2 tumor macrophages, and regulatory T-cells (Tregs) around the tumors. The tumors with non-immunogenic TME are generally termed as “cold” tumors. The presence of immunosuppressive cells in cold tumors can minimize the functionality of conventional immunotherapeutics and result in low patient response rate. Scientists have exploited the advancements in medical nanotechnology to convert the cold tumors into immunogenic “hot” tumors. The adjuvant nanotherapeutics has been developed to trigger the conversion of cold TME to hot TME. PTT is one of the efficient synergistic strategies to convert the cold TME to hot TME. The tumor lysates released post PTT can function as tumor antigens, which can be sensed by the dendritic cells (DC) and assist in development of antigen presenting cells (APC). These APCs drain into the lymph node and activate the CD4+ T-helper cells. The CD4+ T-helper cells further release the pro-inflammatory cytokines and activate cytotoxic CD8+ T-cells and natural killer cells, which will home at the tumor site and initiate the conversion of cold TME into hot TME (Nam et al. 2019). In addition, the PTT can itself result in tumor destruction, which could be beneficial for enhancing the therapeutic efficiency of immunotherapeutics (Chu and Dupuy 2014).

Graphene has been used as a platform for loading the immunoactive agents and assisting in immunotherapy (Zhang et al. 2017b). Wang et al. demonstrated that the heat generated during PTT could itself trigger immune activation and prevent metastatic tumor growth. They hybridized Fe_3O_4 nanoparticles with rGO and further PEGylated the nanocomposite to facilitate photothermal immunotherapy in metastatic breast cancer animal model. Their results demonstrate that animals post PTT triggered the activation of DC in the lymph node and assisted in efficient tumor reduction and improved the median survival time in 4 T1 metastatic animal models. In addition, the presence of Fe_3O_4 in the composition resulted in T2 MRI contrast imaging of the developed nanocomposite (Wang et al. 2019). Scientists have used graphene as a PTT adjuvant for developing novel immunotherapeutics. Tao et al. used GO along with cytosine-phosphate-guanine (CpG), an immunostimulatory oligonucleotide, to enhance the TLR9 induced immunogenicity and facilitate dual photothermal cancer immunotherapy (Tao et al. 2014). GO were dually functionalized with polyethylene glycol (PEG) and positively charged polyethylenimine to not only enhance their bioavailability but also facilitate interactions with negatively charged CpG. The CpG loaded GO resulted in heat generation and improved the secretion of TNF-α and IL-6 pro-inflammatory cytokines from RAW 264.7 macrophage cells. In addition, the tumor bearing mice demonstrated 91% tumor reduction after treating with CpG-GO nanocomposite, with negligible damage to the major organs. Recently, rGO was loaded with indoleamine-2,3-dioxygenase (IDO) inhibitor and used them as photothermal immunotherapeutic to promote immune activation and mitigate the tumor metastases (Yan et al. 2019). IDO is expressed in most of the tumor cells and plays a predominant role in catalysis of tryptophan to kynurenine. The reduced tryptophan levels in the cancer cells can induce immunotolerance and prevent the homing of T-cells at the tumor site. The intravenously injected PEG and folic acid modified rGO-IDO inhibitor nanocomplex resulted in enhanced tumor accumulation, facilitated NIR laser induced PTT, improved DC maturation, and assisted in homing CD8+ cytotoxic T-cells at the tumor site (Fig. 4) (Yan et al. 2019). The results from In vivo abscopal animal models (animal model that mimics the tumor metastases) demonstrate effective tumor reduction, immune activation, and homing of pro-inflammatory immune cells at both primary and distal tumors. In addition, the injection of anti-PD-L1 monoclonal antibodies along with PTT and IDO inhibitor therapy improved the

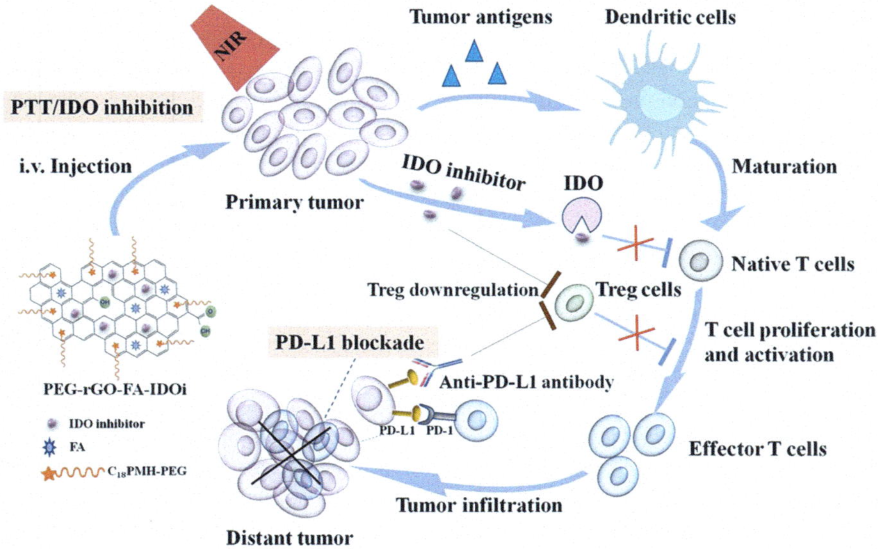

Fig. 4 Schematic illustration of PEG and folic acid modified rGO-IDO inhibitor nanocomplex induce combined PPT and immunotherapy (Adapted with permission from (Yan et al. 2019). Copyright 2019 American Chemical Society)

activity of CD8+ T-cells and resulted in enhanced tumor reduction compared to PTT and IDO inhibitor therapy. These reports indicate that graphene could be a promising candidate to develop novel photothermal immunotherapeutics.

5 Image-Guided Photothermal Therapy

Bioimaging is a vital tool that plays a predominant role in determining the activity of cell/tissue/organs in the body. Bioimaging has helped researchers, clinical doctors, and scientists to diagnose a disease and can also determine the efficiency of the developed therapeutics during the treatment. Graphene allotropes, in particular graphene quantum dots (GQDs), with their intrinsic fluorescence properties have attracted researchers to use them for fluorescence-assisted bioimaging (Kumawat et al. 2019). In addition, GQDs can overcome the issues related to photobleaching and toxicology issues. Furthermore, GQDs and graphene allotropes have been coated with metallic and transition metal oxide nanomaterials and have been used for developing biosensors for disease prognosis.

Computer tomography is an imaging procedure that can provide detailed pictures of targeted tissues/areas in the body. During CT imaging, a beam of X-rays is pointed at the patient's body and the produced X-ray signals are sent to a computer, which are then processed to construct a 3-dimensional(3D) images of tissues/organs. Although CT is a best technique to provide information about hard tissues like bone, it is difficult to distinguish soft tissues with similar mass densities. Nanoparticles, with their ability to demonstrate efficient X-ray attenuation coefficients, have gained significance as new CT contrast agents. Gold and bismuth have been used as contrast agents for CT imaging. Badrigilan et al. fabricated GQD coated bismuth nanoparticles as dual photoacoustic and photothermal adjuvant for improved computer tomography (CT) imaging and PTT (Badrigilan et al. 2018). Their results demonstrate that GQD-BiNPs demonstrated 1.3 times higher contrasting properties compared to convention iodine-based CT agents. With synergistic PTT and contrast imaging properties the developed GQD-BiNPs demonstrated enhanced PTT induced cell death and CT imaging. In another study, Jin et al. fabricated a poly lactic acid microbubble loaded with gold nanoparticles and modified their surface with GO and used them as theranostic agent with ultrasound/CT contrast imaging and PTT. Compared to the control PLA/AU, the GO coated composites demonstrated enhanced ultrasound contrast imaging. In addition, the X-ray attenuation coefficient of the GO coated PLA/AU was 4 times higher than the iodine-based CT agents. With PTT-assisted cancer cell death, these multimodal imaging systems can be useful for clinical applications.

Photoacoustic (PA) imaging involves in the use of light as an excitation source, which cause thermoelastic expansion of the tissues and assist in the generation of ultrasonic (US) waves. These US waves are received and reconstructed to provide real-time images for determining disease progression. Jun et al. fabricated folic acid functionalized chitosan coated graphene oxide nanomaterials for synergistic PA imaging and photothermal chemotherapy (Jun et al. 2019). The folic acid (FA) can assist in targeting the FA receptors on the cancer cells and result in active delivery of therapeutic drugs at the cancer site. They have demonstrated that GO can itself function as a PA contrast agent providing better PA imaging after intravenous administration of the nanocomposite. The synergistic chemo and PTT resulted in complete tumor destruction in the tumor bearing mice. In another study Hu et al. demonstrated that coating the surface of GO with pDOPA and indocyanine green resulted in ~20 times stronger PA signals compared to pristine GO (Hu et al. 2016). Gold/graphene nanohybrids have been widely investigated for synergistic PA imaging and PTT (Fig. 5) (Moon et al. 2015). The ability of Au nanomaterials to absorb NIR radiations and generate PTT has complimented the heating profiles of GO and resulted in generating elevated temperatures at low laser power irradiations (0.25–1 W/cm^2) (Moon et al. 2015). Gao et al. coated the surface of GO with AuNPs and further labeled with matrix metalloprotease 14 (MMP-14) cleavable Cy5.5 peptide (GAC) (Gao et al. 2016). The GAC can selectively sense the tumor microenvironments and result in fluorescence from Cy 5.5 in the presence of MMP which are abundant in the tumors. This presence of GO enhanced the PA imaging and improved the heating profiles of GAC. The PTT activity resulted in complete tumor destruction and no tumor relapse in tumor bearing mice after 30 days. Other nanomaterials like bismuth selenide (Zhang et al. 2017c), bismuth oxide (Badrigilan et al. 2019), and silver sulfide (Song et al. 2018) have been coated over GO/rGO and have been used for dual PA imaging and PTT.

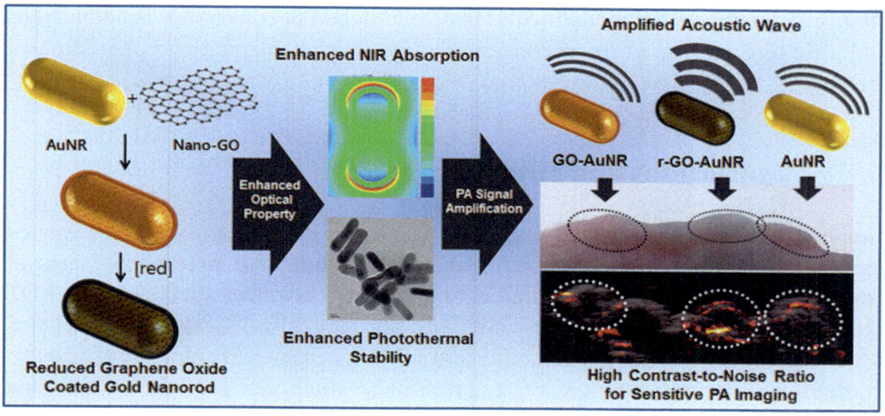

Fig. 5 Schematic illustration of Gold/graphene nanohybrids have been widely investigated for synergistic PA imaging and PTT (Adapted with permission from (Moon et al. 2015). Copyright 2015 American Chemical Society)

Magnetic resonance imaging (MRI) is a noninvasive imaging technique that uses alternating magnetic fields to develop images of organs/tissues in the body. Gadolinium, manganese oxide nanomaterials are used as T1 MRI contrast imaging agents while iron oxide nanoparticles have been used as T2 MRI contrast imaging agents. In addition, the ability of the contrast agents to respond to the magnetic field has assisted in the development of magnetic field-assisted delivery (magneto-targeting) (Lu et al. 2018; Pramanik et al. 2019). Graphene has been integrated with these MRI contrast agents to develop MRI image guided therapeutics. Chen et al. developed a new pulsed-laser-ablation-in liquid method graphene-based magneto-responsive nanohybrids (Chen et al. 2018a). Here, an Fe-target is pulsed with laser in GO-PEG aqueous solution to grow γ-Fe_2O_3 nanoparticles over Go-PEG. The developed nanohybrids demonstrated T2 MRI contrast and photothermal imaging under NIR laser irradiations. Apart from facilitating T2 MRI tumor imaging, the heat released during PTT resulted in complete tumor destruction in H22 tumor bearing mice. Badrigilan et al. conjugated iron oxide/bismuth oxide nanocomposites over GQDs and used them as "three-in-one" MRI/PA imaging and PTT agent for cancer therapy (Badrigilan et al. 2019).The presence of Bi_2O_3 in the nanocomposite resulted in 175% enhancement in X-ray attenuation compared to clinical used dotarem. Furthermore, the developed nanocomposite displayed strong T2 MRI contrast imaging with $r^2 = 62.34\ \text{mM}^{-1}\ \text{s}^{-1}$). The presence of both Bi_2O_3 and GQD in the nanocomposite resulted in enhanced PTT with a photothermal conversion efficiency of 31.8%. Qian et al. fabricated rGO-manganese ferrite nanocomposite, functionalized with PEG and used them for dual T1/T2 MRI contrast imaging and PTT (Qian et al. 2019). The presence of both Mn and Fe_2O_4 resulted in both T1 and T2 contrast imaging with r^1 and r^2 of 11.74 and 295.48 $\text{mM}^{-1}\ \text{S}^{-1}$, respectively. In addition, these composites have been labeled with $^{125-}$I for single photon emission computed tomography. Furthermore, graphene allotropes have been integrated with lanthanide doped upconversion nanomaterials and have also been used for upconversion fluorescence imaging and photothermal therapy.

6 Conclusions and Future Prospects

Graphene derivatives, especially GO and rGO, have been extensively investigated for PTT. In addition, the presence of functional groups with large surface area of graphene can assist in loading/conjugating various chemotherapeutic drugs, PDT agents, metallic nanoparticles, organic dyes, and targeting ligands over graphene, which can improve their bioavailability and achieve enhanced therapeutic efficiency. Furthermore, by integrating with nanomaterials which can provide bio imaging functionality, we can develop theranostic graphene nanocomposites for multifunctional applications. Although graphene is a promising material for developing combinatorial therapeutics with enhanced anti-cancer PTT, several issues need to be addressed before translating them to clinical settings. First, heat generated during PTT can cause changes in the TME, which could either be beneficial or toxic to the biological system. Therefore, the effect of PTT by graphene over the tumor niche needs to be investigated to achieve enhanced therapeutic efficiency in clinical settings. Second, solubility of rGO and GO has been critical in the biological environments. Although PEGylation can enhance their solubility, they can affect the photothermal heat conversion efficiency of the nanomaterials. In addition, the toxicology profiles of composite nanomaterials which are developed for enhancing the heating profiles from graphene derivatives need to be addressed before using them for clinical applications. Third, the densely packed cancer cells result in high interstitial fluid pressures around the tumors, which prevent the entry of cancer cells into the tumor cells. Therefore, it is important to develop strategies which can facilitate deep tumor drug penetration. Fourth, most of the graphene derivatives have enhanced PTT activity in NIR-1 bio-window with wavelengths <980 nm. This could prevent the treatment of deep tumors. Therefore, it would be better to develop strategies which could overcome the issues related to the laser penetration for deep tumor targeting. Fifth, the applications of PTT for developing hybrid immunotherapeutics are at infancy. Although PTT could assist in generating tumor lysate that could be used to activate immune system, the feedback activation of signaling cascade in the neighboring immune cells can result in recruitment of immunosuppressive cells like MDSC, which can limit the immunotherapeutic activity of the developed nanomaterials. However, by using multiple immunotherapeutics along with PTT, we can overcome this hurdle and use graphenebased PTT as an adjuvant for immunotherapy. Therefore, it is important to develop numerous graphene-based novel strategies with multiple functionalities as well a thorough analysis of the systemic toxicology profiles of the developed nanomaterials needs to be investigated before translating graphene-based photothermal therapeutics for clinical applications. We hope graphene derivatives can address the issues associated with PTT and replace the currently used clinical PTT agents in the near future.

Acknowledgments This research was graciously supported by National Research Foundation of Korea (no. 2018R1D1A1A09083269, 2019R1A4A1024116, 2021R1A2C2095113, 2021R1A6A1A03046418). This work was also supported by Korea National University of Transportation in 2018.

References

Akhavan O, Ghaderi E, Aghayee S, Fereydooni Y, Talebi A (2012a) The use of a glucose-reduced graphene oxide suspension for photothermal cancer therapy. J Mater Chem 22(27):13773. https://doi.org/10.1039/c2jm31396k

Akhavan O, Ghaderi E, Emamy H (2012b) Nontoxic concentrations of PEGylated graphene nanoribbons for selective cancer cell imaging and photothermal therapy. J Mater Chem 22(38):20626. https://doi.org/10.1039/c2jm34330d

Assali A, Akhavan O, Adeli M, Razzazan S, Dinarvand R, Zanganeh S, Soleimani M, Dinarvand M, Atyabi F (2018) Multifunctional core-shell nanoplatforms (gold@graphene oxide) with mediated NIR thermal therapy to promote miRNA delivery. Nanomedicine 14(6): 1891–1903. https://doi.org/10.1016/j.nano.2018.05.016

Badrigilan S, Shaabani B, Aghaji NG, Mesbahi A (2018) Graphene quantum dots-coated bismuth nanoparticles for improved CT imaging and photothermal performance. Int J Nano 1850043 https://doi.org/10.1142/S0219581X18500436

Badrigilan S, Shaabani B, Gharehaghaji N, Mesbahi A (2019) Iron oxide/bismuth oxide nanocomposites coated by graphene quantum dots: "three-in-one" theranostic agents for simultaneous CT/MR imaging-guided in vitro photothermal therapy. Photodiagn Photodyn Ther 25: 504–514. https://doi.org/10.1016/j.pdpdt.2018.10.021

Balapanuru J, Yang J-X, Xiao S, Bao Q, Jahan M, Polavarapu L, Wei J, Xu Q-H, Loh KP (2010) A graphene oxide-organic dye ionic complex with DNA-sensing and optical-limiting properties. Angew Chem Int Ed 49(37):6549–6553. https://doi.org/10.1002/anie.201001004

Chen M-L, Gao Z-W, Chen X-M, Pang S-C, Zhang Y (2018a) Laser-assisted in situ synthesis of graphene-based magnetic-responsive hybrids for multimodal imaging-guided chemo/ photothermal synergistic therapy. Talanta 182:433–442. https://doi.org/10.1016/j.talanta.2018.02.030

Chen W, Lv G, Hu W, Li D, Chen S, Dai Z (2018b) Synthesis and applications of graphene quantum dots: a review. Nanotechnol Rev 7(2):157–185. https://doi.org/10.1515/ntrev-2017-0199

Cheng L, Wang C, Feng L, Yang K, Liu Z (2014) Functional nanomaterials for phototherapies of cancer. Chem Rev 114(21):10869–10939. https://doi.org/10.1021/cr400532z

Chu KF, Dupuy DE (2014) Thermal ablation of tumours: biological mechanisms and advances in therapy. Nat Rev Cancer 14(3):199–208. https://doi.org/10.1038/nrc3672

Cui L, Lin X, Lin N, Song Y, Zhu Z, Chen X, Yang CJ (2012) Graphene oxide-protected DNA probes for multiplex microRNA analysis in complex biological samples based on a cyclic enzymatic amplification method. Chem Commun 48(2):194–196. https://doi.org/10.1039/C1CC15412E

Datta NR, Ordóñez SG, Gaipl US, Paulides MM, Crezee H, Gellermann J, Marder D, Puric E, Bodis S (2015) Local hyperthermia combined with radiotherapy and−/or chemotherapy: recent advances and promises for the future. Cancer Treat Rev 41(9):742–753. https://doi.org/10.1016/j.ctrv.2015.05.009

de Melo-Diogo D, Pais-Silva C, Dias DR, Moreira AF, Correia IJ (2017) Strategies to improve cancer Photothermal therapy mediated by nanomaterials. Adv Healthc Mater 6(10):1700073. https://doi.org/10.1002/adhm.201700073

de Melo-Diogo D, Lima-Sousa R, Alves CG, Costa EC, Louro RO, Correia IJ (2018) Functionalization of graphene family nanomaterials for application in cancer therapy. Colloids Surf B Biointerfaces 171:260–275. https://doi.org/10.1016/j.colsurfb.2018.07.030

Ding YH, Zhang P, Zhuo Q, Ren HM, Yang ZM, Jiang Y (2011) A green approach to the synthesis of reduced graphene oxide nanosheets under UV irradiation. Nanotechnology 22(21):215601. https://doi.org/10.1088/0957-4484/22/21/215601

Feng L, Yang X, Shi X, Tan X, Peng R, Wang J, Liu Z (2013) Polyethylene glycol and polyethylenimine dual-functionalized Nano-graphene oxide for photothermally enhanced gene delivery. Small 9(11):1989–1997. https://doi.org/10.1002/smll.201202538

Gao S, Zhang L, Wang G, Yang K, Chen M, Tian R, Ma Q, Zhu L (2016) Hybrid graphene/au activatable theranostic agent for multimodalities imaging guided enhanced photothermal therapy. Biomaterials 79:36–45. https://doi.org/10.1016/j.biomaterials.2015.11.041

Ge J, Lan M, Zhou B, Liu W, Guo L, Wang H, Jia Q, Niu G, Huang X, Zhou H, Meng X, Wang P, Lee C-S, Zhang W, Han X (2014) A graphene quantum dot photodynamic therapy agent with high singlet oxygen generation. Nat Commun 5:4596. https://doi.org/10.1038/ncomms5596

Guo M, Huang J, Deng Y, Shen H, Ma Y, Zhang M, Zhu A, Li Y, Hui H, Wang Y, Yang X, Zhang Z, Chen H (2015) pH-responsive cyanine-grafted graphene oxide for fluorescence resonance energy transfer-enhanced photothermal therapy. Adv Funct Mater 25(1):59–67. https://doi.org/10.1002/adfm.201402762

He S, Song B, Li D, Zhu C, Qi W, Wen Y, Wang L, Song S, Fang H, Fan C (2010) A graphene Nanoprobe for rapid, sensitive, and multicolor fluorescent DNA analysis. Adv Funct Mater 20(3):453–459. https://doi.org/10.1002/adfm.200901639

Hu Z, Li J, Huang Y, Chen L, Li Z (2015) Functionalized graphene/C 60 nanohybrid for targeting photothermally enhanced photodynamic therapy. RSC Adv 5(1):654–664. https://doi.org/10.1039/C4RA13427C

Hu D, Zhang J, Gao G, Sheng Z, Cui H, Cai L (2016) Indocyanine green-loaded polydopamine-reduced graphene oxide nanocomposites with amplifying photoacoustic and photothermal effects for cancer theranostics. Theranostics 6(7):1043–1052. https://doi.org/10.7150/thno.14566

Hu Z, Wang C, Zhao F, Xu X, Wang S, Yu L, Zhang D, Huang Y (2017) Fabrication of a graphene/C 60 nanohybrid via γ-cyclodextrin host–guest chemistry for photodynamic and photothermal therapy. Nanoscale 9(25):8825–8833. https://doi.org/10.1039/C7NR02922E

Huang H, Song W, Rieffel J, Lovell JF (2015) Emerging applications of porphyrins in photomedicine. Front Phys 3. https://doi.org/10.3389/fphy.2015.00023

Huang C, Hu X, Hou Z, Ji J, Li Z, Luan Y (2019) Tailored graphene oxide-doxorubicin nanovehicles via near-infrared dye-lactobionic acid conjugates for chemo-photothermal therapy. J Colloid Interface Sci 545:172–183. https://doi.org/10.1016/j.jcis.2019.03.019

Hummers WS, Offeman RE (1958) Preparation of graphitic oxide. J Am Chem Soc 80(6):1339–1339. https://doi.org/10.1021/ja01539a017

Jha S, Sharma PK, Malviya R (2016) Hyperthermia: role and risk factor for cancer treatment. Achievements Life Sci 10(2):161–167. https://doi.org/10.1016/j.als.2016.11.004

Ji P, Zhang W, Ai S, Zhang Y, Liu J, Liu J, He P, Li Y (2019) Hybridization of graphene oxide into nanogels to acquire higher photothermal effects for therapeutic delivery. Nanotechnology 30(11):115701. https://doi.org/10.1088/1361-6528/aaf8e4

Jun SW, Manivasagan P, Kwon J, Nguyen VT, Mondal S, Ly CD, Lee J, Kang Y-H, Kim C-S, Oh J (2019) Folic acid–conjugated chitosan-functionalized graphene oxide for highly efficient photoacoustic imaging-guided tumor-targeted photothermal therapy. Int J Biol Macromol. https://doi.org/10.1016/j.ijbiomac.2019.11.055

Kalluru P, Vankayala R, Chiang CS, Hwang KC (2016) Nano-graphene oxide-mediated in vivo fluorescence imaging and bimodal photodynamic and photothermal destruction of tumors. Biomaterials 95:1–10. https://doi.org/10.1016/j.biomaterials.2016.04.006

Kim H, Kim WJ (2014) Photothermally controlled gene delivery by reduced graphene oxide-Polyethylenimine nanocomposite. Small 10(1):117–126. https://doi.org/10.1002/smll.201202636

Kim H, Lee D, Kim J, Kim T, Kim WJ (2013) Photothermally triggered cytosolic drug delivery via endosome disruption using a functionalized reduced graphene oxide. ACS Nano 7(8): 6735–6746. https://doi.org/10.1021/nn403096s

Kim Y-K, Na H-K, Kim S, Jang H, Chang S-J, Min D-H (2015) One-pot synthesis of multifunctional au@graphene oxide Nanocolloid Core@Shell nanoparticles for Raman bioimaging, Photothermal, and photodynamic therapy. Small 11(21):2527–2535. https://doi.org/10.1002/smll.201402269

Koppens FHL, Chang DE, García de Abajo FJ (2011) Graphene Plasmonics: A platform for strong light–matter interactions. Nano Lett 11(8):3370–3377. https://doi.org/10.1021/nl201771h

Kosynkin DV, Higginbotham AL, Sinitskii A, Lomeda JR, Dimiev A, Price BK, Tour JM (2009) Longitudinal unzipping of carbon nanotubes to form graphene nanoribbons. Nature 458(7240): 872–876. https://doi.org/10.1038/nature07872

Kumawat MK, Thakur M, Bahadur R, Kaku T, R.S., P., Ninawe, A., & Srivastava, R. (2019) Preparation of graphene oxide-graphene quantum dots hybrid and its application in cancer theranostics. Mater Sci Eng C 103:109774. https://doi.org/10.1016/j.msec.2019.109774

Li F, Park S-J, Ling D, Park W, Han JY, Na K, Char K (2013) Hyaluronic acid-conjugated graphene oxide/photosensitizer nanohybrids for cancer targeted photodynamic therapy. J Mater Chem B 1(12):1678. https://doi.org/10.1039/c3tb00506b

Li Q, Hong L, Li H, Liu C (2017) Graphene oxide-fullerene C60 (GO-C60) hybrid for photodynamic and photothermal therapy triggered by near-infrared light. Biosens Bioelectron 89:477–482. https://doi.org/10.1016/j.bios.2016.03.072

Li J, Huang X, Huang R, Jiang J, Wang Y, Zhang J, Jiang H, Xiang X, Chen W, Nie X, Gui R (2019) Erythrocyte membrane camouflaged graphene oxide for tumor-targeted photothermal-chemotherapy. Carbon 146:660–670. https://doi.org/10.1016/j.carbon.2019.02.056

Liang J, Chen B, Hu J, Huang Q, Zhang D, Wan J, Hu Z, Wang B (2019) pH and thermal dual-responsive graphene oxide Nanocomplexes for targeted drug delivery and photothermal-chemo/photodynamic synergetic therapy. ACS Appl Bio Mater:acsabm.9b00835. https://doi.org/10.1021/acsabm.9b00835

Lima-Sousa R, de Melo-Diogo D, Alves CG, Costa EC, Ferreira P, Louro RO, Correia IJ (2018) Hyaluronic acid functionalized green reduced graphene oxide for targeted cancer photothermal therapy. Carbohydr Polym 200:93–99. https://doi.org/10.1016/j.carbpol.2018.07.066

Liu Z, Zhang J, Tian Y, Zhang L, Han X, Wang Q, Cheng W (2018) Targeted delivery of reduced graphene oxide nanosheets using multifunctional ultrasound nanobubbles for visualization and enhanced photothermal therapy. Int J Nanomedicine 13:7859–7872. https://doi.org/10.2147/IJN.S181268

Liu W, Zhang X, Zhou L, Shang L, Su Z (2019) Reduced graphene oxide (rGO) hybridized hydrogel as a near-infrared (NIR)/pH dual-responsive platform for combined chemo-photothermal therapy. J Colloid Interface Sci 536:160–170. https://doi.org/10.1016/j.jcis.2018.10.050

Lu C-H, Yang H-H, Zhu C-L, Chen X, Chen G-N (2009) A graphene platform for sensing biomolecules. Angew Chem Int Ed 48(26):4785–4787. https://doi.org/10.1002/anie.200901479

Lu Y-J, Lin P-Y, Huang P-H, Kuo C-Y, Shalumon KT, Chen M-Y, Chen J-P (2018) Magnetic graphene oxide for dual targeted delivery of doxorubicin and photothermal therapy. Nano 8(4): 193. https://doi.org/10.3390/nano8040193

Luo S, Yang Z, Tan X, Wang Y, Zeng Y, Wang Y, Li C, Li R, Shi C (2016) Multifunctional photosensitizer grafted on polyethylene glycol and polyethylenimine dual-functionalized nanographene oxide for cancer-targeted near-infrared imaging and synergistic phototherapy. ACS Appl Mater Interfaces 8(27):17176–17186. https://doi.org/10.1021/acsami.6b05383

Ma W, Hu Y, Yang H, Zhang Y, Ding J, Chen L (2019) Au-aided reduced graphene oxide-based nanohybrids for photo-chemotherapy. Mater Sci Eng C 95:256–263. https://doi.org/10.1016/j.msec.2018.10.072

Marcano DC, Kosynkin DV, Berlin JM, Sinitskii A, Sun Z, Slesarev A, Alemany LB, Lu W, Tour JM (2010) Improved synthesis of graphene oxide. ACS Nano 4(8):4806–4814. https://doi.org/10.1021/nn1006368

Martin SJ, Henry CM, Cullen SP (2012) A perspective on mammalian caspases as positive and negative regulators of inflammation. Mol Cell 46(4):387–397. https://doi.org/10.1016/j.molcel.2012.04.026

Mocan T, Matea CT, Cojocaru I, Ilie I, Tabaran FA, Zaharie F, Iancu C, Bartos D, Mocan L (2014) Photothermal treatment of human pancreatic cancer using PEGylated multi-walled carbon nanotubes induces apoptosis by triggering mitochondrial membrane depolarization mechanism. J Cancer 5(8):679–688. https://doi.org/10.7150/jca.9481

Moon H, Kumar D, Kim H, Sim C, Chang J-H, Kim J-M, Kim H, Lim D-K (2015) Amplified photoacoustic performance and enhanced photothermal stability of reduced graphene oxide coated gold nanorods for sensitive photoacoustic imaging. ACS Nano 9(3):2711–2719. https://doi.org/10.1021/nn506516p

Nafiujjaman M, Revuri V, Park H-K, Kwon IK, Cho K-J, Lee Y (2016) Enhanced photodynamic properties of graphene quantum dot conjugated Ce6 nanoparticles for targeted cancer therapy and imaging. Chem Lett 45(8):997–999. https://doi.org/10.1246/cl.160388

Nakano H, Tetsuka H, Spencer MJS, Morishita T (2018) Chemical modification of group IV graphene analogs. Sci Technol Adv Mater 19(1):76–100. https://doi.org/10.1080/14686996.2017.1422224

Nam J, Son S, Park KS, Zou W, Shea LD, Moon JJ (2019) Cancer nanomedicine for combination cancer immunotherapy. Nat Rev Mater 4(6):398–414. https://doi.org/10.1038/s41578-019-0108-1

Pramanik N, Ranganathan S, Rao S, Suneet K, Jain S, Rangarajan A, Jhunjhunwala S (2019) A composite of hyaluronic acid-modified graphene oxide and iron oxide nanoparticles for targeted drug delivery and magnetothermal therapy. ACS Omega 4(5):9284–9293. https://doi.org/10.1021/acsomega.9b00870

Qian R, Maiti D, Zhong J, Xiong S, Zhou H, Zhu R, Wan J, Yang K (2019) Multifunctional nano-graphene based nanocomposites for multimodal imaging guided combined radioisotope therapy and chemotherapy. Carbon 149:55–62. https://doi.org/10.1016/j.carbon.2019.04.046

Riley RS, June CH, Langer R, Mitchell MJ (2019) Delivery technologies for cancer immunotherapy. Nat Rev Drug Discov 18(3):175–196. https://doi.org/10.1038/s41573-018-0006-z

Robinson JT, Tabakman SM, Liang Y, Wang H, Sanchez Casalongue H, Vinh D, Dai H (2011) Ultrasmall reduced graphene oxide with high near-infrared absorbance for photothermal therapy. J Am Chem Soc 133(17):6825–6831. https://doi.org/10.1021/ja2010175

Roy S, Sarkar A, Jaiswal A (2019) Poly(allylamine hydrochloride)-functionalized reduced graphene oxide for synergistic chemo-photothermal therapy. Nanomedicine 14(3):255–274. https://doi.org/10.2217/nnm-2018-0320

Song J, Yang X, Jacobson O, Lin L, Huang P, Niu G, Ma Q, Chen X (2015) Sequential drug release and enhanced photothermal and photoacoustic effect of hybrid reduced graphene oxide-loaded ultrasmall gold Nanorod vesicles for cancer therapy. ACS Nano 9(9):9199–9209. https://doi.org/10.1021/acsnano.5b03804

Song Y-Y, Li C, Yang X-Q, An J, Cheng K, Xuan Y, Shi X-M, Gao M-J, Song X-L, Zhao Y-D, Chen W (2018) Graphene oxide coating core–shell silver sulfide@mesoporous silica for active targeted dual-mode imaging and chemo-photothermal synergistic therapy against tumors. J Mater Chem B 6(29):4808–4820. https://doi.org/10.1039/C8TB00940F

Stankovich S, Piner RD, Chen X, Wu N, Nguyen ST, Ruoff RS (2006) Stable aqueous dispersions of graphitic nanoplatelets via the reduction of exfoliated graphite oxide in the presence of poly (sodium 4-styrenesulfonate). J Mater Chem 16(2):155–158. https://doi.org/10.1039/B512799H

Tao Y, Ju E, Ren J, Qu X (2014) Immunostimulatory oligonucleotides-loaded cationic graphene oxide with photothermally enhanced immunogenicity for photothermal/immune cancer therapy. Biomaterials 35(37):9963–9971. https://doi.org/10.1016/j.biomaterials.2014.08.036

Tetsuka H, Asahi R, Nagoya A, Okamoto K, Tajima I, Ohta R, Okamoto A (2012) Optically Tunable amino-functionalized graphene quantum dots. Adv Mater 24(39):5333–5338. https://doi.org/10.1002/adma.201201930

Thakur M, Kumawat MK, Srivastava R (2017) Multifunctional graphene quantum dots for combined photothermal and photodynamic therapy coupled with cancer cell tracking applications. RSC Adv 7(9):5251–5261. https://doi.org/10.1039/C6RA25976F

Thapa RK, Soe ZC, Ou W, Poudel K, Jeong J-H, Jin SG, Ku SK, Choi H-G, Lee YM, Yong CS, Kim JO (2018) Palladium nanoparticle-decorated 2-D graphene oxide for effective photodynamic and photothermal therapy of prostate solid tumors. Colloids Surf B: Biointerfaces 169: 429–437. https://doi.org/10.1016/j.colsurfb.2018.05.051

Tian B, Wang C, Zhang S, Feng L, Liu Z (2011a) Photothermally enhanced photodynamic therapy delivered by nano-graphene oxide. ACS Nano 5(9):7000–7009. https://doi.org/10.1021/nn201560b

Tian B, Wang C, Zhang S, Feng L, Liu Z (2011b) Photothermally enhanced photodynamic therapy delivered by Nano-graphene oxide. ACS Nano 5(9):7000–7009. https://doi.org/10.1021/nn201560b

Tian B, Wang C, Zhang S, Feng L, Liu Z (2011c) Photothermally enhanced photodynamic therapy delivered by Nano-graphene oxide. ACS Nano 5(9):7000–7009. https://doi.org/10.1021/nn201560b

Tung VC, Allen MJ, Yang Y, Kaner RB (2009) High-throughput solution processing of large-scale graphene. Nat Nanotechnol 4(1):25–29. https://doi.org/10.1038/nnano.2008.329

van Straten D, Mashayekhi V, de Bruijn H, Oliveira S, Robinson D (2017) Oncologic photodynamic therapy: basic principles, current clinical status and future directions. Cancers 9(12):19. https://doi.org/10.3390/cancers9020019

Wang L, Yan J (2019) Superficial synthesis of photoactive copper sulfide quantum dots loaded nano-graphene oxide sheets combined with near infrared (NIR) laser for enhanced photothermal therapy on breast cancer in nursing care management. J Photochem Photobiol B Biol 192:68–73. https://doi.org/10.1016/j.jphotobiol.2018.12.024

Wang G, Yang J, Park J, Gou X, Wang B, Liu H, Yao J (2008a) Facile synthesis and characterization of graphene Nanosheets. J Phys Chem C 112(22):8192–8195. https://doi.org/10.1021/jp710931h

Wang S, Chia P-J, Chua L-L, Zhao L-H, Png R-Q, Sivaramakrishnan S, Zhou M, Goh RG-S, Friend RH, Wee AT-S, Ho PK-H (2008b) Band-like transport in surface-functionalized highly solution-Processable graphene Nanosheets. Adv Mater 20(18):3440–3446. https://doi.org/10.1002/adma.200800279

Wang Y, Wang H, Liu D, Song S, Wang X, Zhang H (2013) Graphene oxide covalently grafted upconversion nanoparticles for combined NIR mediated imaging and photothermal/photodynamic cancer therapy. Biomaterials 34(31):7715–7724. https://doi.org/10.1016/j.biomaterials.2013.06.045

Wang L, Wang M, Zhou B, Zhou F, Murray C, Towner RA, Smith N, Saunders D, Xie G, Chen WR (2019) PEGylated reduced-graphene oxide hybridized with Fe_3O_4 nanoparticles for cancer photothermal-immunotherapy. J Mater Chem B. https://doi.org/10.1039/C9TB00630C

Xing R, Jiao T, Liu Y, Ma K, Zou Q, Ma G, Yan X (2016) Co-assembly of graphene oxide and albumin/photosensitizer Nanohybrids towards enhanced photodynamic therapy. Polymers 8(5): 181. https://doi.org/10.3390/polym8050181

Yan M, Liu Y, Zhu X, Wang X, Liu L, Sun H, Wang C, Kong D, Ma G (2019) Nanoscale reduced graphene oxide-mediated Photothermal therapy together with IDO inhibition and PD-L1 blockade synergistically promote antitumor immunity. ACS Appl Mater Interfaces 11(2):1876–1885. https://doi.org/10.1021/acsami.8b18751

Yin F, Hu K, Chen Y, Yu M, Wang D, Wang Q, Yong K-T, Lu F, Liang Y, Li Z (2017) SiRNA delivery with PEGylated graphene oxide Nanosheets for combined Photothermal and Genetherapy for pancreatic cancer. Theranostics 7(5):1133–1148. https://doi.org/10.7150/thno.17841

Zeng Y, Yang Z, Luo S, Li H, Liu C, Hao Y, Liu J, Wang W, Li R (2015) Fast and facile preparation of PEGylated graphene from graphene oxide by lysosome targeting delivery of photosensitizer to efficiently enhance photodynamic therapy. RSC Adv 5(71):57725–57734. https://doi.org/10.1039/C5RA07535A

Zhang H, Chen J (2018) Current status and future directions of cancer immunotherapy. J Cancer 9(10):1773–1781. https://doi.org/10.7150/jca.24577

Zhang D-Y, Zheng Y, Tan C-P, Sun J-H, Zhang W, Ji L-N, Mao Z-W (2017a) Graphene oxide decorated with Ru(II)–polyethylene glycol complex for lysosome-targeted imaging and photodynamic/photothermal therapy. ACS Appl Mater Interfaces 9(8):6761–6771. https://doi.org/10.1021/acsami.6b13808

Zhang H, Yan T, Xu S, Feng S, Huang D, Fujita M, Gao X-D (2017b) Graphene oxide-chitosan nanocomposites for intracellular delivery of immunostimulatory CpG oligodeoxynucleotides. Mater Sci Eng C 73:144–151. https://doi.org/10.1016/j.msec.2016.12.072

Zhang Y, Zhang H, Wang Y, Wu H, Zeng B, Zhang Y, Tian Q, Yang S (2017c) Hydrophilic graphene oxide/bismuth selenide nanocomposites for CT imaging, photoacoustic imaging, and photothermal therapy. J Mater Chem B 5(9):1846–1855. https://doi.org/10.1039/C6TB02137A

Zhang W, Dai J, Zhang G, Zhang Y, Li S, Nie D (2018a) Photothermal/pH dual-responsive drug delivery system of amino-terminated HBP-modified rGO and the chemo-photothermal therapy on tumor cells. Nanoscale Res Lett 13(1):379. https://doi.org/10.1186/s11671-018-2787-8

Zhang Z, Shi J, Song Z, Zhu X, Zhu Y, Cao S (2018b) A synergistically enhanced photothermal transition effect from mesoporous silica nanoparticles with gold nanorods wrapped in reduced graphene oxide. J Mater Sci 53(3):1810–1823. https://doi.org/10.1007/s10853-017-1628-y

Zhang H, Li Y, Pan Z, Chen Y, Fan Z, Tian H, Zhou S, Zhang Y, Shang J, Jiang B, Wang F, Luo F, Hou Z (2019) Multifunctional Nanosystem based on graphene oxide for synergistic multistage tumor-targeting and combined chemo-photothermal therapy. Mol Pharm 16(5):1982–1998. https://doi.org/10.1021/acs.molpharmaceut.8b01335

Zhen SJ, Wang TT, Liu YX, Wu ZL, Zou HY, Huang CZ (2018) Reduced graphene oxide coated cu 2−x se nanoparticles for targeted chemo-photothermal therapy. J Photochem Photobiol B Biol 180:9–16. https://doi.org/10.1016/j.jphotobiol.2018.01.020

Zou L, Wang H, He B, Zeng L, Tan T, Cao H, He X, Zhang Z, Guo S, Li Y (2016) Current approaches of photothermal therapy in treating cancer metastasis with nanotherapeutics. Theranostics 6(6):762–772. https://doi.org/10.7150/thno.14988

Part IV

Graphene for Nanobiosensors, Biochips, and Antibacterial Agents

Graphene for Nanobiosensors and Nanobiochips

Mijeong Kang and Seunghun Lee

Abstract

Graphene family nanomaterials have interesting electronic structures which determine their electrical, optical, and mechanical properties. Especially, their unique chemical properties enable interactions with biological substances and chemical reagents, and the interactions have further an influence on the observable properties of the graphene family nanomaterials. Such aspects render graphene family nanomaterials versatile for various types of biosensing as a target recognition unit and a recognition-to-signal transduction unit. In this chapter, we look over the recent progress on the graphene-based biosensors, which is categorized in terms of (1) the role of graphene family nanomaterials (target recognition, signal transduction), (2) the sensing mechanisms and modes (electrochemical, electrical, fluorescent, Raman scattering), and (3) the formats of sensing devices (paper, lab-on-a-chip, wearable devices).

Keywords

Graphene family nanomaterials · Materials properties · Biosensors · Sensing mechanisms · Measurement modes · Sensing devices

M. Kang (✉)
Department of Cogno-Mechatronics Engineering, Pusan National University, Busan, South Korea
e-mail: mkang@pusan.ac.kr

S. Lee (✉)
Department of Physics, Pukyong National University, Busan, South Korea
e-mail: seunghun@pknu.ac.kr

D.-W. Han, S. W. Hong (eds.), *Multifaceted Biomedical Applications of Graphene*, Advances in Experimental Medicine and Biology 1351,
https://doi.org/10.1007/978-981-16-4923-3_10

1 Introduction

Biosensors are devices that convert the chemical and/or physical information on biological or medical samples to readable signals. In general, they are composed of the element (e.g., antibody, aptamer) to recognize a specific target substance (e.g., neurotransmitter, cytokine, pathogen) in a sample (e.g., blood, urine) and the element to convert an event of target recognition to a certain physical signal (e.g., electrochemical, electrical, optical signals) as illustrated in Fig. 1a. Biosensors must meet the major requirements with respect to sensitivity and selectivity of target detection and other ones such as low device-to-device variation, reusability, long-term stability, or low fabrication cost for practical usages. Recently, there are growing demands for other device properties for future applications, such as miniaturizability for point-of-care diagnosis, wireless operation for the integration into internet-of-things, and mechanical flexibility and optical transparency for wearable platform.

Graphene and graphene-derivatives such as graphene oxide (GO), reduced graphene oxide (rGO), heteroatom-doped graphene (e.g., doped with N, B, S, or F), and graphene quantum dot (GQD) have been extensively studied and employed in biosensors. Pristine graphene is an ultrathin (0.34 nm-thick, optical transparency of >95%) sheet of sp^2-bonded carbon atoms organized in a perfect honeycomb lattice, across which π electrons are delocalized (Fig. 1b). This confers unique (1) electrical (e.g., electrical conductivity of 1.42×10^6 S/m, sheet resistance of

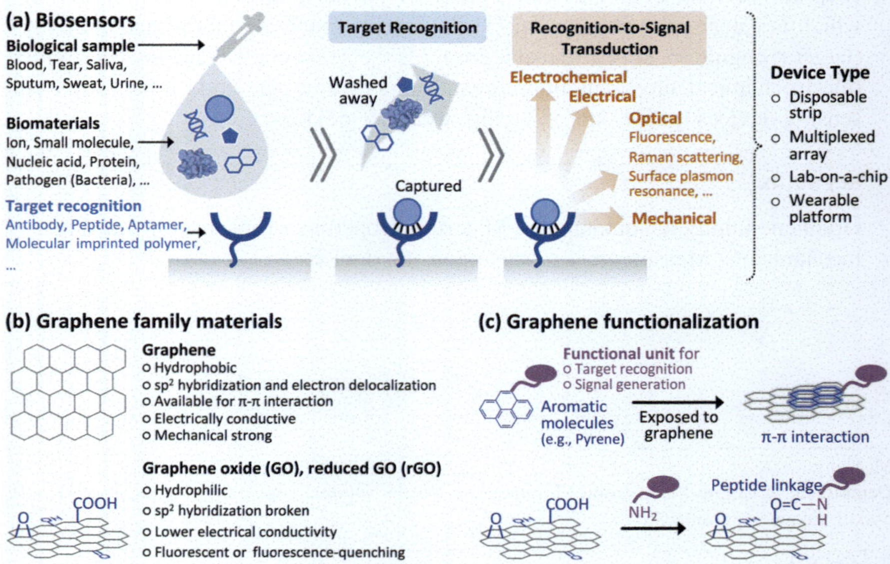

Fig. 1 Graphene-based biosensors. (**a**) General layout of biosensors. (**b**) Graphene family materials and their intrinsic properties utilized in biosensors. (**c**) Chemical properties of graphene family materials for their functionalization with other sensor components

125 Ω/sq), (2) chemical (e.g., availability of π-π interaction), (3) mechanical (e.g., flexible, Young's modulus of ~1 TPa, intrinsic strength of 130 GPa) properties on graphene, and when the sp^2-hybridization is perturbed (e.g., by introducing oxygen-containing functional groups to the edges of graphene) the graphene-derivatives can have (4) particular optical properties (e.g., fluorescence). In addition to these intrinsic properties, graphene family materials can be equipped with other properties required for biosensors via surface functionalization with functional components (Fig. 1c, see details in the following section). As a versatile platform for biosensor component in terms of physical properties and chemical reactivities, graphene (and its derivatives) has been one of the top priority materials for biosensors.

2 Graphene Family Materials as Functional Components of Biosensors

2.1 Graphene-Assisted Target Recognition

Graphene derivatives have several chemical features that facilitate the recognition of target substances. First, π electrons at the basal plane of graphene provides direct interaction with various biomolecules containing aromatic groups (e.g., dopamine, ascorbic acid, uric acids) via π-π interaction. The strength of π-π interaction is dependent on the chemical structure of the biomolecules. This allows even unmodified graphene to form distinct interfaces with different biomolecules, which can be observed by the interface-sensitive measurements (e.g., kinetics of electron transfer through the graphene-solution interface) (Ping et al. 2012). Second, the graphene derivatives that have negative charges attract the positively charged target substances and repel the negatively charged interferants (Wongkaew et al. 2019). Third, more sophisticatedly, graphene can be functionalized with the specific target recognition components via non-covalent or covalent bonding. A molecule possessing aromatic rings are used as a non-covalent linker tethered to the target recognition components, which closely places the components to graphene via π-π stacking. Pyrene, for instance, consists of four fused benzene rings, and has been used as a linker (Fig. 1c, upper). In the cases of graphene-derivatives, electrostatic interaction and hydrogen boning are also used for surface functionalization (Wang et al. 2020a). For the graphene-derivatives that have reactive functional groups (e.g., –COOH), the recognition components labeled with the other conjugation pair (e.g., $–NH_3$) can be coupled to the graphene derivatives via covalent bond formation (e.g., amide bonding,) (Fig. 1c, lower). The functionalization of graphene is of great significance for biosensing applications and new functional modification methods are being developed for further improvement of graphene-based biosensors.

2.1.1 Graphene as a Signal Transducer

Graphene-Based Electrode for Electrochemical Sensor

In electrochemical biosensors, the signal transduction relies on the electron transfer between an electrode and the substances that can be reduced or oxidized at mild electrical potential (within ± few V vs. Ag/AgCl reference electrode), which is called redox-active, in a sample solution. As simply depicted in Fig. 2a, once the redox-active substance diffuses in the solution and reaches the electrode-solution interface, it can either receive electrons from the electrode (resulting in reduction current) or give electrons to the electrode (resulting in oxidation current) if the electrode is appropriately polarized. Many redox-active biomolecules and redox-probe molecules have their own standard reduction potentials so that they generate reduction or oxidation current around specific potentials, which gives the specificity to electrochemical biosensors.

Various materials including noble metals, semiconductors, and carbon materials have been used as electrodes of electrochemical biosensors (Wongkaew et al. 2019). Noble metals, especially gold, are used due to high electrical conductivity, fast electron transfer kinetics, and availability for chemical functionalization via strong

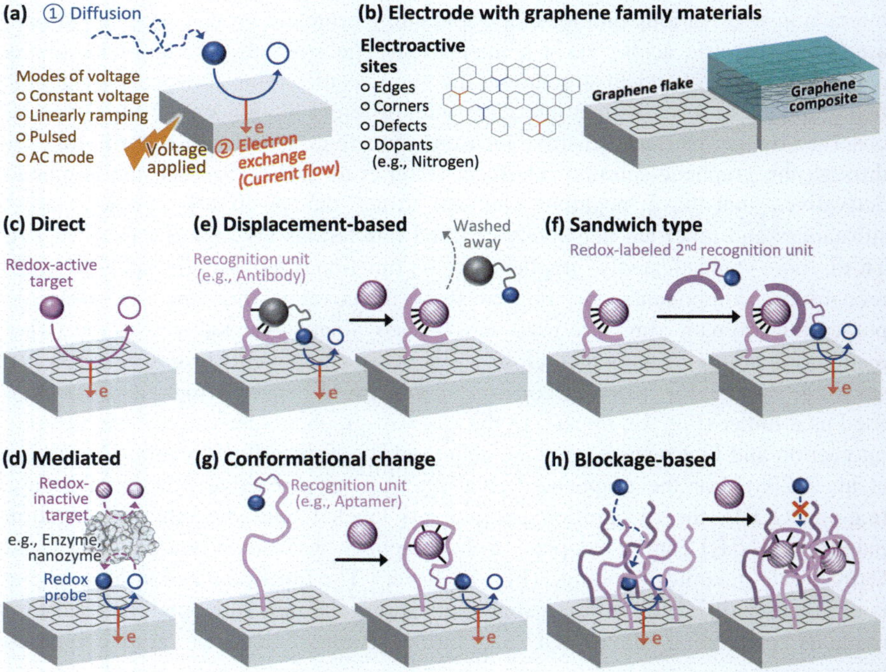

Fig. 2 Graphene-based electrochemical biosensors. (**a**) Two fundamental steps in an electrochemical reaction. (**b**) Electrode functionalized with graphene family materials. (**c–h**) Representative detection mechanisms used in electrochemical biosensors

metal-sulfur covalent bond, although there remains the issue of high costs. Semiconductors (e.g., indium tin oxide) have attracted attention as electrochemical sensor materials because of their lower costs compared with noble metals; however, their usages are limited due to slower electron transfer kinetics. Carbon-based materials, in the graphitic form (e.g., glassy carbon, highly oriented pyrolytic graphite, graphene), have established themselves as the most practical electrode materials in electrochemical sensors. They are composed of sp^2-hybridized carbon atoms and their π electrons contribute to high electrical conductivity. Graphene, in particular, enables fast electron transfer process at its edges, corners, and defects (Fig. 2b). Since graphene has large surface area, it provides a number of such electroactive sites and therefore can increase electrochemical signals and improve the detection sensitivity. Graphene and its derivatives can be used as an individual sheet attached on a supporting electrode or as nanocomposites, where a large number of graphene sheets are entrapped and dispersed, deposited on the electrode (Fig. 2b).

Electrochemical biosensors can be classified in terms of measurement techniques. In amperometry, a constant potential is applied to an electrode for a certain time, and current flow as a result of reduction or oxidation of redox-active substances is simultaneously recorded. This method is simple and robust so that it has been adopted in commercial blood glucose sensor. Voltammetric techniques are also based on the current measurement, but featured by the variation of potential. Potential can be linearly ramped over time (linear sweep voltammetry, LSV; cyclic voltammetry, CV), which is used to find at which potential the substances of interest react; or it can be pulsed in various ways (e.g., differential pulsed voltammetry, DPV; square wave voltammetry, SWV), which largely removes the background signal that can hide the (mostly) small sensing signals, and therefore such pulsed voltammetries can improve detection sensitivity. Oscillating potentials can be applied to the electrode at different oscillation frequencies and the impedance, a frequency-dependent resistance, can be measured by electrochemical impedance spectroscopy (EIS). The impedance is highly sensitive to the small changes in the electrode-solution interface so that EIS has been used to detect the binding events on the electrode. Graphene-based materials can affect the sensitivities of these measurement techniques. For instance, an electrical impedance sensor for cancer diagnosis can have more comprehensive cell signals by coating the electrode with graphene (Fig. 3) (Wang et al. 2018a). In this sensor, a cancer cell was interfaced with the electrode and its electrical activities associated with biophysical behaviors (e.g., proliferation, metastasis) were monitored. To increase the measurement sensitivity, three-dimensional polymeric structure at microscale was fabricated to maximize the interfacial area between the electrode and the cancer cell, and here graphene was coated to the structure allowing it to act as an electrode (Fig. 3, left). Due to the small dimension, high electrical conductivity, and, importantly, mechanical flexibility, graphene formed efficient electrical bio-interface (Fig. 3, middle) and yielded the electrical signals about two times stronger than those measured from the two-dimensional electrode (Fig. 3, right). This three-dimensional graphene-coated electrode can serve as an improved tool for the dynamic studies on the clinical cell-associated electrical signals.

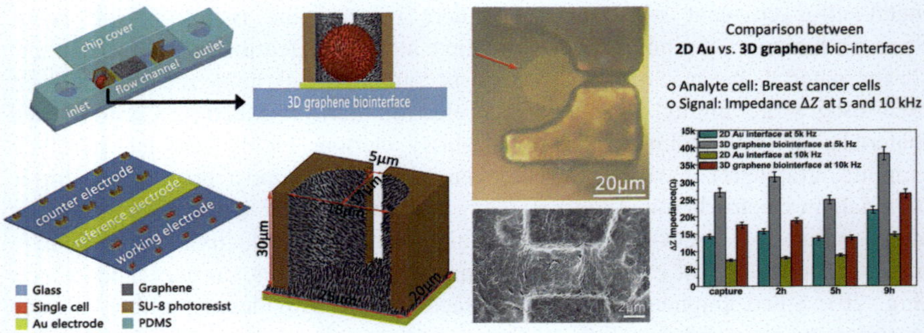

Fig. 3 Graphene-coated electrode sensor for cancer cell analysis. (Left) Three-dimensional structure coated with graphene for the electrode-cell interface with improved topographical interaction. (Middle) Bio-interface between graphene-coated electrode and a cancer call. (Right) Electrical signals (changes in impedance, ΔZ) of a cancer cell measured by three-dimensional graphene-electrodes and conventional two-dimensional electrode. Reproduced from Wang et al. (2018a)

Another classification of the electrochemical biosensors is based on the mechanism of converting the target recognition event to the electrochemical signals. For detecting the materials that are intrinsically redox-active, graphene-functionalized electrodes can be "directly" used without further conjugation with recognition components (Fig. 2c). In this case, however, it is not easy to secure the detection specificity since in real biological and clinical samples (e.g., blood) there are high concentrations of other redox-active materials, and their electrochemical signals can appear within the potential window of the target signal. This interference can be minimized if graphene strongly interacts with either the target substance or the interfering species, subsequently facilitating the electron transfer process of either one of them. For instance, the electrode with graphene composite coat was used for therapeutic monitoring of a redox-active antipsychotic drug, clozapine, in clinical blood serum samples where typical interfering species such as uric acid are present at higher concentration than the drug molecule (Fig. 4a) (Kang et al. 2017). Many interfering biomolecules intrinsically have slow electron transfer kinetics and show less-defined oxidation current peaks over a wide potential range when measured by the electrode without graphene coat (i.e., bare electrode, electrode with graphene-free composite), resulting in the overlap with the drug signal. Graphene provided detection specificity by displacing the interfering signals out of the range of the target signal; close contact between graphene and uric acid via π-π interaction catalyzed the interferants oxidation.

Many biologically and clinically significant substances do not have the redox properties that are readily accessible to typical electrochemical methods. To detect such substances, electrochemical sensors need the target-recognizing component and the signal-generating component. There are smart detection mechanisms that orchestrate the operations of such functional components. As illustrated in Fig. 3d, a recognition unit such as enzyme and nanozyme changes the chemical properties

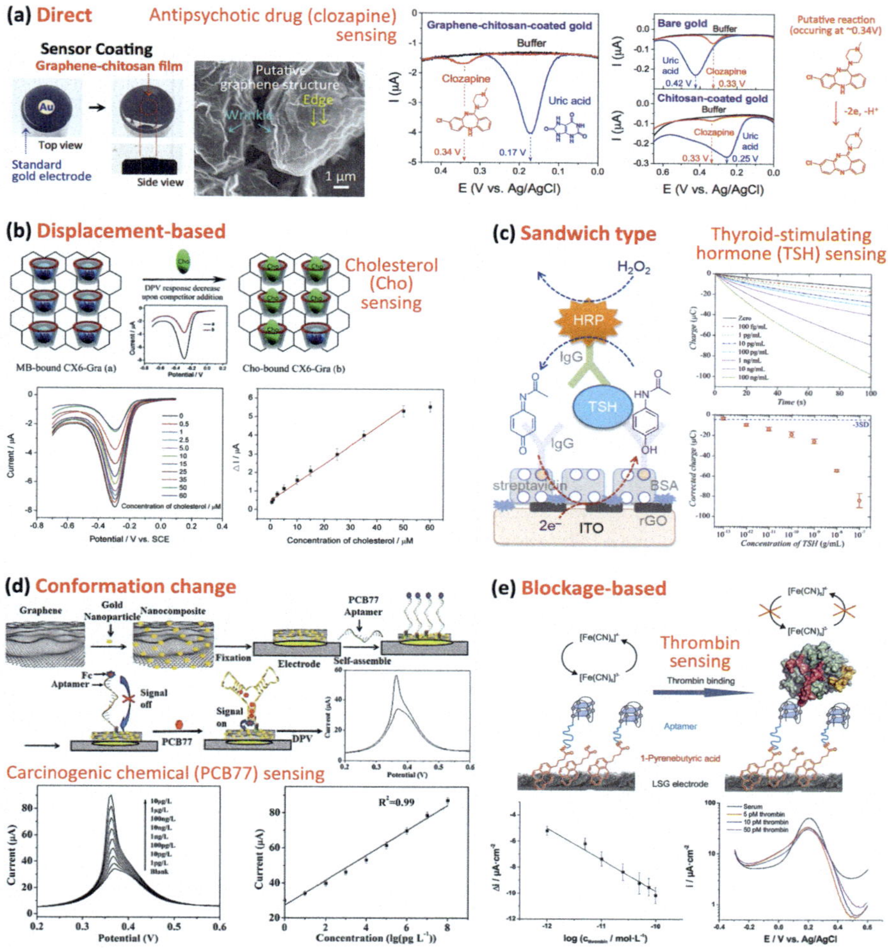

Fig. 4 (**a**) Graphene-composite-deposited electrode for direct monitoring of the blood serum level of antipsychotic drug (clozapine). Reproduced from Kang et al. (2017) (**b**) Reduced graphene oxide complexed with supramolecule (calixarene, with internal volume pre-occupied by pseudo-target) for displacement-based detection of small target biomolecule (cholesterol). Reproduced from Yang et al. (2016) (**c**) Electrochemical enzyme-linked immunosorbent assay formed on reduced graphene oxide, where one graphene-tethered antibody captures target (thyroid-stimulating hormone, TSH) and the other anti-TSH antibody generates electrochemical signals. Reproduced from Yan et al. (2020)) (**d**) Reduced graphene oxide electrode functionalized with the redox-labeled aptamer that varies secondary structure in response to the target (carcinogenic compound, PCB77). Reproduced from Wu et al. (2017) (e) Aptamer-functionalized graphene electrode for the turn-off electrochemical signal by the blockage of electrode surface upon aptamer-target (thrombin) binding. Reproduced from Fenzl et al. (2017)

(e.g., intra- or intermolecular covalent bond, oxidation state) of the target, of which enzymatic reactions are mediated by a small redox-active molecule (redox probe) and quantified by the measurement of redox probe's signal. For sensing the redox-inactive targets that do not have the corresponding enzyme or nanozyme generating electrochemical signals, various detection mechanisms have been developed employing some redox-labeled components as described below.

In "displacement type" electrochemical sensors (Fig. 3e), a recognition unit (e.g., antibody, supramolecule) is anchored to the graphene-coated electrode and, prior to detection, complexed with a redox-labeled pseudo-target or an intrinsically redox-active pseudo-target, which has moderate affinity to the recognition unit. Due to the close proximity of the pseudo-target to the electrode, strong electrochemical signal is measured. In the presence of targets, the pseudo-target is displaced by the target due to the stronger binding between the recognition unit and the target substance, removed from the solution, and the electrochemical signal decreases in proportion to the target concentration. This assay is effective in detecting the small target substances with a single binding site. Figure 4b shows an example detecting a principal sterol molecule, cholesterol (Cho), using a GO electrode functionalized with cyclic oligomers, calix[6]arenes (Yang et al. 2016). Calix[6]arenes have strong host–guest interaction with cholesterol and relatively weaker interaction with redox-active methylene blue (MB). The functionalized GO was first exposed to the MB solution, and then a strong MB signal was measured. When Cho was added to this solution, the MB molecule was displaced by Cho, leading to a turn-off signal.

Figure 2f illustrates the detection process in "sandwich type" electrochemical sensors. This method is applied to detect a large target substance that has multiple binding sites and therefore can be complexed with more than two recognition components (e.g., antibody, aptamer). One of them is covalently linked to graphene and captures the target. The other recognition components labeled with redox probe are bound to the target and generate turn-on electrochemical signals, which is proportional to the target concentration. An electrochemical enzyme-linked immunosorbent assay, that is electrochemical ELISA, is based on the sandwich type detection mechanism and Fig. 4c shows an example (Yan et al. 2020). An indium tin oxide (ITO) electrode was partially modified with rGO, and the antibody specifically capturing thyroid-stimulating hormone (TSH, consisting of two peptide chains, molecular weight: ~28,000 Da) was tethered to the electrode. TSH in the analyte solution was captured by the antibody, followed by the second interaction with another anti-TSH antibody that was linked to a redox-signal-generating enzyme, horseradish peroxidase (HRP). HRP is one of the most popular enzymes catalyzing the reaction that turns on the redox activity of a substrate (acetaminophen in Fig. 4c). As illustrated in Fig. 4c, the electrode surface is highly crowded with many components in the immunosensing layer and likely to lose the electrocatalytic activity to some extent. To minimize such deterioration, rGO was introduced to the ITO surface and maintain the high electrocatalytic activity of the sensor.

Certain types of recognition components (e.g., aptamer) can transform into different three-dimensional structure by the event of target binding. If these recognition components are labeled with a redox probe, generally at one end, and linked to the graphene-coated electrode at the other end, the binding event changes the redox probe's proximity to the electrode (either getting away from or coming close to the electrode) and affects the electrochemical signal accordingly (Fig. 2g). Aptamers are fit for this assay because they have highly versatile and flexible secondary structures (e.g., single-stranded and double-stranded segments, hairpin stem/loops, multi-branched loops) and have reactive groups for the terminal labeling with redox probes (e.g., methylene blue, ferrocene) (Xu et al. 2009). In the aptamer-based detection of highly toxic industrial compound (PCB77) that induces various harmful effects including carcinogenesis, the anti-PCB77 aptamer was labeled with ferrocene (Fc) at one end and with thiol (–SH) at the other end (Fig. 4d) (Wu et al. 2017). This aptamer was covalently attached via Au–S bond to the Au surface of the Au nanoparticles-rGO composite electrode. Before capturing PCB77, aptamers remained unfolded and kept Fc away from the electrode surface. In the presence of PCB77, the aptamer and PCB77 were complexed in a configuration that pushed Fc to be in the proximity of the electrode, generating an electrochemical signal. In this sensor, the presence of rGO in the electrode coat improved the measurement sensitivity and repeatability in comparison with the Au nanoparticle-coated electrode without rGO.

Lastly, Fig. 2h describes the "blockage-based" format of electrochemical assay. The surface of graphene-coated electrode is firstly functionalized with recognition components, while enabling a diffusible redox probe in the solution to encounter the electrode surface and then generate an electrochemical signal. When this electrode is exposed to the analyte solution that contains target substance, the target can form a complex with the recognition unit and partially block the electrode surface, resulting in the turn-off signal of the redox probe. For example, the detection of blood coagulation factor thrombin was performed by using the anti-thrombin aptamer as a recognition component, which was anchored to the graphene electrode through pyrene moiety (Fig. 4e) (Fenzl et al. 2017). This electrode allowed the diffusion of a small redox probe, ferricyanide, to the electrode surface and its electrochemical signal. When thrombin was captured by the aptamer, ferricyanide diffusion to electrode was hindered, and consequently the signal decreased. Graphene in this sensor offered high sensing signal due to its large electrochemically active surface area (18-fold larger than the physical area) and high electron transfer rate ($k_0 = 0.0044 \pm 0.0003$ cm·s^{-1}, which is higher than the k_0 of other carbon materials (e.g., 0.00033 cm·s^{-1} of the basal-plane pyrolytic graphite)).

Graphene Channel in Field Effect Transistor Biosensors

In electrical sensing, target molecules are measured in field-effect transistor (FET) structures. FET-based devices, in general, have metal-oxide-semiconductor structures. When the potential of metal layer changes by any means, the electric field induces band bending of a semiconductor layer, which leads to accumulation or depletion of charge carriers, and changes electrical properties subsequently (e.g., conductivity or I-V characteristic) (Syu et al. 2018). Applying gate voltage is a way to induce such changes, but solution potential (e.g., pH value) or change of electro-static circumstance, including biomolecules nearby, is expected to bring similar

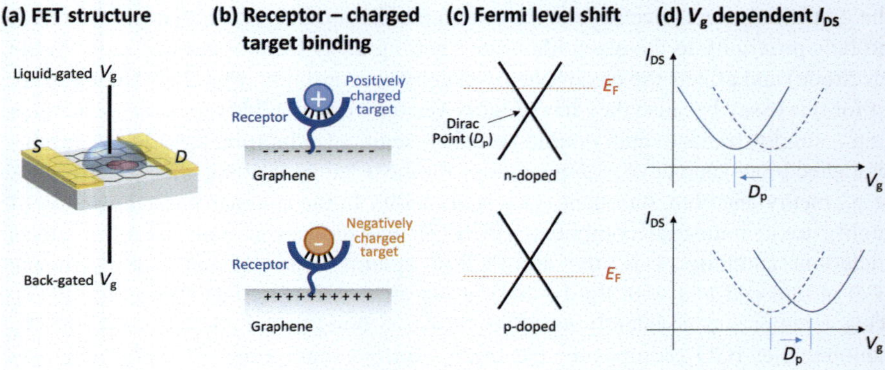

Fig. 5 Sensing principle of graphene-based field-effect transistor (GFET). (**a**) Scheme of general GFET biosensors. (**b**) Receptor and electrically charged target binding. Positively (or negatively) charged target molecules induce n-type (or p-type) doping to the surface of graphene due to the electrostatic gating effect. Charged target (or ion) directly absorbed on the surface of graphene is expected to bring the same electrostatic effect. It results in Fermi level shift (**c**) and changes gate-voltage dependent source-drain current (**d**). Direct charge transfer from analytes to graphene yields a similar but opposite effect

effects (Syu et al. 2018; Nehra and Singh 2015). Compared to other biosensing methods, FET-based biosensing provides many advantages such as high sensitivity, fast detection, mass production, low cost, and label-free real-time sensing (Syu et al. 2018).

Graphene has been a promising material for future electronics because of its high electron/hole mobility due to Dirac band structure, high transparency and mechanical strength (Geim 2009). In sensing application-wise, the surface of graphene consisting of carbon atoms is potentially highly sensitive to any changes in its surrounding environment (Zhang et al. 2020). Such unique structure and properties render graphene an ideal material for FET structure-based biosensor application. The basic structure of graphene-based FET biosensor and working principle are depicted in Fig. 5. Receptors (or probes) are placed and immobilized on the surface of graphene, and they take electrically negative or positive biomolecules (Zhang et al. 2020; Zubiarrain-Laserna and Kruse 2020). Eventually, this process makes electrostatic interaction between targeted molecules and the graphene surface and results in the shift of Dirac point and the subsequent change in I_{DS}-V_G (gate-voltage dependent source-drain current) characteristics (e.g., resistance under a constant V_G). Because most biomolecules, such as antibodies, DNA, peptides, proteins, and lipids, are inherently charged, their detection with FET-based biosensors could be achieved label-free. Even for less-charged biomolecules, charge variation can be made by receptors and chemical reaction design (Hao et al. 2017). The gate voltage can be applied and regulated by a back-gating or liquid-gating manner.

Since Ohno et al. have reported electrolyte-gated GFET for electrical detecting pH and protein absorptions (Ohno et al. 2009), much progress has been made in the development of GFET-based biosensors. Many studies in GFET-based biosensors have devoted to (1) functionalizing the surface of graphene (Xie et al. 2016; Hao et al. 2020; Wu et al. 2020; Soikkeli et al. 2016; Kim et al. 2013a), (2) sensing of new biomolecules (Seo et al. 2020), as well as (3) fabricating practical devices such as wearable and portable POC devices (Xiang et al. 2016; Wang et al. 2020b; Gao et al. 2020; Tehrani et al. 2020; Wang et al. 2019). Functionalization of the graphene surface, including surface passivation, is an important step for immobilizing desired receptors or linkers, and avoiding electrostatic interaction with not desired molecules, which eventually leads to improvement of sensitivity and limit of detection (LoD) (Wu et al. 2020). In addition, Debye screening which hinders electrostatic interaction between targeted biomolecules and the surface of graphene is a fundamental obstacle to improve sensitivity and applicability. Several studies have suggested possible routes to overcome Debye screening effect. For instance, Hwang et al. have demonstrated that structurally deformed (i.e., crumpled) graphene forms "electrical hot spots" in the sensing channel, which reduce the charge screening at the concave regions (Fig. 6a–c) (Hwang et al. 2020). Strategies to maximize the electrostatic effect on the graphene surface have also been demonstrated through unique structures, novel probe and biochemical reaction design (Fig. 6d–g) (Hwang et al. 2018; Gao et al. 2018; Hao et al. 2017; Kanai et al. 2020; Liu et al. 2018; Kim et al. 2017b; Li et al. 2017; Chen et al. 2016; Xu et al. 2014). Understanding the exact sensing mechanism as well as designing reproducible and reliable sensing routes with high sensitivity are still challenging. Low-cost and straightforward device fabrication methods, including the preparation of high quality and large area graphene (Xu et al. 2018; Zheng et al. 2015), are indeed required for practicality. Recent reports on GFET-based biosensors are summarized in Table 1 with brief notes highlighting their significance. Comprehensive reviews can also be found elsewhere (Zubiarrain-Laserna and Kruse 2020; Yan et al. 2014; Fu et al. 2017; Green and Norton 2015; Park et al. 2016; Suvarnaphaet and Pechprasarn 2017).

Graphene as Fluorophore or Fluorescence Quencher in Fluorescent Biosensor

In fluorescence-based biosensors, fluorophores—small molecules of which electrons are excited by light and relaxed by emitting the light in the visible range—are used to indicate the presence of target substances. Graphene family materials are extensively used as fluorescent-quenchers that can turn off the signals of the fluorophores adsorbed on graphene-based materials and then turn them on in response to target substances as illustrated in Fig. 7a. A fluorophore can be covalently linked to a free recognition component and produces strong fluorescence when present alone. Once it is mixed with graphene and forms complex, its fluorescence is quenched by either one of the two pathways: energy transfer pathway or electron transfer pathway. When the fluorophore in the complex gains energy by light and subsequently its electron is excited, the energy can be transferred to graphene by dipole-dipole interaction; this is called fluorescence resonance energy transfer. In this case, the excited electron of the fluorophore is relaxed non-radiatively (i.e., yields no fluorescence). If the fluorophore and graphene are in close contact (within few nm) and their electronic structures lie appropriately (i.e., the energy level of excited state of

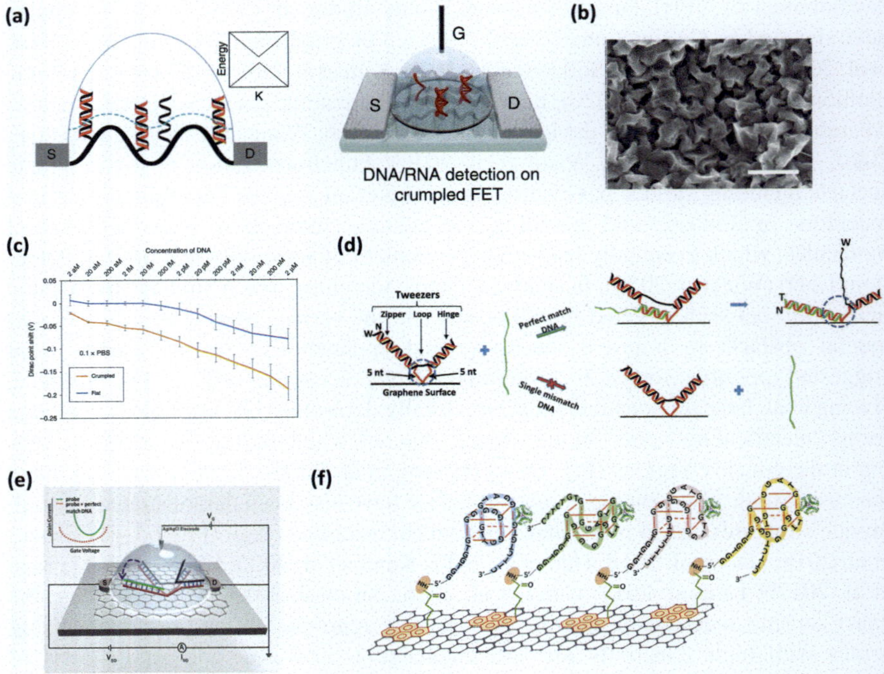

Fig. 6 (**a**) Scheme of crumpled graphene FET DNA sensor. (**b**) SEM images of crumpled graphene. (**c**) Dirac point shifts of the flat and crumpled graphene FET sensor plotted as a function of target DNA concentration. Reproduced from Hwang et al. (2020). (**d**) DNA nanotweezers probe design. The probe was horizontally laid down on the graphene surface, placing the longer DNA sequence of charge accumulation part (blue dotted circle) in close proximity to the graphene surface. (**e**) Schematic of graphene FET sensor with DNA tweezer probe. Reproduced from Hwang et al. (2018). (**f**) Aptameric GFET-based insulin sensor. Guanine-rich IGA3, negatively charged aptamer folds into a compact and stable antiparallel or parallel G-quadruplex conformation upon binding with insulin, resulting in closer interaction with the graphene surface. Reproduced from Hao et al. (2017)

fluorophore is higher than that of graphene), the excited electrons of the fluorophore can be directly transferred to graphene. This electron transfer pathway also results in no fluorescence from the fluorophore. If the recognition component has higher binding affinity to the target substance than to graphene, it can be complexed with the target and the complex can get detached from graphene. By this event, the fluorescence of the fluorophore is recovered. Fig. 8a shows the sensing mechanism where GO serves as fluorescence quencher for quantifying the tumor-derived exosome (Li et al. 2020). Exosomes, the nanometer-sized extracellular vesicles originated from cells, have surface proteins that are specific to the origin cell and thus have been largely utilized as disease biomarkers. In the biosensor to detect the exosome from prostate cancer cells, aptamers recognizing the surface protein of the

Table 1 Recent reports of graphene field effect transistor-based biosensor

Channel	Target	Receptor	Detection limit	Note	Ref.
GFET	IL-6, insulin	Aptamer	618 fM (IL-6), 766 fM (insulin)	E-field assisted functionalization	Hao et al. (2020)
GFET	TNF-α, IFN-γ	Aptamer	2.75 pM (TNF-α), 2.89 pM (FN-γ)	Wearable and deformable device	Wang et al. (2020b)
rGO FET	HepG2-MVs	Aptamer	84 particles/μL	Au decorated and dual aptamer	Wu et al. (2020)
GFET	ssDNA		8×10^{-21} M	Crumpled graphene and DNA amplification detection	Ganguli et al. (2020)
GFET	Staphylococcus aureus, Acinetobacter baumannii	Peptide	10^4 cells/mL	Dielectrophoresis and E-field assisted binding	Kumar et al. (2020)
GFET	SARS-CoV-2	Antibody	1 fg/mL	For COVID-19 detection	Seo et al. (2020)
GFET	miRNA	ssDNA	10 fM	On flexible polyimide substrate	Gao et al. (2020)
GFET	pH		62 mV pH^{-1}	With D-melanin, printable pH sensor	Tehrani et al. (2020)
GFET	Peptide of bone gla protein	Antibody	100 fg/mL	Open-sandwich immunoassay	Kanai et al. (2020)
GFET	TNF-α	Aptamer	5×10^{-12} M	Ultraflexible and stretchable	Wang et al. (2019)
rGO FET	Glutamate	mGluR	10 fM		Li et al. (2019)
rGO FET	Exosome	Antibody	33 particles/μL		Yu et al. (2019)
GFET	VSV enveloped HIV-1, MLV	Antibody	47.8 aM	Liquid coplanar-gate GFET on plastic	Kim et al. (2019)
GFET	DNA	DNA	25 aM	DNA hybridization detection	Campos et al. (2019)

(continued)

Table 1 (continued)

Channel	Target	Receptor	Detection limit	Note	Ref.
GFET	IL-6	Aptamer	12 pM	Portable sensing	Hao et al. (2019)
GFET	AFP	Antibody	0.1 ng/mL		Kim et al. (2018)
rGO FET	TSH	Antibody	0.2×10^{-15} M	For serum	Andoy et al. (2018)
GFET	p24, cTn1 and CCP	Antibody	100 fg/mL (p24), 10 fg/mL (cTn1, CCP)	Amine functionalized	Islam et al. (2019)
GFET	HBsAg	Aptamer	10 aM	With carboxylic polypyrrole nanowires	Cho et al. (2018)
GFET	DNA	DNA-tweezer	100×10^{-12} M	Part of V shape dsDNA replaced by target DNA	Hwang et al. (2018)
GFET	BPA	DNA	10 ng/mL	Destroying DNA attached to Au NP on GFET / reusable sensor	Liu et al. (2018)
GFET	DNA	Hairpin-structured probe DNA	5 fM	Target recycling and hybridization chain reaction for amplification	Gao et al. (2018)
GFET	Phospholipase D, cholera toxin B, amyloid beta 40			With supported lipid bilayer to monitor cell membrane - protein interaction	Kuo et al. (2018)
GFET	Insulin	Aptamer	35 pM	Formation of G-quadruplexes, which brings negatively charged insulin and DNA strands to the close vicinity of the graphene surface	Hao et al. (2017)
rGO FET	Dopamine		100×10^{-18} M	Nonenzymatic and organic dopamine sensor/Pt decorated	Oh et al. (2017)
GFET	Streptavidin	Biotin, peptide		Monomolecular self-assembly of designed peptide protein receptors to enhance the coupling between graphene and charged molecules	Kim et al. (2017b)

GFET	Pb^{2+}	G-rich ssDNA	163.7 ng/L	Changing G-quadruplex structures by Pb^{2+} ion	Li et al. (2017)
GFET	Kanamycin A	Aptamer	11.5×10^{-9} M	Employing high κ material for gating	Wang et al. (2016a)
rGO FET	NO	Iron−porphyrin	1 pM	Porphyrin-functionalized graphene	Xie et al. (2016)
GFET	Norovirus virus like particle	Antibody	0.1 μg/mL	Inkjet-printed flexible biosensor	Xiang et al. (2016)
GFET	CEA	Antibody	100 pg/mL	Non-covalent modification	Zhou et al. (2017)
rGO FET	Matrilysin activity		10 ng/mL (400 pM)	Matrilysin cutting polypeptide	Chen et al. (2016)
GFET	Peptides	HFBI + peptides		Receptor modules for improving sensing function and immobilization simultaneously	Soikkeli et al. (2016)
rGO FET	D-dimer	Anti-D-dimer TiO_2	10 pg/mL	Photocatalysis-induced renewable transistor achieved by rGO encapsulated TiO_2	Zhang et al. (2016)
rGO FET	DNA	DNA	1 nM	Au decorated graphene / scalable production of sensor arrays	Gao et al. (2016)
rGO FET	C6 protein	Antibody	1×10^{-15} M	Self-assembled platform that combines a natural cellular material (Lycopodium clavatum pollen spores) with rGO	Wang et al. (2016b)
GFET	DNA	Peptides	10 fM	Directional transfer technique based on CVD-grown graphene	Zheng et al. (2015)
GFET	Topoisomerase I	DNA	300 pM	Binding kinetics of the essential human enzyme	Zuccaro et al. (2015)
GFET	PSA–ACT	Antibody	100 fg/mL	ZnO NR, TiO_2 structures	Kim et al. (2013b)
GFET	Glucose	Glucose oxidase	0.1 mM	Silk protein for enzyme immobilization	You and Pak (2014)

(continued)

Table 1 (continued)

Channel	Target	Receptor	Detection limit	Note	Ref.
GFET	K^+, Pb^{2+}	Aptamer	100 μM (K^+), 10 μM (Pb^{2+})	Aptamer terminated methylene blue molecule for better sensitivity	Xu et al. (2014)
GFET	Alamethicin, gramicidin A	Lipid bilayer		Individual ion channel activity	Wang et al. (2014b)
GFET	Naltrexone	μ-Opioid receptor	10 pg/mL	G protein-coupled receptors	Lerner et al. (2014)
GFET	Streptavidin	Biotin, peptides	50 ng/mL		Khatayevich et al. (2014)
GFET	PA	Aptamer	1.2 aM	Additional signal enhancement obtained by the secondary aptamer-conjugated au NP	Kim et al. (2013a)
rGO FET	H_2O_2		100 pM	Graphene-polypyrrole-nanotube composites	Park et al. (2014)
rGO FET	DNA	PNA	100 fM		Cai et al. (2014)
GFET	IgG	Antibody	13 pM	Vertically oriented graphene with au NP	Mao et al. (2013)
GFET	Carboxypeptidase B		1 μM	Au NPs and polypeptides. NP dissemble by the enzyme interaction with substrate (proteolysis)	Myung et al. (2012)
rGO (NP) FET	HER2, EGFR	Antibody	1 pM (HER2), 100 pM (EGFR)	Graphene encapsulated NP	Myung et al. (2011)

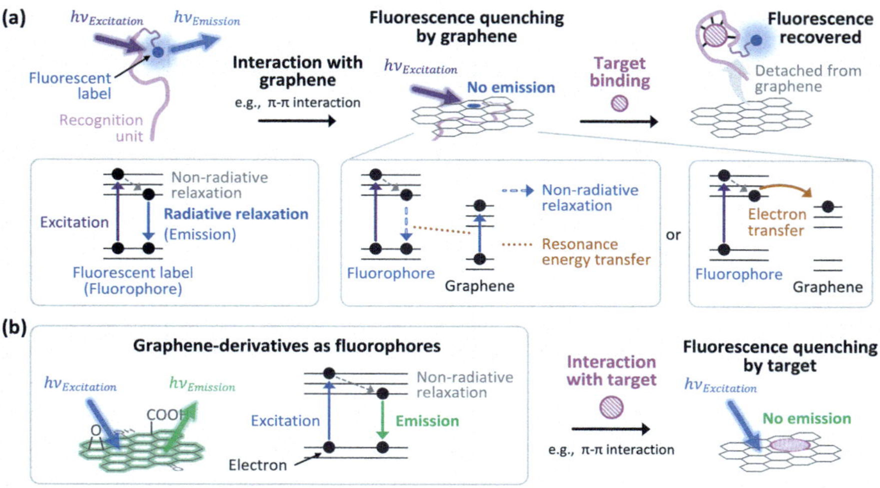

Fig. 7 General detection mechanism of graphene-based fluorescence biosensors. Graphene-based materials are used as (**a**) fluorescence quenchers or (**b**) fluorophores

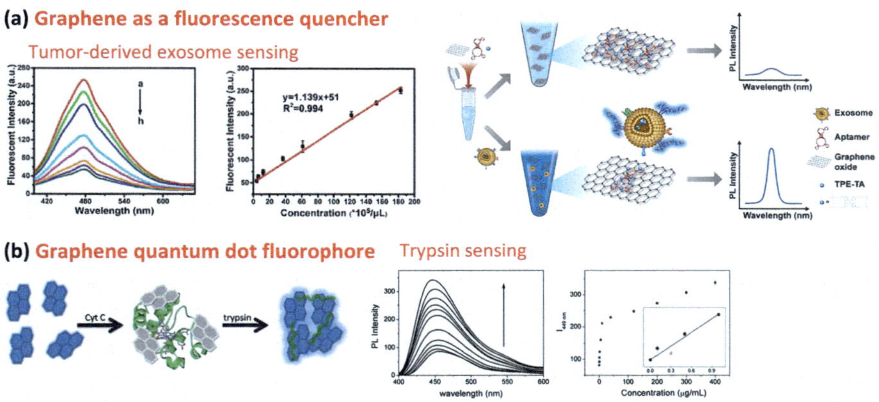

Fig. 8 (**a**) Complex of the fluorescence-quenching graphene and the fluorophore-aptamer that has affinity to the tumor-derived exosome. Fluorescence is quenched in the absence of target exosome, while it is restored by the aptamer-exosome binding that releases the fluorophore-aptamer from graphene. Reproduced from Li et al. (2020) (**b**) Fluorescent graphene quantum dot for trypsin detection. Fluorescence is quenched by cytochrome c (Cyt c) prior to measurement and restored by trypsin that destroys Cyt c. Reproduced from Li et al. (2013)

exosome, aggregation-induced fluorescence probes (TPE-TAs), and GO (as a fluorescence quencher) were used and complexed prior to detection. The complex was formed by the electrostatic attraction between the negatively charged aptamer and the positively charged TPE-TAs, which can generate the strong fluorescence signal of the aggregated TPE-TAs in the complex, and then by the hydrogen bond and π-π stacking between the aptamer and GO, which ultimately quenches the fluorescence signal. In the presence of the target exosomes, the aptamer preferentially interacted with the exosome protein and, subsequently, the aptamer-exosome complex containing TPE-TAs left the GO surface, which recovered the TPE-TAs fluorescence. Because of the excellent biocompatibility, hydrophilicity, and fluorescent quenching ability, GO served as a promising component involved in signal generation.

When the sp^2 hybridization network of carbon atoms in a tiny piece of graphene-based materials is perturbed by the presence of sp^3-hybridized carbon atoms, usually found at the edge and bonded with oxygen-containing functional groups, the electronic structures of the materials are modified to have a band gap (Wang et al. 2014a). Nanoscale GQDs, prepared by cutting GO or rGO, are such materials and show intrinsically strong and stable fluorescence, which allows them to perform as small fluorophores. For instance, the suppression and restoration of GQD fluorescence can be used to detect trypsin, an enzyme associated with diseases such as pancreatitis (Fig. 8b) (Li et al. 2013). GQDs were first complexed with cytochrome c (Cyt c), a well-known electron transfer protein, and the GQD fluorescence was quenched subsequently. When trypsin was added to the complex, it cleaved Cyt c into smaller fragments making them ineffective fluorescence quenchers. As a result, increase in fluorescence signal was observed with increasing the trypsin concentration.

Graphene as a Substrate for Raman Scattering-Based Biosensor

Raman spectroscopy is a technique commonly used to observe the vibrational modes of molecules, of which signal can be regarded as a signature fingerprint of the molecules. In this measurement, monochromatic light is incident upon a molecule and induces the transition of the molecule to the vibrationally excited state ending up with inelastic scattering, that is, Raman scattering (Fig. 9a). The energy of the scattered light is highly sensitive to the functional groups of a molecule, and this makes Raman spectroscopy a very powerful tool for molecular identification. Application of Raman spectroscopy as biosensing technology, however, is limited, because of the low cross-section of Raman scattering (i.e., intrinsically low Raman scattering signal). Figure 9b shows one way to overcome this limitation by introducing the materials that can generate strong electric field over their surface and then placing the Raman-active molecule close to the surface. Raman signal intensity increases with increasing the strength of electric field of the surface as the fourth power of field strength; this is termed surface-enhanced Raman scattering (SERS). Graphene family materials can serve as a good SERS substrate due to its abilities (1) to reduce the interference from the molecule's fluorescence (i.e., fluorescent background), (2) to generate strong electric field, especially from oxygen species (Yu et al. 2011), and (3) to transfer charges to the adsorbed molecules (Wang et al. 2018b). In addition, π-electrons of graphene attract the target substances that contains aromatic groups to graphene surface, intensifying the graphene-target interaction, beneficial to the target Raman signal.

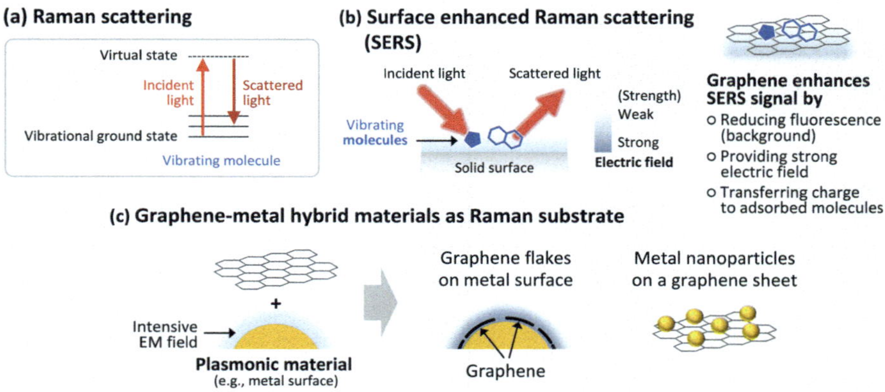

Fig. 9 Graphene as a surface enhanced Raman scattering (SERS) substrate. (**a**) Vibrational transition of a molecule by incident light of a certain photon energy and the subsequent scattered light of different energy (i.e., usually a portion of photon energy is absorbed by the molecule), which is termed Raman scattering. (**b**) Enhancement of Raman signals by the close contact between the Raman-active molecules and the solid surface that has strong electric field. (**c**) Graphene-metal hybrid materials as SERS substrates

When graphene is complexed with the metal surface of nanoscale roughness (e.g., metallic nanoparticle aggregates), the hybrid materials perform as highly sensitive SERS substrates (Fig. 9c). The aggregates of coinage metal (e.g., Au, Ag, Cu) nanoparticles produce intensive electric field so that they have been extensively used in SERS-based biosensors. However, it is difficult to fabricate the reproducible nanoparticle aggregates or to protect the metallic surface from oxidation (in the case of Ag (Lu et al. 2018) and Cu (Xu et al. 2015)). Graphene can mitigate the former issue by dispersing and attaching the nanoparticles on its flat surface (Lin et al. 2015). Graphene is also used to delay the surface oxidation of metal nanoparticles; the nanoparticles are coated with graphene and obtain long-term stability of their SERS-activities. Fig. 10 clearly shows the advantage of integrating GO and Au nanostars (GNSs) for SERS-based label-free detection of a biomolecule in diluted biofluid (Pan et al. 2019). The hybrid material was prepared by functionalizing GO sheets with dispersant (polyvinylpyrrolidone, PVP) and positively charged polymer (polydiallyldimethyl ammonium chloride, PDDA) and then decorated with the negatively charged GNSs via electrostatic attraction. This substrate sensitively measured the SERS signal of bilirubin in diluted serum, a hydrophobic metabolite (consisting of four pyrrolic rings) associated with hepatic dysfunction, mental disorders, brain damage, etc. GO in the SERS substrate improved the sensitivity by quenching the fluorescence background and preconcentrating bilirubin via π-π interactions.

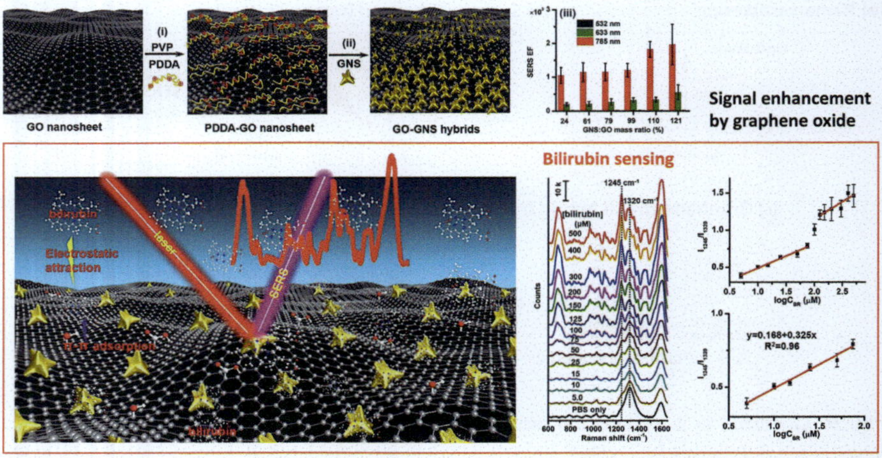

Fig. 10 Graphene oxide and Au nanostar hybrid as a SERS substrate to measure the level of bilirubin in diluted serum. Reproduced from Pan et al. (2019)

3 Device Formats of Graphene-Based Point-of-Care Biosensors

Point-of-care (POC) biosensors are small portable devices that can be operated in real time by any users whenever and wherever they are needed. They are highly demanded these days with the desire to monitor one's health conditions continuously or periodically for routine health care as well as for a timely information that assists clinical decision (e.g., for personalized medication) (Huang et al. 2019). The World Health Organization (WHO) has issued ASSURED (Affordable, Sensitive, Specific, User-friendly, Rapid and robust, Equipment-free, and Delivered to the end-users) criteria for POT devices. Graphene family materials have emerged as promising POC components of target recognition and/or signal generation in terms of their material properties as well as the ease of incorporation into the POC devices.

3.1 Disposable Paper-Based Devices

Paper-based sensing is a highly promising format that meets ASSURED criteria. There are lateral-flow and paper-folding types, which have different ways of solution processing (Baharfar et al. 2020; Nguyen et al. 2020). In the lateral-flow type sensors, the solution runs from one end of the paper sensor to the other by capillary force. While traveling the sensor paper, the solution is processed by the functional components locally deposited on the paper. Fig. 11a shows an example where a paper-based lateral-flow strip for pathogen detection has a test line containing the

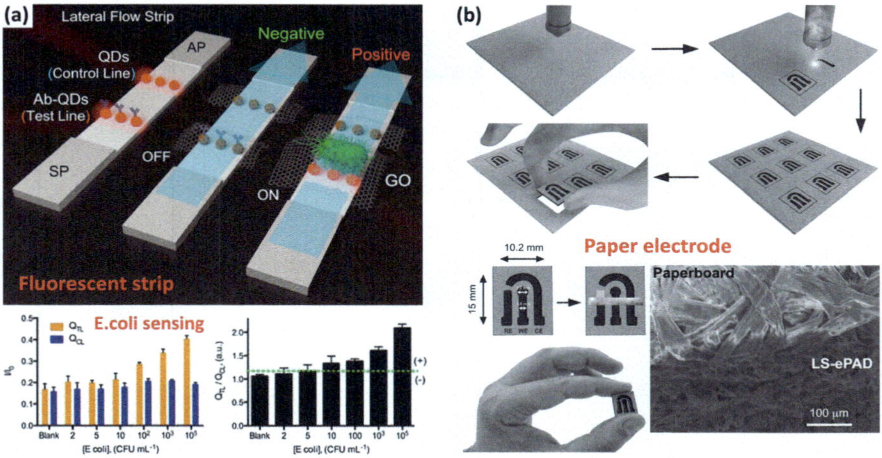

Fig. 11 Graphene-on-paper-based biosensor. (**a**) Fluorescent lateral-flow strip for pathogen detection. Reproduced from Morales-Narvaez et al. (2015). (**b**) One-step fabrication of graphene-patterned paper electrodes using a laser scribing technique. Reproduced from De Araujo et al. (2017)

fluorescent quantum dots (QDs) coated with anti-pathogen antibody (Ab-QDs) and a control line containing bare QDs, and the sample solution flows from the sample pad (SP) to the absorbent pad (AP) (Morales-Narvaez et al. 2015). If target pathogens are present in the solution, they are captured by Ab-QDs, which is revealed by the addition of fluorescence-quenching GOs and the subsequent measurement of the fluorescence from the two lines; the fluorescence from the test line is kept by the pathogen that separates GOs and Ab-QDs while the fluorescence from the control line is always quenched regardless of the presence of pathogen, and the fluorescence ratio is dependent on the pathogen concentration. In the paper-folding type (also called origami type) sensors, multiple functional components (e.g., target recognizer, signal generator) of a sensor are separated on a single sheet (reducing fabrication cost), and easily and quickly (<1 min) assembled just by folding the paper (providing user-friendly operation).

One critical factor for the success of this type of sensor in POC applications is the cost-effective production of the single-use paper sensors to secure the affordability requirement. Various technologies (e.g., drop-casting, screen printing, ink printing, electro-spraying, electro-spinning) have been developed to fabricate paper-based sensors, and graphene family materials and their composites (with dispersing agents) are highly compatible with such technologies (Kudr et al. 2020; Cinti and Arduini 2017). In addition, since graphene can be pyrolytically synthesized from carbonaceous matters, graphene-patterned paper sensors can be simply fabricated by the local pyrolysis of a paper board using laser light (Fig. 11b) (De Araujo et al. 2017). This scalable fabrication method is highly promising for the development of POC sensing devices.

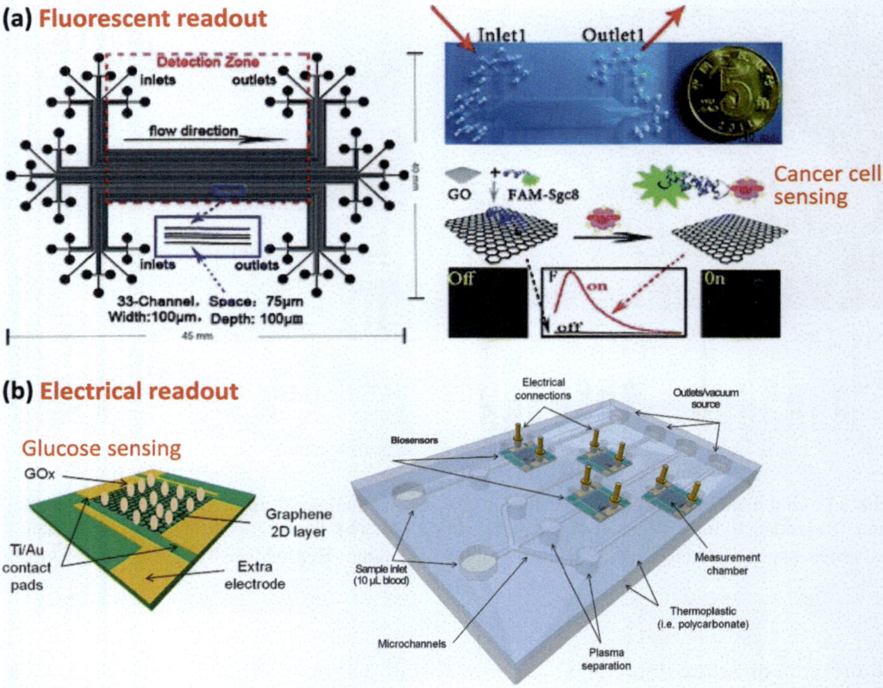

Fig. 12 Graphene-incorporating analytical lab-on-a-chips (LOCs). (**a**) Fluorescence-based multichannel LOC for cancer cell detection. Graphene oxide and fluorophore-labeled aptamer generate the fluorescence signal in response to the presence of target cancer cells. Reproduced from Cao et al. (2012). (**b**) Design for standardized multiplex LOCs composed of multiple graphene-based electronics (e.g., graphene FET sensor). Reproduced from Viswanathan et al. (2015)

3.2 Lab-on-a-Chips

Lab-on-a-chips (LOCs) are small analytical devices (< few cm^2) that perform multiple laboratory techniques (e.g., fluid transport, extraction, recognition, signal-generation), typically integrated with an external fluid flow controller. LOCs facilitate POC analysis by shortening the analysis time, reducing the sample consumption, and providing portability. The adoption of graphene-based materials in LOCs is in an infant stage, but the materials show potentials as powerful functional components of LOCs because of their thickness, electrical conductivity, and optical transparency (Morales-Narvaez et al. 2017). For instance, GO can be incorporated into the multichannel LOC as a fluorescence quencher for the detection of cancer cells as shown in Fig. 12a (Cao et al. 2012). In this sensor, the fluorophore-labeled aptamer (FAM-Sgc8) for the T-cell acute lymphoblastic leukemia cell is attached to GO (via π-π stacking) that quenches the fluorescence of the fluorophore. The target cells injected to the microfluidic channels of the LOC interact with FMA-Sgc8 strongly and release it from GO, by which the fluorescence from the aptamer is recovered. Multiple graphene-electronics with different target recognition components (e.g., graphene-FET as glucose sensor, Fig. 12b) can be integrated into LOC for multiplex detections (Viswanathan et al. 2015).

3.3 Wearable Biosensors

Real continuous POC monitoring of health condition or disease status can be realized by the biocompatible sensors mounted on human tissue (e.g., skin tissue). Such wearable or implantable biosensors can greatly improve the quality of health care managements. One critical concern about wearable sensors is the mechanical mismatch between the soft tissue and solid sensors. This requires mechanical flexibility and stretchability of the sensors to conform to the non-planar structure and the movement of human tissue to minimize irritation and motion artifacts (Huang et al. 2019). Graphene and its derivatives meet these requirements with their high Young's modulus and remarkable mechanical flexibility. There have been great developments of wearable biosensors using graphene-based materials in various forms such as contact lenses (Kim et al. 2017a) (Fig. 13a), microneedles (Jin et al. 2019) (Fig. 13b), patches (Xuan et al. 2018) (Fig. 13c), tattoo (Ameri et al. 2017), and so on. On the surface of the contact lens shown in Fig. 13a, graphene-based electronics for wireless detection of glucose in tears was integrated. In this sensor, the graphene composite (with Ag nanowire) provided sufficient optical transparency and mechanical stretchability, which provided users comfort and unobstructed vision. Figure 13b shows a minimally invasive transdermal biosensor composed of the microneedles coated with functional rGO composite (with Pt nanoparticle). At the microneedle inserted into skin tissue, transdermal hydrogen peroxide can be detected, where Pt nanoparticle catalyzed hydrogen peroxide oxidation and rGO offered fast electron transfer through the composite coat. Patch-type flexible biosensors for the target substances in sweat can be fabricated with graphene family materials. For instance, micropatterned rGO composite containing AuPt alloy nanoparticles, chitosan, and glucose oxidase successfully measured glucose while attached on human wrist. rGO played an important role in immobilizing glucose oxidase, which specifically reacts with glucose and generates redox-active hydrogen peroxide.

4 Conclusions

Graphene and its derivatives have gained increasing interests, especially from the biosensor community because of their exceptional chemical and physical properties. The lateral size, oxidation level, and defects of graphene-based materials affect their electronic structure and modulate the electrical, optical, and mechanical properties, which determines the roles of graphene as biosensor components. π-electrons and oxygen-containing moieties of graphene-based materials allow various

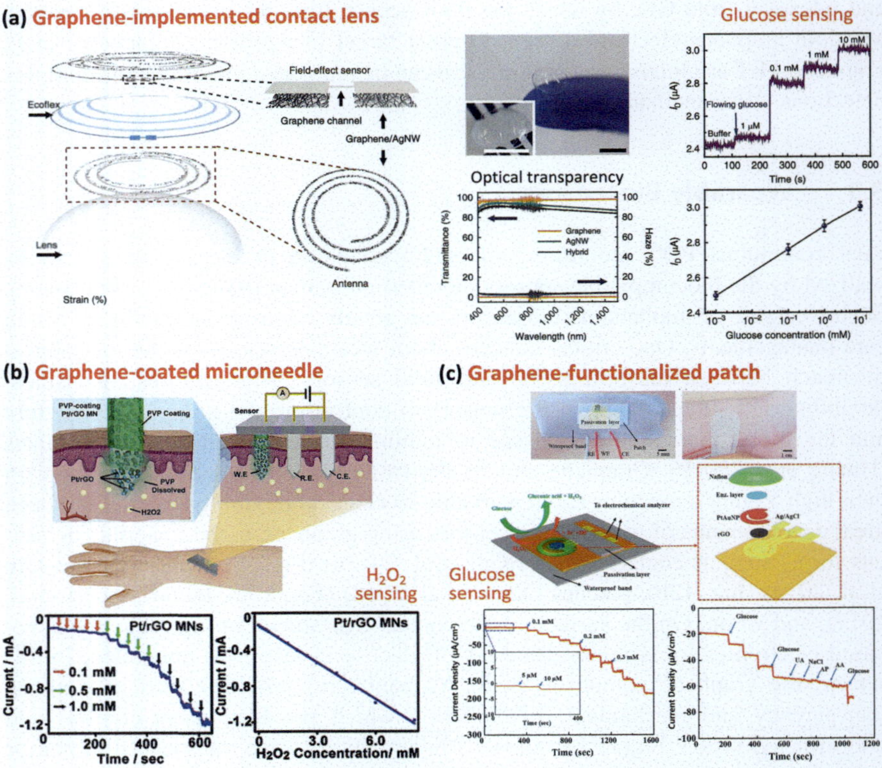

Fig. 13 Various formats of wearable biosensors functionalized with graphene-based materials. (**a**) Graphene-electronics-coated contact lens for the detection of glucose in tears. Reproduced from Kim et al. (2017a). (**b**) Microneedle electrode coated with graphene oxide composite for detecting transdermal hydrogen peroxide. Reproduced from Jin et al. (2019). (**c**) Graphene-patterned flexible patch to detect glucose in sweat. Reproduced from Xuan et al. (2018)

functionalization for advanced sensing performances. The possibility that graphene can be utilized with various detection methods—electrochemical, electrical, optical (e.g., fluorescent, Raman scattering), and so on—is of great advantage and there have been developed various detection mechanisms to utilize the properties of graphene to the greatest extent. In addition to the excellent sensing performance with high sensitivity, economical perspective of the fabrication of graphene-based sensing devices as well as their aptitude for the advanced point-of-care applications accelerate the development of graphene-based biosensors. It is encouraged to establish the standards associated with material properties and sensing performances.

Acknowledgment This work was supported by Pusan National University Research Grant, 2020 and the National Research Foundation of Korea Grant funded by the Korean Government (NRF-2020R1A4A4078780).

References

Ameri SK, Ho R, Jang HW, Tao L, Wang YH, Wang L, Schnyer DM, Akinwande D, Lu NS (2017) Graphene electronic tattoo sensors. ACS Nano 11:7634–7641

Andoy NM, Filipiak MS, Vetter D, Gutierrez-Sanz O, Tarasov A (2018) Graphene-based electronic immunosensor with femtomolar detection limit in whole serum. Adv Mater Technol 2018:3

Baharfar M, Rahbar M, Tajik M, Liu G (2020) Engineering strategies for enhancing the performance of electrochemical paper-based analytical devices. Biosens Bioelectron 167:112506

Cai BJ, Wang ST, Huang L, Ning Y, Zhang ZY, Zhang GJ (2014) Ultrasensitive label-free detection of PNA-DNA hybridization by reduced graphene oxide field-effect transistor biosensor. ACS Nano 8:2632–2638

Campos R, Borme J, Guerreiro JR, Machado G, Cerqueira MF, Petrovykh DY, Alpuim P (2019) Attomolar label-free detection of DNA hybridization with electrolyte-gated graphene field-effect transistors. Acs Sens 4:286–293

Cao LL, Cheng LW, Zhang ZY, Wang Y, Zhang XX, Chen H, Liu BH, Zhang S, Kong JL (2012) Visual and high-throughput detection of cancer cells using a graphene oxide-based FRET aptasensing microfluidic chip. Lab Chip 12:4864–4869

Chen H, Chen P, Huang JF, Selegard R, Platt M, Palaniappan A, Aili D, Tok AIY, Liedberg B (2016) Detection of Matrilysin activity using polypeptide functionalized reduced graphene oxide field-effect transistor sensor. Anal Chem 88:2994–2998

Cho KH, Shin DH, Oh J, An JH, Lee JS, Jang J (2018) Multidimensional conductive nanofilm-based flexible aptasensor for ultrasensitive and selective HBsAg detection. ACS Appl Mater Interfaces 10:28412–28419

Cinti S, Arduini F (2017) Graphene-based screen-printed electrochemical (bio)sensors and their applications: efforts and criticisms. Biosens Bioelectron 89:107–122

De Araujo WR, Frasson CMR, Ameku WA, Silva JR, Angnes L, Paixao TRLC (2017) Single-step reagentless laser scribing fabrication of electrochemical paper-based analytical devices. Angewandte Chem-Int 56:15113–15117

Fenzl C, Nayak P, Hirsch T, Wolfbeis OS, Alshareef HN, Baeumner AJ (2017) Laser-scribed graphene electrodes for aptamer-based biosensing. Acs Sens 2:616–620

Fu WY, Jiang L, Van Geest EP, Lima LMC, Schneider GF (2017) Sensing at the surface of graphene field-effect transistors. Adv Mater 29:6

Ganguli A, Faramarzi V, Mostafa A, Hwang MT, You S, Bashir R (2020) High sensitivity graphene field effect transistor-based detection of DNA amplification. Adv Funct Mater 30:28

Gao J, Gao Y, Han Y, Pang J, Wang C, Wang Y, Liu H, Zhang Y, Han L (2020) Ultrasensitive label-free MiRNA sensing based on a flexible graphene field-effect transistor without functionalization. ACS Appl Electron Mater 2:1090–1098

Gao ZL, Kang H, Naylor CH, Streller F, Ducos P, Serrano MD, Ping JL, Zauberman J, Rajesh Carpick RW, Wang YJ, Park YW, Luo ZT, Ren L, Johnson ATC (2016) Scalable production of sensor arrays based on high-mobility hybrid graphene field effect transistors. ACS Appl Mater Interfaces 8:27546–27552

Gao ZL, Xia H, Zauberman J, Tomaiuolo M, Ping JL, Zhang QC, Ducos P, Ye HC, Wang S, Yang XP, Lubna F, Luo ZT, Ren L, Johnson ATC (2018) Detection of sub-fM DNA with target recycling and self-assembly amplification on graphene field-effect biosensors. Nano Lett 18: 3509–3515

Geim AK (2009) Graphene: status and prospects. Science 324:1530–1534

Green NS, Norton ML (2015) Interactions of DNA with graphene and sensing applications of graphene field-effect transistor devices: A review. Anal Chim Acta 853:127–142

Hao Z, Pan YL, Huang C, Wang ZR, Lin Q, Zhao XZ, Liu SQ (2020) Modulating the linker immobilization density on aptameric graphene field effect transistors using an electric field. Acs Sens 5:2503–2513

Hao Z, Pan YL, Shao WW, Lin Q, Zhao XZ (2019) Graphene-based fully integrated portable nanosensing system for on-line detection of cytokine biomarkers in saliva. Biosens Bioelectron 134:16–23

Hao Z, Zhu YB, Wang XJ, Rotti PG, Dimarco C, Tyler SR, Zhao XZ, Engelhardt JF, Hone J, Lin Q (2017) Real-time monitoring of insulin using a graphene field-effect transistor aptameric Nanosensor. ACS Appl Mater Interfaces 9:27504–27511

Huang HZ, Su S, Wu N, Wan H, Wan S, Bi HC, Sun LT (2019) Graphene-based sensors for human health monitoring. Front Chem 2019:7

Hwang MT, Heiranian M, Kim Y, You S, Leem J, Taqieddin A, Faramarzi V, Jing YH, Park I, Van Der Zande AM, Nam S, Aluru NR, Bashir R (2020) Ultrasensitive detection of nucleic acids using deformed graphene channel field effect biosensors. Nat Commun 2020:11

Hwang MT, Wang ZJ, Ping JL, Ban DK, Shiah ZC, Antonschmidt L, Lee J, Liu YS, Karkisaval AG, Johnson ATC, Fan CH, Glinsky G, Lal R (2018) DNA nanotweezers and graphene transistor enable label-free genotyping. Adv Mater 2018:30

Islam S, Shukla S, Bajpai VK, Han YK, Huh YS, Kumar A, Ghosh A, Gandhi S (2019) A smart nanosensor for the detection of human immunodeficiency virus and associated cardiovascular and arthritis diseases using functionalized graphene-based transistors. Biosens Bioelectron 126: 792–799

Jin QC, Chen HJ, Li XL, Huang XS, Wu QN, He G, Hang T, Yang CD, Jiang Z, Li EL, Zhang AH, Lin ZH, Liu FM, Xie X (2019) Reduced graphene oxide nanohybrid-assembled microneedles as mini-invasive electrodes for real-time transdermal biosensing. Small 5:15

Kanai Y, Ohmuro-Matsuyama Y, Tanioku M, Ushiba S, Ono T, Inoue K, Kitaguchi T, Kimura M, Ueda H, Matsumoto K (2020) Graphene field effect transistor-based Immunosensor for ultra-sensitive noncompetitive detection of small antigens. Acs Sens 5:24–28

Kang M, Kim E, Winkler TE, Banis G, Liu Y, Kitchen CA, Kelly DL, Ghodssi R, Payne GF (2017) Reliable clinical serum analysis with reusable electrochemical sensor: toward point-of-care measurement of the antipsychotic medication clozapine. Biosens Bioelectron 95:55–59

Khatayevich D, Page T, Gresswell C, Hayamizu Y, Grady W, Sarikaya M (2014) Selective detection of target proteins by peptide-enabled graphene biosensor. Small 10:1505–1513

Kim DH, Oh HG, Park WH, Jeon DC, Lim KM, Kim HJ, Jang BK, Song KS (2018) Detection of alpha-fetoprotein in hepatocellular carcinoma patient plasma with graphene field-effect transistor. Sensors 18:4032

Kim DJ, Park HC, Sohn IY, Jung JH, Yoon OJ, Park JS, Yoon MY, Lee NE (2013a) Electrical graphene Aptasensor for ultra-sensitive detection of anthrax toxin with amplified signal transduction. Small 9:3352–3360

Kim DJ, Sohn IY, Jung JH, Yoon OJ, Lee NE, Park JS (2013b) Reduced graphene oxide field-effect transistor for label-free femtomolar protein detection. Biosens Bioelectron 41:621–626

Kim J, Kim M, Lee MS, Kim K, Ji S, Kim YT, Park J, Na K, Bae KH, Kim HK, Bien F, Lee CY, Park JU (2017a) Wearable smart sensor systems integrated on soft contact lenses for wireless ocular diagnostics. Nat Commun 27:8

Kim JE, No YH, Kim JN, Shin YS, Kang WT, Kim YR, Kim KN, Kim YH, Yu WJ (2017b) Highly sensitive graphene biosensor by monomolecular self-assembly of receptors on graphene surface. Appl Phys Lett 110:20

Kim JW, Kim S, Jang YH, Lim KI, Lee WH (2019) Attomolar detection of virus by liquid coplanar-gate graphene transistor on plastic. Nanotechnology 2019:30

Kudr J, Zhao L, Nguyen EP, Arola H, Nevanen TK, Adam V, Zitka O, Merkoci A (2020) Inkjet-printed electrochemically reduced graphene oxide microelectrode as a platform for HT-2 mycotoxin immunoenzymatic biosensing. Biosens Bioelectron 2020:156

Kumar N, Wang WJ, Ortiz-Marquez JC, Catalano M, Gray M, Biglari N, Hikari K, Ling X, Gao JM, Van Opijnen T, Burch KS (2020) Dielectrophoresis assisted rapid, selective and single cell detection of antibiotic resistant bacteria with G-FETs. Biosens Bioelectron 2020:156

Kuo CJ, Chiang HC, Tseng CA, Chang CF, Ulaganathan RK, Ling TT, Chang YJ, Chen CC, Chen YR, Chen YT (2018) Lipid-modified graphene-transistor biosensor for monitoring amyloid-beta aggregation. ACS Appl Mater Interfaces 10:12311–12316

Lerner MB, Matsunaga F, Han GH, Hong SJ, Xi J, Crook A, Perez-Guilar JM, Park YW, Saven JG, Liu RY, Johnson ATC (2014) Scalable production of highly sensitive nanosensors based on graphene functionalized with a designed G protein-coupled receptor. Nano Lett 14:2709–2714

Li B, Liu CC, Pan WL, Shen JL, Guo JY, Luo TT, Feng JJ, Situ B, An TX, Zhang Y, Zheng L (2020) Facile fluorescent aptasensor using aggregation-induced emission luminogens for exosomal proteins profiling towards liquid biopsy. Biosens Bioelectron 2020:168

Li X, Zhu SJ, Xu B, Ma K, Zhang JH, Yang B, Tian WJ (2013) Self-assembled graphene quantum dots induced by cytochrome c: a novel biosensor for trypsin with remarkable fluorescence enhancement. Nanoscale 5:7776–7779

Li YJ, Wang C, Zhu YB, Zhou XH, Xiang Y, He M, Zeng SY (2017) Fully integrated graphene electronic biosensor for label-free detection of lead (II) ion based on G-quadruplex structure-switching. Biosens Bioelectron 89:758–763

Li YT, Jin X, Tang LN, Lv WL, Xiao MM, Zhang ZY, Gao C, Zhang GJ (2019) Receptor-mediated field effect transistor biosensor for real-time monitoring of glutamate release from primary hippocampal neurons. Anal Chem 91:8229–8236

Lin TW, Wu HY, Tasi TT, Lai YH, Shen HH (2015) Surface-enhanced Raman spectroscopy for DNA detection by the self-assembly of ag nanoparticles onto ag nanoparticle-graphene oxide nanocomposites. Phys Chem Chem Phys 17:18443–18448

Liu SH, Fu Y, Xiong C, Liu ZK, Zheng L, Yan F (2018) Detection of bisphenol A using DNA-functionalized graphene field effect transistors integrated in microfluidic systems. ACS Appl Mater Interfaces 10:23522–23528

Lu ZY, Liu YJ, Wang MH, Zhang C, Li Z, Huo YY, Li Z, Xu SC, Man BY, Jiang SZ (2018) A novel natural surface-enhanced Raman spectroscopy (SERS) substrate based on graphene oxide-ag nanoparticles-Mytilus coruscus hybrid system. Sens Actuat B-Chem 261:1–10

Mao S, Yu KH, Chang JB, Steeber DA, Ocola LE, Chen JH (2013) Direct growth of vertically-oriented graphene for field-effect transistor biosensor. Sci Rep 2013:3

Morales-Narvaez E, Baptista-Pires L, Zamora-Galvez A, Merkoci A (2017) Graphene-based biosensors: going simple. Adv Mater 29:7

Morales-Narvaez E, Naghdi T, Zor E, Merkoci A (2015) Photoluminescent lateral-flow immunoassay revealed by graphene oxide: highly sensitive paper-based pathogen detection. Anal Chem 87:8573–8577

Myung S, Solanki A, Kim C, Park J, Kim KS, Lee KB (2011) Graphene-encapsulated nanoparticle-based biosensor for the selective detection of cancer biomarkers. Adv Mater 23:2221

Myung S, Yin PT, Kim C, Park J, Solanki A, Reyes PI, Lu YC, Kim KS, Lee KB (2012) Label-free polypeptide-based enzyme detection using a graphene-nanoparticle hybrid sensor. Adv Mater 24:6081

Nehra A, Singh KP (2015) Current trends in nanomaterial embedded field effect transistor-based biosensor. Biosens Bioelectron 74:731–743

Nguyen VT, Song S, Park S, Joo C (2020) Recent advances in high-sensitivity detection methods for paper-based lateral-flow assay. Biosens Bioelectron 152:112015

Oh J, Lee JS, Jun J, Kim SG, Jang J (2017) Ultrasensitive and selective organic FET-type nonenzymatic dopamine sensor based on platinum nanoparticles-decorated reduced graphene oxide. ACS Appl Mater Interfaces 9:39526–39533

Ohno Y, Maehashi K, Yamashiro Y, Matsumoto K (2009) Electrolyte-gated graphene field-effect transistors for detecting pH protein adsorption. Nano Lett 9:3318–3322

Pan X, Li L, Lin H, Tan J, Wang H, Liao M, Chen C, Shan B, Chen Y, LI M (2019) A graphene oxide-gold nanostar hybrid based-paper biosensor for label-free SERS detection of serum bilirubin for diagnosis of jaundice. Biosens Bioelectron 145:111713

Park CS, Yoon H, Kwon OS (2016) Graphene-based nanoelectronic biosensors. J Ind Eng Chem 38:13–22

Park JW, Park SJ, Kwon OS, Lee C, Jang J (2014) Polypyrrole nanotube embedded reduced graphene oxide transducer for field-effect transistor-type H2O2 biosensor. Anal Chem 86: 1822–1828

Ping JF, Wu J, Wang YX, Ying YB (2012) Simultaneous determination of ascorbic acid, dopamine and uric acid using high-performance screen-printed graphene electrode. Biosens Bioelectron 34:70–76

Seo G, Lee G, Kim MJ, Baek SH, Choi M, Ku KB, Lee CS, Jun S, Park D, Kim HG, Kim SJ, Lee JO, Kim BT, Park EC, Kim SI (2020) Rapid detection of COVID-19 causative virus (SARS-CoV-2) in human nasopharyngeal swab specimens using field-effect transistor-based biosensor. ACS Nano 14:12257–12258

Soikkeli M, Kurppa K, Kainlauri M, Arpiainen S, Paananen A, Gunnarsson D, Joensuu JJ, Laaksonen P, Prunnila M, Linder MB, Ahopelto J (2016) Graphene biosensor programming with genetically engineered fusion protein monolayers. ACS Appl Mater Interfaces 8:8257–8264

Suvarnaphaet P, Pechprasarn S (2017) Graphene-based materials for biosensors: a review. Sensors 2017:17

Syu YC, Hsu WE, Lin CT (2018) Review-field-effect transistor biosensing: devices and clinical applications. Ecs J Solid State Sci Technol 7:Q3196–Q3207

Tehrani Z, Whelan SP, Mostert AB, PAULIN JV, ALI MM, AHMADI ED, GRAEFF CFO, GUY OJ, GETHIN DT (2020) Printable and flexible graphene pH sensors utilising thin film melanin for physiological applications. 2D Mater 2020:7

Viswanathan S, Narayanan TN, Aran K, Fink KD, Paredes J, Ajayan PM, Filipek S, Miszta P, Tekin HC, Inci F, Demirci U, Li P, Bolotin KI, Liepmann D, Renugopalakrishanan V (2015) Graphene-protein field effect biosensors: glucose sensing. Mater Today 18:513–522

Wang C, Li YJ, Zhu YB, Zhou XH, Lin Q, He M (2016a) High-kappa solid-gate transistor configured graphene biosensor with fully integrated structure and enhanced sensitivity. Adv Funct Mater 26:7668–7678

Wang L, Zhu SJ, Wang HY, Qu SN, Zhang YL, Zhang JH, Chen QD, Xu HL, Han W, Yang B, Sun HB (2014a) Common origin of Green luminescence in carbon Nanodots and graphene quantum dots. ACS Nano 8:2541–2547

Wang LL, Ng WB, Jackman JA, Cho NJ (2016b) Graphene-functionalized natural microcapsules: modular building blocks for ultrahigh sensitivity bioelectronic platforms. Adv Funct Mater 26: 2097–2103

Wang XH, Liu AP, Xing Y, Duan HW, Xu WZ, Zhou Q, Wu HP, Chen C, Chen BY (2018a) Three-dimensional graphene biointerface with extremely high sensitivity to single cancer cell monitoring. Biosens Bioelectron 105:22–28

Wang Y, Li SS, Yang HY, Luo J (2020a) Progress in the functional modification of graphene/graphene oxide: a review. RSC Adv 10:15328–15345

Wang YY, Pham TD, Zand K, Li JF, Burke PJ (2014b) Charging the quantum capacitance of graphene with a single biological ion channel. ACS Nano 8:4228–4238

Wang ZQ, Wu SS, Ciacchi LC, Wei G (2018b) Graphene-based nanoplatforms for surface-enhanced Raman scattering sensing. Analyst 143:5074–5089

Wang ZR, Hao Z, Yu SF, De Moraes CG, Suh LH, Zhao XZ, Lin Q (2019) An ultraflexible and stretchable aptameric graphene nanosensor for biomarker detection and monitoring. Adv Funct Mater 29:44

Wang ZR, Hao Z, Yu SF, Huang C, Pan YL, Zhao XZ (2020b) A wearable and deformable graphene-based affinity nanosensor for monitoring of cytokines in biofluids. Nanomaterials 10: 8

Wongkaew N, Simsek M, Griesche C, Baeumner AJ (2019) Functional nanomaterials and nanostructures enhancing electrochemical biosensors and lab-on-a-Chip performances: recent progress, applications, and future perspective. Chem Rev 119:120–194

Wu D, Yu Y, Jin D, Xiao MM, Zhang ZY, Zhang GJ (2020) Dual-aptamer modified graphene field-effect transistor nanosensor for label-free and specific detection of hepatocellular carcinoma-derived microvesicles. Anal Chem 92:4006–4015

Wu LD, Lu XB, Fu XC, Wu LX, Liu H (2017) Gold nanoparticles dotted reduction graphene oxide nanocomposite based electrochemical Aptasensor for selective, rapid, sensitive and congener-specific PCB77 detection. Sci Rep 7:1–7

Xiang LC, Wang Z, Liu ZH, Weigum SNE, Yu, Q. K. & Chen, M. Y. (2016) Inkjet-printed flexible biosensor based on graphene field effect transistor. IEEE Sensors J 16:8359–8364

Xie H, Li YT, Lei YM, Liu YL, Xiao MM, Gao C, Pang DW, Huang WH, Zhang ZY, Zhang GJ (2016) Real-time monitoring of nitric oxide at single-cell level with porphyrin-functionalized graphene field-effect transistor biosensor. Anal Chem 88:11115–11122

Xu K, Meshik X, Nichols BM, Zakar E, Dutta M, Stroscio MA (2014) Graphene- and aptamer-based electrochemical biosensor. Nanotechnology 25:205501

Xu SC, Jiang SZ, Zhang C, Yue WW, Zou Y, Wang GY, Liu HL, Zhang XM, Li MZ, Zhu ZS, Wang JH (2018) Ultrasensitive label-free detection of DNA hybridization by sapphire-based graphene field-effect transistor biosensor. Appl Surf Sci 427:1114–1119

Xu SC, Man BY, Jiang SZ, Wang JH, Wei J, Xu SD, Liu HP, Gao SB, Liu HL, Li ZH, Li HS, Qiu HW (2015) Graphene/cu nanoparticle hybrids fabricated by chemical vapor deposition as surface-enhanced Raman scattering substrate for label-free detection of adenosine. ACS Appl Mater Interfaces 7:10977–10987

Xu Y, Cheng GF, He PG, Fang YZ (2009) A review: electrochemical Aptasensors with various detection strategies. Electroanalysis 21:1251–1259

Xuan X, Yoon HS, Park JY (2018) A wearable electrochemical glucose sensor based on simple and low-cost fabrication supported micro-patterned reduced graphene oxide nanocomposite electrode on flexible substrate. Biosens Bioelectron 109:75–82

Yan F, Zhang M, Li, J. H. (2014) Solution-gated graphene transistors for chemical and biological sensors. Adv Healthc Mater 3:313–331

Yan K, Haque AMJ, Nandhakumar P, Bhatia A, Lee NS, Yoon YH, Yang H (2020) Boosting electrochemical immunosensing performance by employing acetaminophen as a peroxidase substrate. Biosens Bioelectron 5:165

Yang L, Zhao H, Li YC, Ran X, Deng GG, Zhang YQ, Ye HZ, Zhao GF, Li CP (2016) Indicator displacement assay for cholesterol electrochemical sensing using a calix[6]arene functionalized graphene-modified electrode. Analyst 141:270–278

You XQ, Pak JJ (2014) Graphene-based field effect transistor enzymatic glucose biosensor using silk protein for enzyme immobilization and device substrate. Sens Actuat B-Chem 202:1357–1365

Yu XX, Cai HB, Zhang WH, Li XJ, Pan N, Luo Y, Wang XP, Hou JG (2011) Tuning chemical enhancement of SERS by controlling the chemical reduction of graphene oxide Nanosheets. ACS Nano 5:952–958

Yu Y, Li YT, Jin D, Yang F, Wu D, Xiao MM, Zhang H, Zhang ZY, Zhang GJ (2019) Electrical and label-free quantification of exosomes with a reduced graphene oxide field effect transistor biosensor. Anal Chem 91:10679–10686

Zhang C, Xu JQ, Li YT, Huang L, Pang DW, Ning Y, Huang WH, Zhang ZY, Zhang GJ (2016) Photocatalysis-induced renewable field-effect transistor for protein detection. Anal Chem 88: 4048–4054

Zhang XY, Jing QS, Ao S, Schneider GF, Kireev D, Zhang ZJ, Fu WY (2020) Ultrasensitive field-effect biosensors enabled by the unique electronic properties of graphene. Small 16:e1902820

Zheng C, Huang L, Zhang H, Sun ZY, Zhang Z, Zhang GJ (2015) Fabrication of ultrasensitive field-effect transistor DNA biosensors by a directional transfer technique based on CVD-grown graphene. ACS Appl Mater Interfaces 7:16953–16959

Zhou L, Mao HJ, Wu CY, Tang L, Wu ZH, Sun H, Zhang HL, Zhou HB, Jia CP, Jin QH, Chen XF, Zhao JL (2017) Label-free graphene biosensor targeting cancer molecules based on non-covalent modification. Biosens Bioelectron 87:701–707

Zubiarrain-Laserna A, Kruse P (2020) Review-graphene-based water quality sensors. J Electrochem Soc 167:037539

Zuccaro L, Tesauro C, Kurkina T, Fiorani P, Yu HK, Knudsen BR, Kern K, Desideri A, Balasubramanian K (2015) Real-time label-free direct electronic monitoring of topoisomerase enzyme binding kinetics on graphene. ACS Nano 9:11166–11176

Antibacterial Activity of Graphene-Based Nanomaterials

Hongjian Zhou, Fengming Zou, Kwangnak Koh, and Jaebeom Lee

Abstract

Recent research has shown that graphene as a novel "green" antibacterial material possess excellent antibacterial properties with no risk of bacterial resistance for daily life due to its physical damage-based bactericidal mechanism. Therefore, an increasing amount of research has been focused towards evaluating the antibacterial effects of graphene and graphene-based hybrid materials. In this chapter, we reviewed the antibacterial activity and mechanism of graphene-based nanomaterials and highlighted the importance of size, morphology, and composites in the application of antibacterial materials development. Finally, we made a summary and outlook on this research field.

H. Zhou
Key Laboratory of Materials Physics, Centre for Environmental and Energy Nanomaterials, Anhui Key Laboratory of Nanomaterials and Nanotechnology, CAS Center for Excellence in Nanoscience, Institute of Solid State Physics, HFIPS, Chinese Academy of Sciences, Hefei, People's Republic of China
e-mail: hjzhou@issp.ac.cn

F. Zou
Institute of Health & Medical Technology, Hefei Institutes of Physical Science, Chinese Academy of Sciences, Hefei, Anhui, People's Republic of China
e-mail: fmzou@hmfl.ac.cn

K. Koh
Office of General Education, Pusan National University, Busan, South Korea
e-mail: koh@pusan.ac.kr

J. Lee (✉)
Department of Chemistry, Chemical Engineering and Applied Chemistry, Chungnam National University, Daejeon, South Korea
e-mail: nanoleelab@cnu.ac.kr

D.-W. Han, S. W. Hong (eds.), *Multifaceted Biomedical Applications of Graphene*, Advances in Experimental Medicine and Biology 1351,
https://doi.org/10.1007/978-981-16-4923-3_11

Keywords

Graphene · Bacteria · Antibacterial activity · Antibacterial materials · Bacterial resistance

1 Introduction

Bacteria are prokaryotes that can be symbiotic and parasitic with plants and animals. Pathogenic bacteria in plants caused large economic losses in the agricultural production (Dimiev and Tour 2014; Whitman et al. 1998; Dean et al. 2012). Bacteria are also closely related to human life. Interestingly, there are tenfold more bacterial cells in human body than human cells, mainly found in the gut flora and on the skin (Van Der Heijden et al. 2008). Most bacteria in the body are harmless; however, some pathogenic bacteria may cause infectious diseases, such as cholera, anthrax, leprosy, etc. Bacterial respiratory diseases tend to be the most fatal, with tuberculosis alone killing approximately 2 million people per year, mostly in sub-Saharan Africa (Van Der Heijden et al. 2008; Ayele et al. 2004). This still happens even though many different technologies have been applied in the treatment (Boehme et al. 2011; Hanna et al. 1999; Pamme and Manz 2004).

Bacterial infections are often treated using antibiotics, while some bacteria are resistant to commonly used antibiotics (Wright 2005). Antibiotic resistance is a serious threat to global health because of increasingly risky diseases (Ocsoy et al. 2013; Neu 1992). According to the report in 2013 by the Centers for Disease Control and Prevention of United States (U.S.), antibiotic resistant bacteria has caused more than 2 million of infections and 23,000 deaths in the USA (Slayton et al. 2015). Furthermore, the number of approved antibiotics has decreased in the past decade. Thus, there is an urgent need for novel antibacterial treatments (Spellberg et al. 2004; Alanis 2005). Recently, novel antimicrobials for crops, humans, and animals have been developed through nanotechnology (Ocsoy et al. 2013). In particular, polymeric, carbon-based, and metallic nanomaterials are of interest to develop novel antibacterial agents and often functionalized to have enhanced biocompatibility and bioactivity (Han and Atabaev 2020; Han and Chrzanowski 2018; Shin et al. 2018). Recent advances in two-dimensional (2D) nanomaterials proposed their clinical potential including bioimaging, drug delivery, antimicrobial and antibacterial effect, and regenerative therapy (SelestinRaja et al. 2020; Lin et al. 2018; Liu et al. 2013; Lingamdinne et al. 2021; Kang et al. 2021). Especially, graphene and their derivatives are the most highlighted 2D nanomaterial in terms of their exceptional antibacterial property.

Graphene is a two-dimensional, single-atom-thick sheet of hexagonally arranged carbon atoms. Graphene oxide (GO) is a graphene sheet with carboxylic groups at its edges, and phenol hydroxyl and epoxide groups on its basal plane (Dimiev and Tour 2014). Graphene and GO have widespread use in electroconductive films, bio-sensor, bioimaging, and drug delivery due to their extraordinary thermal, mechanical, and electrical properties (Chung et al. 2013; Zhang et al. 2012).

Recently, the antibacterial activity of graphene-based nanomaterials has been reported widely, like GO-polymer, GO-dye, GO-silver, and GO-TiO_2 (Table 1) (Zhou et al. 2013; Mejías Carpio et al. 2012; Kim et al. 2014). The antibacterial activity of GO is attributed to its ability to puncture membranes and induce oxidative stress (Liu et al. 2011a). Many factors are related to the antibacterial activity of graphene-based nanomaterials, such as material properties, surface chemical properties, as well as interaction conditions of graphene and bacterial cells (Dal Lago et al. 2011; El Badawy et al. 2011; Kvítek et al. 2008; El Badawy et al. 2011; Pal et al. 2007; Yang et al. 2012). However, the underlying antibacterial mechanism of graphene-based nanomaterials remains to be elucidated. The field of graphene-related antibacterial activity has seen significant advances in a short time; thus, a comprehensive review of the latest developments is necessary.

Here, we discuss recent advances in graphene-based antibacterial nanomaterials. In particular, we focus on the antibacterial mechanism of graphene-based nanomaterials and the importance of size and morphological control of graphene-based nanomaterials for antibacterial development. We also discuss the major challenges and opportunities for future research.

2 Antibacterial Mechanisms of Graphene-Based Nanomaterials

2.1 Interaction of Bacteria with Graphene-Modified Surfaces

Since the outer layer of bacterial cell is made up of varieties of biological molecules, numerous physicochemical and biological parameters affect attachment of bacteria to graphene surfaces, which can promote or inhibit subsequent biofilm formation (Notley et al. 2013). The chemical property of the substrate can affect the attachment of bacteria and the formation of the biofilm. Bacteria have great affinity for hydrophobic surfaces in terms of its membrane structure (Busscher and Van Der Mei 1997; Rosenberg et al. 1980; Phoenix et al. 2002). The physical structure of the surface is also a critical factor regulating bacterial interaction. Interestingly, graphene has a similar intermediate wettability to graphite, and different preparation methods may result in varying charges. The inhibition of bacterial surface attachment is considered an antibacterial effect; whereas surface contact that effectively destroys bacteria is termed a bactericidal effect. Some chemicals, such as compounds containing quaternary ammonium groups, exhibit strong bactericidal effects by inducing membrane permeability. Furthermore, the surface morphology of graphene can also be used to kill bacteria. Typically, the rough surface has high aspect rations, which can lead to the rupture of bacterial cell wall under surface contact. Therefore, the surface contact plays a vital role in antibacterial activity of graphene-based nanomaterials through the interaction of bacteria with graphene.

Table 1 Graphene-based nanomaterials and nanocomposites used in antibacterial researches

Antibacterial material	Bacteria strains	References
DNA-directed silver nanoparticles on graphene oxide	*Xanthomonas perforans*	Ocsoy et al. (2013)
Ag@polydopamine/graphene oxide sheet nanocomposite	*Escherichia coli*	Zhou et al. (2013)
Polyvinyl-*N*-carbazole–graphene oxide nanocomposite	*Escherichia coli, Cupriavidus metallidurans, Bacillus subtilis, and Rhodococcus opacus.*	Mejías Carpio et al. (2012)
Reduced graphene oxide nanosheets and exfoliated layered titanate nanosheets	*Escherichia coli*	Kim et al. (2014)
Graphite, graphite oxide, graphene oxide, and reduced graphene oxide	*Escherichia coli*	Liu et al. (2011a)
Graphene-based antibacterial paper	*Escherichia coli*	Hu et al. (2010)
Graphene oxide and reduced graphene oxide	*Pseudomonas aeruginosa*	Gurunathan et al. (2012)
Rough surfaced graphene film	*Escherichia coli, mycobacterium smegmatis,* and *Staphylococcus aureus*	Zou et al. (2017)
Graphene-based chitosan films	*Pseudomonas aeruginosa*	Lim et al. (2012)
Graphene nanosheets	*Escherichia coli*	Akhavan et al. (2011)
Graphene and graphene oxide nanowalls	*Escherichia coli*	Akhavan and Ghaderi (2010)
Graphene oxide	*Pseudomonas syringae* and *Xanthomonas campestris pv. Undulosa*	Chen et al. (2014)
Silver nanoparticle and graphene oxide nanosheet composites	*Escherichia coli* and *Staphylococcus aureus*	Bao et al. (2011)
Monodispersed silver nanoparticles on graphene oxide sheets	*Escherichia coli*	Liu et al. (2011b)
Silver nanoparticles in an aqueous suspension of graphene oxide sheets	*Escherichia coli* and *Pseudomonas aeruginosa*	Das et al. (2011)
Carbon nanotube, sliver, and graphene oxide nanocomposites	*Bacillus subtilis, Staphylococcus aureus, Escherichia coli,* and *salmonella typhimurium*	Yun et al. (2013)
Graphene oxide–silver nanocomposite	*Escherichia coli* and *Staphylococcus aureus*	Tang et al. (2013)
Iron oxide nanoparticles and silver nanoparticles on the surface of graphene oxide	*Escherichia coli* and *Staphylococcus aureus*	Tian et al. (2014)
Graphene oxide and molybdenum disulfide nanocomposites	*Escherichia coli*	Kim et al. (2017)
Amino-functionalized graphene oxide	*Staphylococcus aureus* and *Escherichia coli*	Mei et al. (2019)

(continued)

Table 1 (continued)

Antibacterial material	Bacteria strains	References
Ag_2S-MgO/GO nanocomposite	*Escherichia coli*	Wang et al. (2020)
Graphene oxide film	*Escherichia coli, mycobacterium smegmatis,* and *Staphylococcus aureus*	Radhi et al. (2021)
Reduced graphene oxide-silver modified glass ionomer cement	*Streptococcus mutans*	Chen et al. (2020)

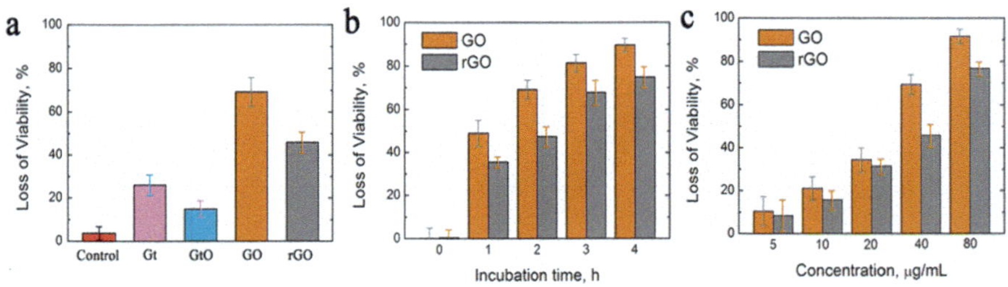

Fig. 1 (**a**) Cell viability measurements after incubation with graphite (Gt), graphite oxide (GtO), graphene oxide (GO), and reduced graphite oxide (rGO) dispersions. (**b**) Time-dependent antibacterial activities of GO and rGO. (**c**) Concentration dependent antibacterial activities of GO and rGO. Copyright ACS and reprinted with permission from (Liu et al. 2011a)

2.2 Antibacterial Mechanisms

Recently, antibacterial mechanisms of graphene have been reported extensively, mainly focused on membrane stress, oxidative stress independent of reactive oxygen species (ROS) generation, and nutrient deprivation imposed by adhesion of graphene-based nanomaterials (Hui et al. 2014). Two sites on graphene have been proposed to induce membrane stress: the sharp edges of graphene, which may puncture cell membranes resulting in the leakage of cellular content (Liu et al. 2011a; Hu et al. 2010), and the basal planes, which may destructively extract membrane lipids (Liu et al. 2012; Tu et al. 2013). It has been suggested that bacterial death is primarily induced by the basal planes, where a variety of antibacterial processes, including membrane stress, oxidative stress, and nutrient deprivation, can occur (Gurunathan et al. 2012). Liu et al. (2011a) investigated the antibacterial mechanisms of four different graphene derivatives: graphite (Gt), graphite oxide (GtO), GO, and reduced graphite oxide (rGO). In this study, they found that GO has the strongest antibacterial activity, followed by rGO, Gt, and GtO (Fig. 1). It demonstrated that graphene nanosheets directly contacted with bacteria and disrupted the cell membrane. Four graphene derivatives were able to oxidize glutathione, which serves as a redox mediator in bacteria, but there is no superoxide anion (O^{2-}) detected. Compared to insulated GO and GtO, conductive ones are highly oxidative. Both membrane disruption and oxidative stress have spurred the antimicrobial activity of graphene-based materials. Moreover, this antimicrobial activity

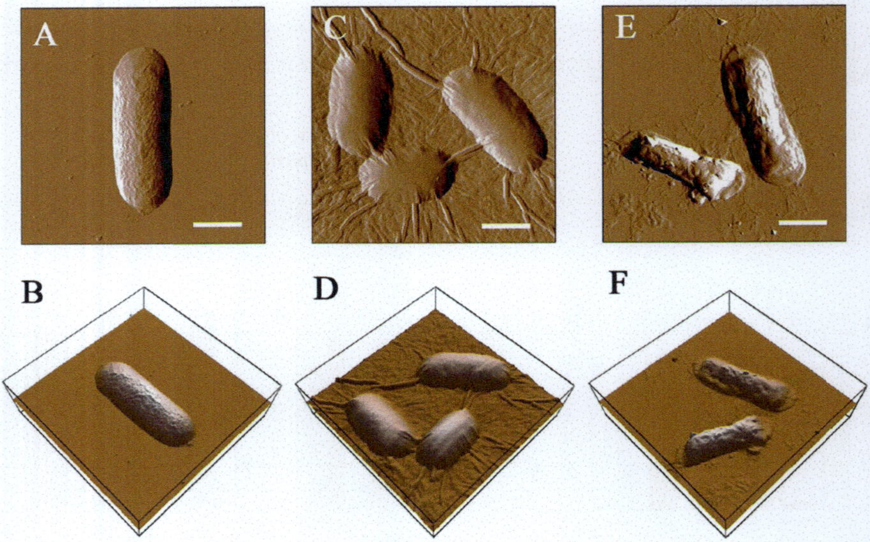

Fig. 2 Atomic force microscopy (AFM) amplitude and three dimensional (3D) images of *E. coli* cells after incubation with graphene oxide (GO) sheets. (**a**, **b**) *E. coli* incubation with deionized water for 2 h, (**c**, **d**) *E. coli* incubation with the 40 μg/mL GO-240 suspension for 2 h. Scale bars = 1 μm. Copyright ACS and reprinted with permission from Liu et al. (2012)

tends to have three steps. First, the cells attach to the graphene-based materials through membrane stress caused by direct contact with sharp nanosheets, and then the superoxide anions were oxidized, independently. It proves that the density, size, and conductivity of graphene-based materials could be precisely to either reduce their threat to health and environment or enhance their beneficial applications.

Literally, size of graphene sheets may be a considerable parameter. Compared with smaller GO particles, larger sheets have stronger antibacterial activity (Liu et al. 2012). Different sizes of GO sheets were obtained by using the modified Hummer's method by changing sonication time. Larger sheets can reduce the viability of *Escherichia coli (E. coli)* obviously through effective isolation and wrapping around the bacterial cells (Fig. 2). The results suggest that the size-dependent antibacterial activity of GO sheets was not attributed to their different aggregation states or oxidation capacity but the different lateral sizes of GO sheets were well dispersed and have a similar oxidation capacity towards glutathione. Finally, large GO sheets cover cells easily, and then isolate their active membrane sites.

Tu et al. (2013) illuminated the mechanism of graphene nanosheet action through experiments and simulations. They showed that the inner and outer cell membranes of *E. coli* were degraded by graphene and GO nanosheets, thereby reducing *E. coli* viability. From the simulation results, it is observed that the graphene nanosheets penetrated into the cell membranes and extracted phospholipids due to the strong dispersion interactions between the graphene and lipid molecules, which is a new explanation for graphene-induced cytotoxicity and antibacterial activity. Hui et al.

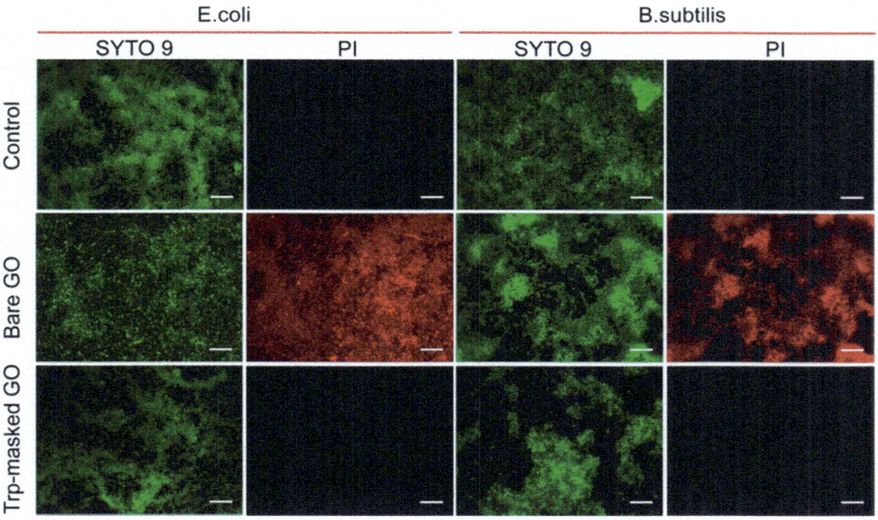

Fig. 3 Representative fluorescence microscopy images of bacterial strains treated with 200 μg/ml bare graphene oxide (GO) sheets or Trp-masked GO (Trp:GO = 12:1), and subsequently stained with SYTO 9 (green) and propidium iodide (PI, red). Scale bar = 20 μm. Copyright ACS and reprinted with permission from Hui et al. (2014)

(2014) reported that GO alone has the potential to kill both bacteria and mammalian cells, whose basal planes equipped with non-covalent adsorption may be deactivating GO cytotoxicity (Fig. 3). These results clearly showed that membrane permeabilization was an important part of antibacterial activity of GO.

Zou et al. (2017) found the relationship between the surface roughness of graphene film and the antibacterial effect to three different bacterial (*Escherichia coli, Mycobacterium smegmatis, and Staphylococcus aureus*) through experiment and simulation. They postulated that bacteria easily interacted with the diameter-matched surface roughness of graphene film, which caused substantial damages to the bacterial cell membrane. Akhavan et al. (2011) proposed that bacterial cells can be blocked within aggregated graphene sheets, biologically isolated from growth medium, and cannot proliferation anymore. Compared with the oxygen-containing functional groups, the bacteria could be better trapped by the graphene sheets with reduced functional groups. Through this mechanism, bacteria proliferation was inhibited (Liu et al. 2012). The physicochemical properties of graphene-based materials can be better adjusted to reduce their risk of side effects or develop their potential antimicrobial applications, including the functional groups, density, size, and conductivity. Sitansu Sekhar Nanda et al studied the mechanism of GO using Raman spectroscopy. They observed that Raman peaks of Adenine and proteins in GO treated *E. coli* and *E. faecalis* were consistent with SEM and Bio-AFM results (Nanda et al. 2016). Alayande et al. (2019) illustrated the role of GO layer morphology using *Pseudomonas aeruginosa*. They found that GO thickness

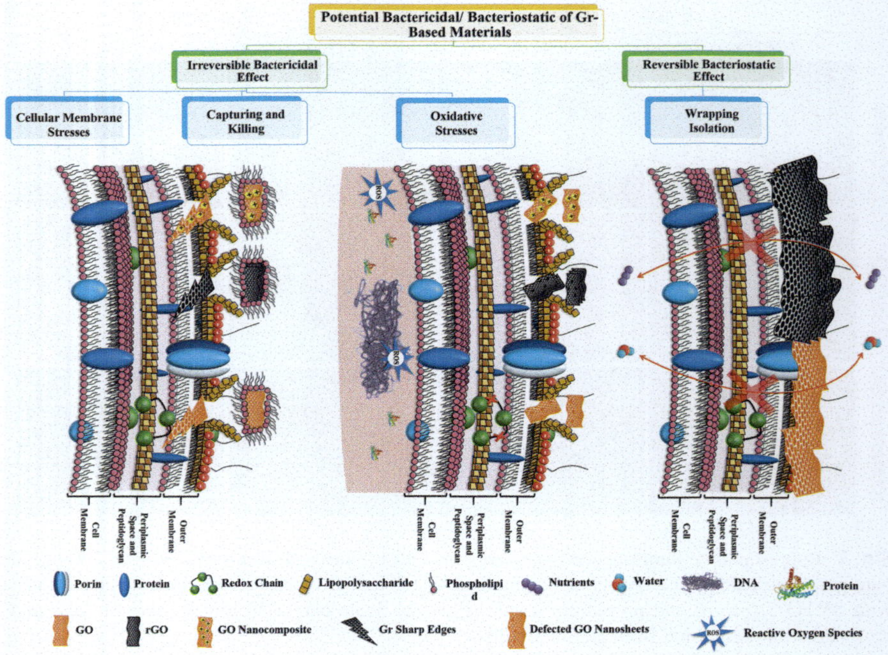

Fig. 4 Permanent and temporary bacterial cell growth inhibition mechanisms by Gr-based materials (A colour version of this figure can be viewed online). Copyright PNAS and reprinted with permission from Lu et al. (2017)

(>1.71429 mg cm^{-2}) has strong adhesion to *P. aeruginosa*, which facilitated to block bacteria cells. Lu et al. (2017) found that when compared with random and horizontal GO nanosheets, vertically arranged GO nanosheets have stronger antibacterial activity due to the increased edge density with a preferential location to membrane damage. In summary, the main mechanisms for graphene-induced antibacterial activity were schematized in Fig. 4 including membrane stress, oxidative stress, and wrapping isolation.

Bacterial cells lose their integrity due to exposure to graphene-based nanomaterials, which was irreversible damage. The direct extraction of phospholipids from lipid membranes was first observed with the help of computer simulations, and was validated by transmission electron microscope (TEM) results (Tu et al. 2013). On the two-dimensional graphene surface, extracted lipid molecules were found due to the redistribution of the hydrophobic tails to maximize hydrophobic interactions with the graphene surface. The destructive lipid extraction was observed in both the outer and inner membranes of *E. coli*, which is induced by graphene nanosheets. The observed graphene insertion and destructive lipid extraction support the notion that graphene nanosheets induce serious membrane stress, thereby significantly reducing cell viability. Furthermore, the resulting antibacterial activity can be also enhanced by increased lateral size and concentration of graphene.

Another key factor of graphene-based nanomaterials induced antibacterial activity is oxidative stress, which caused the oxidation of nucleic acid, proteins, and lipid. Ultimately, bacterial cell membrane was damaged and cellular growth was inhibited via ROS generation. For the wrapping isolation, the interaction between bacteria and the robust surface trap of GO was enhanced, and then caused substantial damages to bacterial cell membrane.

3 Antibacterial Effect of Graphene

Currently, it becomes a hot spot to study graphene-modified surfaces for antibacterial applications. Akhavan and Ghaderi firstly investigated bacterial interactions with a graphene-like surface in 2010 by means of gram-negative (*E. coli*) and gram-positive (*Staphylococcus aureus*, *S. aureus*) (Akhavan and Ghaderi 2010). They deposited single- and multiple-layer of GO and reduced GO onto a stainless steel substrate. The particles were oriented arrangement such that a significant number of edges were exposed. Interestingly, the viability of *E. coli* and *S. aureus* was both decreased (Fig. 5). However, some tendencies were observed from the results. For example, the viability of *S. aureus*, the gram-positive bacteria, was low. Furthermore, the rGO surface was more capable of resisting attachment and killing the bacteria. The membrane damage was induced by contact with the GO and reduced GO particles through measurement of the cytoplasmic milieu efflux, which was consistent with the observation of greater toxicity towards the gram-positive bacteria. In addition, the reduced GO particles have much sharper edges than oxidized GO, which was more potentially for antibacterial applications. However, this theory is not well supported by direct particle imaging or the current literature (Novoselov et al. 2004; Novoselov et al. 2005).

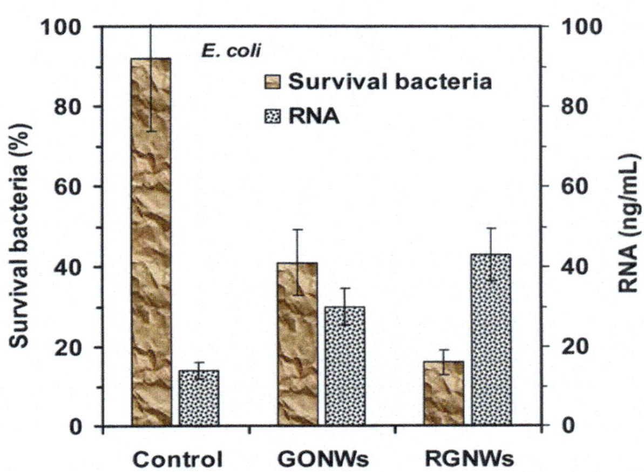

Fig. 5 Cytotoxicity of Graphene Oxide nanowalls (GONWs) and Reduced Graphene nanowalls (RGNWs) in *E. coli*, and concentrations of RNA in the Phosphate Buffer Solution (PBS) of *E. coli* bacteria exposed to the nanowalls. Copyright ACS and reprinted with permission from Akhavan et al. (2011)

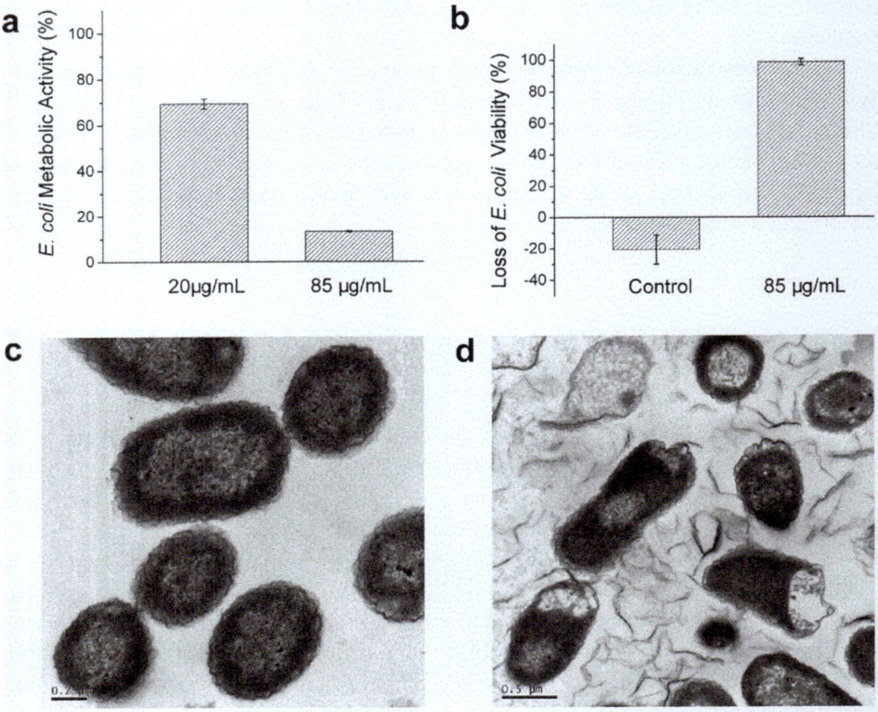

Fig. 6 Antibacterial activity of graphene oxide (GO) nanosheets. (**a**) Metabolic activity of *E. coli* incubation with 20 and 85 μg/mL GO nanosheets at 37 °C for 2 h. (**b**) Antibacterial activity of 85 μg/mL αGO nanosheets against *E. coli* DH5α Cells. (**c** and **d**) TEM images of *E. coli* and *E. coli* exposed to GO nanosheets at 37 °C for 2 h. Copyright ACS and reprinted with permission from Hu et al. (2010)

By vacuum filtration, Hu et al. (2010) made macroscopic freestanding GO and rGO paper which was later discovered possessing strong antibacterial effect (Fig. 6). GO has the potential application in environmental and clinical field in view of the scalability and low cost of the graphene-based antibacterial paper. Gurunathan et al. (2012) investigated the antibacterial activity of GO and rGO triggered by oxidative stress in *Pseudomonas aeruginosa* (*P. aeruginosa*). The survival rate of *P. aeruginosa* was reduced in a dose- and time-dependent manner due to the superoxide radical produced by exposure to GO and rGO. Furthermore, GO and rGO also exhibited dose-dependent antibacterial activity against *P. aeruginosa* cells through the generation of ROS, leading to cell death, which was further confirmed by nuclear fragmentation.

Chen et al. (2014) investigated the mechanism with which GO and typical phytopathogens interact by evaluating the antimicrobial activity of GO against two bacterial (*Pseudomonas syringae* and *Xanthomonas campestris* pv. *undulosa*) and even fungal pathogens (*Fusarium graminearum* and *Fusarium oxysporum*). The results indicated that GO has a significant effect on the reproduction of all four pathogens, killing nearly 90% of the bacteria, repressing 80% of the macroconidia germination, and inducing partial cell swelling and lysis at 500 mg/ml. It hypothesizes that GO envelops the bacterial and fungal spores through GO sheet aggregation, thereby perturbing the cell membrane, decreasing the bacterial membrane potential, and inducing electrolyte leakage from the fungal spores. Thus, GO can interact with pathogens through mechanically wrapping and damaging the positional cell membrane. This will finally lead to cell lysis, which may be the major reason of toxicity. This study suggests that the GO may initiate antibacterial activity against multi-drug resistant bacterial and fungal phytopathogens, and provides useful information regarding the applications of GO in crop diseases.

Currently, all studies about antibacterial activity were done with GO or rGO particles, particularly focusing on surface attachment and cell viability. There remains no consensus regarding the inherent antibacterial properties of these surfaces, while the available information supports the standpoint that cytotoxicity varies depending on the direction of surface contact between the graphene-based materials and the bacteria. Sharp edges of GO and rGO sheets induce cell rupture, whereas studies involving the basal planes have shown no antibacterial activity. As the above, so far, materials used to study the antibacterial activity of graphene-based nanomaterials are quite distinct from "pristine" graphene or epitaxial graphene. Until now, there is no studies about evaluating the antibacterial activity of pure graphene with highly conjugated, defect-free, single layers of sp^2-hybridized carbon.

Recently, numerus studies have demonstrated that metal oxide nanoparticles have good antibacterial activity. Among them, Ag nanoparticles have been used as biological sterilizers against varieties of pathogens (Sondi and Salopek-Sondi 2004), fungi (Kim et al. 2009), and viruses (Zodrow et al. 2009; Elechiguerra et al. 2005). In order to achieve high antibacterial activity, Liu et al. (2011b) and Bao et al. (2011) both reported enhanced antibacterial activity of Ag and GO sheets. They investigated the antibacterial activity of GO–Ag composites in *E. coli*, gram-negative bacteria. Compared with Ag nanoparticles, the antibacterial activity of these composites was markedly enhanced, suggesting that the prepared nanocomposites may have the potential to be an effective antibacterial material.

Das et al. (2011) developed a uniform, water-soluble Ag@rGO nanocomposite without additional reductants, which displayed better antibacterial activity than pure Ag nanoparticles synthesized by microwave irradiation. Furthermore, these nanoparticles showed an equivalent antibacterial effect to the antibiotic ampicillin. Song et al. (2003) synthesized Ag nanoparticles by reducing silver nitrate ($AgNO_3$) in a GO chemical suspension. The gram-negative bacteria *E. coli* and *P. aeruginosa* were used to investigate the antimicrobial activity of these hybrid materials by evaluating bacterial growth kinetics under different conditions. It turned out that *P. aeruginosa* was more sensitive to the Ag nanoparticle–GO suspension than *E. coli*.

Yun et al. (2013) evaluated the antibacterial activity of carbon nanotubes and graphene oxide nanocomposites with silver in gram-negative and gram-positive bacteria. The antibacterial activity of the carbon nanocomposites was different between the gram-positive and gram-negative bacteria from the inhibition to cell

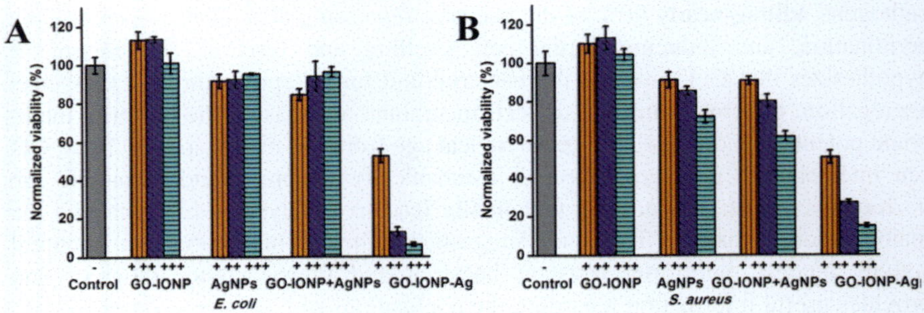

Fig. 7 Cell viability of (**a**) *E. coli* and (**b**) *S. aureus* after treatment with graphene oxide-iron oxide nanoparticles (GO-IONP), Ag nanoparticles, a simple mixture of GO-IONP and Ag nanoparticles, and a GO-IONP-Ag nanocomposite at different concentrations for 2.5 h. Copyright ACS and reprinted with permission from Tian et al. (2014)

growth and the minimum inhibitory concentration (MIC). Tang et al. (2013) developed antibacterial agents where an Ag nanoparticle was anchored to GO (GO−Ag nanocomposites) by using different ratios of Ag nanoparticles. GO−Ag nanocomposites are active antibacterial since they contain more Ag nanoparticles than Ag nanoparticles and even more active at low doses (2.5 μg/ml). Furthermore, the GO−Ag nanocomposite was more toxic against *E. coli* than *S. aureus*.

Ocsoy et al. (2013) demonstrated the use of nanotechnology to control plant disease. They prepared DNA-guided Ag nanoparticles that grow in GO. The viability of *X. perforans* in culture and plants were effectively decreased by the Ag@dsDNA@GO composites. At 16 ppm Ag@dsDNA@GO, the composites exhibit excellent antibacterial quality such as improving stability, enhancing antibacterial activity, and being more absorptive. The 100 ppm Ag@dsDNA@GO to tomato plants can significantly reduce bacterial spot in sample comparison. This is the current gold-standard growth treatment with no phytotoxicity. Kim et al. (2014) found that the antibacterial activity of graphene oxide-titanate hybrid film was much greater than that of the graphene film alone. The improvement of the functionality of graphene film benefitted from the presence of layered metal oxide nanosheets. These freestanding hybrid films composed of rGO-layered titanate exhibited unprecedented high antibacterial properties, eventually the complete sterilization of *E. coli* (100%) was achieved within 15 min (Kim et al. 2014).

Tian et al. (2014) obtained a novel, multifunctional antibacterial material, the GO-iron oxide nanoparticles (IONP)-Ag nanocomposite by co-cultivating IONPs and Ag nanoparticles on the surface of GO (Fig. 7). Compared to the widely used Ag nanoparticles as antibacterial agents, the GO-IONP-Ag nanocomposites exhibited greater antibacterial activity towards both *E. coli* and *S. aureus*. Due to the strong near-infrared (NIR) absorbance, GO-IONP-Ag nanocomposites were also used for photothermal treatment, which turned to be more effective against *S. aureus* at low concentrations. Moreover, these nanoparticles can be easily recycled by magnetic separation and reused with the presence of magnetic IONPs. Given advantages described above as well as its easy preparation and inexpensive attributes, the GO-IONP-Ag nanocomposite could be potentially applied in the healthcare and environmental engineering industries. Kim et al. (2017) prepared graphene oxide (GO) and molybdenum disulfide (MoS_2) nanocomposites, which exhibited enhanced antibacterial activity to *Escherichia coli* with increased glutathione oxidation capacity and partial conductivity. Mei et al. (2019) prepared amino-functionalized graphene oxide ($GO\text{-}NH_2$) with enhanced antibacterial properties through excellent photothermal efficiency.

No conclusions have been drawn regarding the antibacterial mechanism of the graphene-based nanocomposites. However, based on the mechanisms proposed by researchers, the GO sheet itself tends to attach and wrap bacteria spontaneously, enhancing the interaction between the bacteria and GO (Liu et al. 2011a). Eventually, the Ag nanoparticles on GO cause direct and irreversible damage to the cell membrane by denaturing proteins in cell wall and invading the cell during the bacterial cell-wrapping process. The Ag nanoparticles and Ag^+ ions released from the composites can react with thiol, carboxyl, hydroxyl, amino, phosphate, and imidazole groups on and within the cell, thereby inducing cell inactivation and death. The Ag nanoparticles, Ag^+ ions, and Ag-containing compounds have all been extensively used as universal germicides (Sondi and Salopek-Sondi 2004). However, bare Ag nanoparticles have a much lower antibacterial activity than Ag@dsDNA@GO composites, possibly because of their agglomeration when in contact with bacteria through media and the loss of active Ag atoms (Kvítek et al. 2008; Shrivastava et al. 2007). In addition, the polymeric ligands on the surface of Ag nanoparticles protect living organisms from agglomeration, which has resulted in the loss of Ag^+ ions (Ma et al. 2011). In contrast, Ag@dsDNA@GO has more antibacterial activity because of the co-effect of GO and Ag nanoparticles. Therefore, hybridization of functionalized metal nanoparticles with GO could be more effective antibacterial. Novel work to predict antibacterial activity and rationally design effective antimicrobial nanomaterials is currently underway.

Hybridization of graphene into the catalyst materials to endow antibacterial effect is also widely adopted strategy. Wang et al. (2020) introduced Ag_2S-MgO/GO nanocomposite with photocatalytic and antimicrobial activities. The examination of antibacterial and antifungal properties of the nanocomposites were conducted on *Bacillus vallismortis, Escherichia coli, Aspergillus flavus* and *Trichoderma viride*. The synergistic antibacterial and antifungal effect of Ag_2S and GO induced great reduction in the number of bacteria and fungi compared to pristine MgO. On the other hand, the antimicrobial property of graphene is often applied in dentistry due to the disruption of the integrity of bacterial membrane and generation of ROS. Radhi et al. (2021) elucidated mechanism of graphene-based nanomaterials antimicrobial activities for application in dentistry. The incorporation of graphene materials into dental implants significantly enhanced antimicrobial properties, while not compromising their mechanical properties. The results indicated that antimicrobial effect can be altered by various intrinsic and extrinsic factors including size, shape, surface chemistry, external electromagnetic radiation, and substrate materials. Chen et al. (2020) incorporated rGO-Ag nanocomposite into glass ionomer cement (GIC) which is recently developed dental restorative materials. The GIC often suffer from restoration failures due to the poor mechanical properties and secondary caries. The study indicated that rGO-Ag nanocomposite induced significant decrease of Streptococcus mutans which was demonstrated by fewer bacteria population, reduced stack on implant surface, and decreased bacteria viability.

4 Conclusions

The development of novel antibacterial and antivirals has been of interest due to the worldwide outbreak of bacterial and viral diseases. Most studies focused on antibacterial applications of GO and rGO derivatives and their effect on bacterial surface adhesion. These particles were proved to have physiochemical properties through evaluating study of bacteria and graphene interaction. As noted, the unique properties of graphene such as its high electron mobility are significantly diminished upon oxidation and reduction. To fully understand the interactions between bacteria and graphene, exfoliated graphene containing extended conjugation or epitaxial grown graphene should be used in further studies. Graphene and GO nanomaterials have potential applications in the clinical treatment of *S. aureus*, or bacteria with substantially different surface chemistries such as highly hydrophobic bacteria. Such an interdisciplinary approach is complicated; thus, effective collaboration between scientists from different disciplines will be necessary for further development of these antibacterial materials.

Acknowledgment This work was supported by the Korean Health Technology R&D Project, Ministry of Health & Welfare, Republic of Korea (Grant Nos. HI16C1553, HI17C1260), the National Natural Science Foundation of China (Grant No. 51872291 and 51502296), Anhui Provincial Natural Science Foundation (1808085MH268), and the National Research Foundation of Korea grant funded by the Korea government (MS IP) (Grant Nos. 2016R1A2B4012072, 2017R1A41015627, 2019R1A2C2007825). We are also grateful for the support of Hefei leading talent in 2017 for Fengming Zou.

References

Akhavan O, Ghaderi E (2010) Toxicity of graphene and graphene oxide nanowalls against bacteria. ACS Nano 4(10):5731–5736

Akhavan O, Ghaderi E, Esfandiar A (2011) Wrapping bacteria by graphene nanosheets for isolation from environment, reactivation by sonication, and inactivation by near-infrared irradiation. J Phys Chem B 115(19):6279–6288

Alanis AJ (2005) Resistance to antibiotics: are we in the post-antibiotic era? Arch Med Res 36(6): 697–705

Alayande AB, Chae S, Kim IS (2019) Surface morphology-dependent spontaneous bacterial behaviors on graphene oxide membranes. Sep Purif Technol 226:68–74

Ayele WY, Neill SD, Zinsstag J, Weiss MG, Pavlik I (2004) Bovine tuberculosis: an old disease but a new threat to Africa. Int J Tuberc Lung D 8(8):924–937

Bao Q, Zhang D, Qi P (2011) Synthesis and characterization of silver nanoparticle and graphene oxide nanosheet composites as a bactericidal agent for water disinfection. J Colloid Interface Sci 360(2):463–470

Boehme CC, Nicol MP, Nabeta P, Michael JS, Gotuzzo E, Tahirli R, Gler MT, Blakemore R, Worodria W, Gray C, Huang L, Caceres T, Mehdiyev R, Raymond L, Whitelaw A, Sagadevan K, Alexander H, Albert H, Cobelens F, Cox H, Alland D, Perkins MD (2011) Feasibility, diagnostic accuracy, and effectiveness of decentralised use of the Xpert MTB/RIF test for diagnosis of tuberculosis and multidrug resistance: a multicentre implementation study. Lancet 377(9776):1495–1505

Busscher HJ, Van Der Mei HC (1997) Physico-chemical interactions in initial microbial adhesion and relevance for biofilm formation. Adv Dent Res 11(1):24–32

Chen J, Peng H, Wang X, Shao F, Yuan Z, Han H (2014) Graphene oxide exhibits broad-spectrum antimicrobial activity against bacterial phytopathogens and fungal conidia by intertwining and membrane perturbation. Nanoscale 6(3):1879–1889

Chen J, Zhao Q, Peng J, Yang X, Yu D, Zhao W (2020) Antibacterial and mechanical properties of reduced graphene-silver nanoparticle nanocomposite modified glass ionomer cements. J Dent 96:103332–103339

Chung C, Kim Y-K, Shin D, Ryoo S-R, Hong BH, Min D-H (2013) Biomedical applications of graphene and graphene oxide. Acc Chem Res 46(10):2211–2224

Dal Lago V, França de Oliveira L, de Almeida GK, Kobarg J, Borba Cardoso M (2011) Size-selective silver nanoparticles: future of biomedical devices with enhanced bactericidal properties. J Mater Chem 21(33):12267–12273

Das MR, Sarma RK, Saikia R, Kale VS, Shelke MV, Sengupta P (2011) Synthesis of silver nanoparticles in an aqueous suspension of graphene oxide sheets and its antimicrobial activity. Colloids Surf B Biointerfaces 83(1):16–22

Dean RV, Van Kan JAL, Pretorius ZA, Hammond Kosack KE, Di Pietro A, Spanu PD, Rudd JJ, Dickman M, Kahmann R, Ellis J, Fosteroster GD (2012) The top 10 fungal pathogens in molecular plant pathology. Mol Plant Pathol 13(4):414–430

Dimiev AM, Tour JM (2014) Mechanism of graphene oxide formation. ACS Nano 8(3):3060–3068

El Badawy AM, Silva RG, Morris B, Scheckel KG, Suidan MT, Tolaymat TM (2011) Surface charge-dependent toxicity of silver nanoparticles. Environ Sci Technol 45(1):283–287

Elechiguerra JL, Burt JL, Morones JR, Camacho-Bragado A, Gao X, Lara HH, Yacaman MJ (2005) Interaction of silver nanoparticles with HIV-1. J Nanobiotechnol 3(1):6

Gurunathan S, Han JW, Dayem AA, Eppakayala V, Kim J-H (2012) Oxidative stress-mediated antibacterial activity of graphene oxide and reduced graphene oxide in Pseudomonas aeruginosa. Int J Nanomedicine 7:5901–5914

Han D-W, Atabaev TS (2020) The two faces of nanomaterials: toxicity and bioactivity. MDPI

Han D-W, Chrzanowski W (2018) Frontiers in toxicity and functionalization of nanomaterials. MDPI

Hanna BA, Ebrahimzadeh A, Elliott LB, Morgan MA, Novak SM, Rusch-Gerdes S, Acio M, Dunbar DF, Holmes TM, Rexer CH, Savthyakumar C, Vannier AM (1999) Multicenter evaluation of the BACTEC MGIT 960 system for recovery of mycobacteria. J Clin Microbiol 37(3):748–752

Hu W, Peng C, Luo W, Lv M, Li X, Li D, Huang Q, Fan C (2010) Graphene-based antibacterial paper. ACS Nano 4(7):4317–4323

Hui L, Piao J-G, Auletta J, Hu K, Zhu Y, Meyer T, Liu H, Yang L (2014) Availability of the basal planes of graphene oxide determines whether it is antibacterial. ACS Appl Mater Interfaces 6(15):13183–13190

Kang MS, Kang JI, Thi PL, Park KM, Hong SW, Choi YS, Han DW, Park KD (2021) Three-dimensional printable gelatin hydrogels incorporating graphene oxide to enable spontaneous myogenic differentiation. ACS Macro Lett 10(4):426–432

Kim K-J, Sung WS, Suh BK, Moon S-K, Choi J-S, Kim JG, Lee DG (2009) Antifungal activity and mode of action of silver nano-particles on *Candida albicans*. Biometals 22(2):235–242

Kim IY, Park S, Kim H, Park S, Ruoff RS, Hwang S-J (2014) Strongly-coupled freestanding hybrid films of graphene and layered titanate nanosheets: an effective way to tailor the physicochemical and antibacterial properties of graphene film. Adv Funct Mater 24(16):2288–2294

Kim TI, Kwon B, Yoon J, Park I-J, Bang GS, Park Y, Seo Y-S, Choi S-Y (2017) Antibacterial activities of graphene oxide–molybdenum Disulfide nanocomposite films. ACS Appl Mater Interfaces 9(9):7908–7917

Kvítek L, Panáček A, Soukupová J, Kolář M, Večeřová R, Prucek R, Holecová M, Zbořil R (2008) Effect of surfactants and polymers on stability and antibacterial activity of silver nanoparticles (NPs). J Phys Chem C 112(15):5825–5834

Lim H, Huang N, Loo C (2012) Facile preparation of graphene-based chitosan films: enhanced thermal, mechanical and antibacterial properties. J Non-Cryst Solids 358:525–530

Lin J, Huang Y, Huang P (2018) Graphene-based nanomaterials in bioimaging. Biomed Appl Funct Nanomater 2018:247–287

Lingamdinne LP, Lee S, Choi J-S, Lebaka VR, Durbaka VRP, Koduru JR (2021) Potential of the magnetic hollow sphere nanocomposite (graphene oxide-gadolinium oxide) for arsenic removal from real field water and antimicrobial applications. J Hazard Mater 402:123882–123893

Liu S, Zeng TH, Hofmann M, Burcombe E, Wei J, Jiang R, Kong J, Chen Y (2011a) Antibacterial activity of graphite, graphite oxide, graphene oxide, and reduced graphene oxide: membrane and oxidative stress. ACS Nano 5(9):6971–6980

Liu L, Liu J, Wang Y, Yan X, Sun DD (2011b) Facile synthesis of monodispersed silver nanoparticles on graphene oxide sheets with enhanced antibacterial activity. New J Chem 35(7):1418–1423

Liu S, Hu M, Zeng TH, Wu R, Jiang R, Wei J, Wang L, Kong J, Chen Y (2012) Lateral dimension-dependent antibacterial activity of graphene oxide sheets. Langmuir 28(33):12364–12372

Liu J, Liang C, Dusan L (2013) Graphene and graphene oxide as new nanocarriers for drug delivery applications. Acta Biomater 9(12):9243–9257

Lu X, Feng X, Werber JR, Chu C, Zucker I, Kim J-H, Osuji CO, Elimelech M (2017) Enhanced antibacterial activity through the controlled alignment of graphene oxide nanosheets. Proc Natl Acad Sci 114(46):E9793–E9801

Ma J, Zhang J, Xiong Z, Yong Y, Zhao XS (2011) Preparation, characterization and antibacterial properties of silver-modified graphene oxide. J Mater Chem 21(10):3350–3352

Mei L, Lin C, Cao F, Yang D, Jia X, Hu S, Miao X, Wu P (2019) Amino-functionalized graphene oxide for the capture and photothermal inhibition of bacteria. ACS Appl Nano Mater 2(5): 2902–2908

Mejías Carpio IE, Santos CM, Wei X, Rodrigues DF (2012) Toxicity of a polymer–graphene oxide composite against bacterial planktonic cells, biofilms, and mammalian cells. Nanoscale 4(15): 4746–4756

Nanda SS, Yi DK, Kim K (2016) Study of antibacterial mechanism of graphene oxide using Raman spectroscopy. Sci Rep 6(1):28443–28454

Neu HC (1992) The crisis in antibiotic resistance. Science 257(5073):1064–1073

Notley SM, Crawford RJ, Ivanova EP (2013) Bacterial interaction with graphene particles and surfaces. In advances in graphene science and technology. Adv Graphene Sci:99–272

Novoselov KS, Geim AK, Morozov SV, Jiang D, Zhang Y, Dubonos SV, Grigorieva IV, Firsov AA (2004) Electric field effect in atomically thin carbon films. Science 306(5696):666–669

Novoselov KS, Jiang D, Schedin F, Booth TJ, Khotkevich VV, Morozov SV, Geim AK (2005) Two-dimensional atomic crystals. Proc Natl Acad Sci 102(30):10451–10453

Ocsoy I, Paret ML, Ocsoy MA, Kunwar S, Chen T, You M, Tan W (2013) Nanotechnology in plant disease management: DNA-directed silver nanoparticles on graphene oxide as an antibacterial against *Xanthomonas perforans*. ACS Nano 7(10):8972–8980

Pal S, Tak YK, Song JM (2007) Does the antibacterial activity of silver nanoparticles depend on the shape of the nanoparticle? A study of the gram-negative bacterium. Appl Environ Microbiol 73(6):1712–1720

Pamme N, Manz A (2004) On-Chip free-flow magnetophoresis: continuous flow separation of magnetic particles and agglomerates. Anal Chem 76(24):7250–7256
Phoenix VR, Martinez RE, Konhauser KO, Ferris FG (2002) Characterization and implications of the cell surface reactivity of sp. strain KC97. Appl Environ Microbiol 68(10):4827–4834
Radhi A, Mohamad D, Rahman FSA, Abdullah AM, Hasan H (2021) Mechanism and factors influence of graphene-based nanomaterials antimicrobial activities and application in dentistry. J Mater Res Technol 11:1290–1307
Rosenberg M, Gutnick D, Rosenberg E (1980) Adherence of bacteria to hydrocarbons: a simple method for measuring cell-surface hydrophobicity. FEMS Microbiol Lett 9(1):29–33
SelestinRaja I, Lee JH, Hong SW, Shin D-M, Lee JH, Han D-W (2020) A critical review on genotoxicity potential of low dimensional nanomaterials. J Hazard Mater 409:124915–124930
Shin YC, Lee JH, Lee I-S, Han D-W (2018) Biosafety of carbon-based nanoparticles and nanocomposites. Fab Biomed Appl 2018:431–458
Shrivastava S, Bera T, Roy A, Singh G, Ramachandrarao P, Dash D (2007) Characterization of enhanced antibacterial effects of novel silver nanoparticles. Nanotechnology 18(22): 225103–225112
Slayton RB, Toth D, Lee BY, Tanner W, Bartsch SM, Khader K, Wong K, Brown K, McKinnell JA, Ray W, Miller LG, Rubin M, Kim DS, Adler F, Cao C, Avery L, Stone NTB, Kallen A, Samore M, Huang SS, Fridkin S, Jernigan JA (2015) Vital signs: estimated effects of a coordinated approach for action to reduce antibiotic-resistant infections in health care facilities - United States. MMWR Morb Mortal Wkly Rep 64(30):826–831
Sondi I, Salopek-Sondi B (2004) Silver nanoparticles as antimicrobial agent: a case study on *E. coli* as a model for Gram-negative bacteria. J Colloid Interface Sci 275(1):177–182
Song R, Kim Y-S, Lee CO, Sohn YS (2003) Synthesis and antitumor activity of DNA binding cationic porphyrin–platinum(II) complexes. Tetrahedron Lett 44(8):1537–1540
Spellberg B, Powers JH, Brass EP, Miller LG, Edwards JE (2004) Trends in antimicrobial drug development: implications for the future. Clin Infect Dis 38(9):1279–1286
Tang J, Chen Q, Xu L, Zhang S, Feng L, Cheng L, Xu H, Liu Z, Peng R (2013) Graphene oxide–silver nanocomposite as a highly effective antibacterial agent with species-specific mechanisms. ACS Appl Mater Interfaces 5(9):3867–3874
Tian T, Shi X, Cheng L, Luo Y, Dong Z, Gong H, Xu L, Zhong Z, Peng R, Liu Z (2014) Graphene-based nanocomposite as an effective, multifunctional, and recyclable antibacterial agent. ACS Appl Mater Interfaces 6(11):8542–8548
Tu Y, Lv M, Xiu P, Huynh T, Zhang M, Castelli M, Liu Z, Huang Q, Fan C, Fang H, Zhou R (2013) Destructive extraction of phospholipids from Escherichia coli membranes by graphene nanosheets. Nat Nanotechnol 8:594–602
Van Der Heijden MGA, Bardgett RD, Van Straalen NM (2008) The unseen majority: soil microbes as drivers of plant diversity and productivity in terrestrial ecosystems. Ecol Lett 11(3):296–310
Wang H, Li G, Fakhri A (2020) Fabrication and structural of the Ag_2S-MgO/graphene oxide nanocomposites with high photocatalysis and antimicrobial activities. J Photochem Photobiol B Biol 207:111882–111888
Whitman WB, Coleman DC, Wiebe WJ (1998) Prokaryotes: the unseen majority. Proc Natl Acad Sci 95(12):6578–6583
Wright GD (2005) Bacterial resistance to antibiotics: enzymatic degradation and modification. Adv Drug Deliv Rev 57(10):1451–1470
Yang X, Gondikas AP, Marinakos SM, Auffan M, Liu J, Hsu-Kim H, Meyer JN (2012) Mechanism of silver nanoparticle toxicity is dependent on dissolved silver and surface coating in *Caenorhabditis elegans*. Environ Sci Technol 46(2):1119–1127
Yun H, Kim JD, Choi HC, Chul-Won L (2013) Antibacterial activity of CNT-Ag and GO-Ag nanocomposites against gram-negative and gram-positive bacteria. Bull Kor Chem Soc 34(11): 3261

Zhang Y, Nayak TR, Hong H, Cai W (2012) Graphene: a versatile nanoplatform for biomedical applications. Nanoscale 4(13):3833–3842

Zhou H, Liu Y, Chi W, Yu C, Yu Y (2013) Preparation and antibacterial properties of Ag@polydopamine/graphene oxide sheet nanocomposite. Appl Surf Sci 282:181–185

Zodrow K, Brunet L, Mahendra S, Li D, Zhang A, Li Q, Alvarez PJJ (2009) Polysulfone ultrafiltration membranes impregnated with silver nanoparticles show improved biofouling resistance and virus removal. Water Res 43(3):715–723

Zou F, Zhou H, Jeong DY, Kwon J, Eom SU, Park TJ, Hong SW, Lee J (2017) Wrinkled surface-mediated antibacterial activity of graphene oxide Nanosheets. ACS Appl Mater Interfaces 9(2): 1343–1351

Part V

Conclusions

Reflections and Outlook on Multifaceted Biomedical Applications of Graphene

Iruthayapandi Selestin Raja, Suck Won Hong, and Dong-Wook Han

Abstract

Two-dimensional nanomaterials have been widely explored by researchers due to their nanosized thickness and quantum size effect. They were layered double hydroxides, transition metal dichalcogenides, transition metal oxides, and synthetic silicate clays. Among the 2D nanomaterials, graphene and their derivatives were investigated extensively at first as they exhibited exceptional conductivity and a zero-band gap semimetal nature. Though graphene family nanomaterials (GFNs) were utilized for several physicochemical applications, including electronic, electric, mechanic, photonic, magnetic, and catalytic devices, their biomedical applications are still meritorious. Biosensor, bioimaging, drug delivery, tumor ablation, and tissue regeneration are some of them. The outlook of the present book chapters encompasses the preparation of GFNs, physicochemical properties, biomedical applications, biosafety, and their future directions.

Keywords

2D nanomaterials · Graphene family nanomaterials · Biomedical applications · Tissue regeneration · Drug delivery · Cancer treatment · Biosafety

I. S. Raja
BIO-IT Fusion Technology Research Institute, Pusan National University, Busan, South Korea

S. W. Hong · D.-W. Han (✉)
Department of Cogno-Mechatronics Engineering, College of Nanoscience and Nanotechnology, Pusan National University, Busan, South Korea
e-mail: swhong@pusan.ac.kr; nanohan@pusan.ac.kr

D.-W. Han, S. W. Hong (eds.), *Multifaceted Biomedical Applications of Graphene*, Advances in Experimental Medicine and Biology 1351,
https://doi.org/10.1007/978-981-16-4923-3_12

1 Two-Dimensional Nanomaterials

Two-dimensional (2D) nanomaterials have received great attention among researchers due to their nanosized thickness, quantum size effect, and related physicochemical properties (Khan et al. 2020). With a wide selection of elemental composition, the family of 2D materials includes metals, semimetals, semiconductors, and insulators with direct and indirect bandgaps ranging from ultraviolet to infrared regions (Zeng et al. 2018). In the past few years, layered double hydroxides (Zhang et al. 2020a, b), transition metal dichalcogenides (Meng et al. 2020), transition metal oxides (Jia et al. 2020), synthetic silicate clays (Khatoon et al. 2020), and other types of 2D nanomaterials have been reported widely (Fig. 1) (Chhowalla et al. 2013; Chimene et al. 2015). 2D nanomaterials are confined to the nanometer length scale of <100 nm (Jones and Grainger 2009). Because of extreme thinness in shape, they possess high surface-to-volume ratios and anisotropic properties compared to 3D nanomaterials (Smith et al. 2014; Pattnaik and Kumar Sutar 2019). The diverse physicochemical properties of 2D nanomaterials make them suitable for a wide range of biomedical applications such as bioimaging, biosensor, drug delivery, tumor ablation, and tissue regeneration (Aillon et al. 2009; Andreescu et al. 2012; Palombo et al. 2014; Lee et al. 2016a, b; Kim et al. 2017a, b; Ryoo et al. 2018; Park et al. 2020). They achieve a high level of surface interaction with biological macromolecules on a small scale. They can load large numbers of drug molecules and control their release kinetic behavior (Zhang et al. 2020a, b). They improve the mechanical properties of the nanocomposites even at low concentrations owing to the high modulus values (Dewapriya et al. 2020). The thinner surface of nanomaterials accesses them to respond rapidly to the external

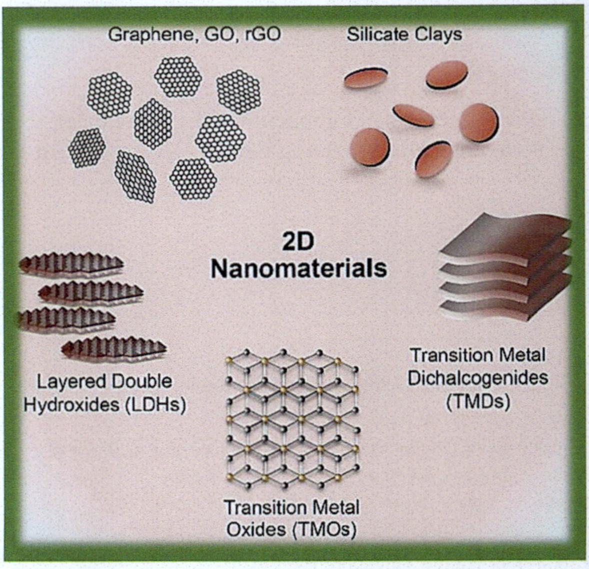

Fig. 1 Two-dimensional nanomaterials, including graphene, silicate clays, transition metal dichalcogenides, transition metal oxides, and layered double hydroxides (Reproduced from Chimene et al. 2015, Copyright (2015) John Wiley and Sons)

light source, which has led to an effective material in photothermal therapy and photodynamic therapy applications (Wang and Cheng 2019; Liu et al. 2020). The safety information for 2D nanomaterials is still practically nonexistent compared to other types of nanomaterials (Zhou et al. 2020).

2 Graphene-Based Materials

Graphene is known as the first widely explored 2D material. The electrons in the lattice of graphene are highly mobile, exhibiting exceptional conductivity as the nanomaterial possesses a zero-band gap semimetal nature (Chen et al. 2013; Mu et al. 2015). Due to novel intrinsic properties, such as electronic, mechanical, and thermal transport, graphene-based materials have found enormous numbers of biomedical applications and achievements, as shown in Fig. 2 (Lee et al. 2012; Butler et al. 2013; Shin et al. 2017a, b).

Graphene, an allotrope of carbon, is composed of a honeycomb lattice with the sp^2-hybridized 2D structural arrangement of carbon atoms (Lakshmanan and Maulik 2018). They are the primary starting material to produce fullerenes and carbon

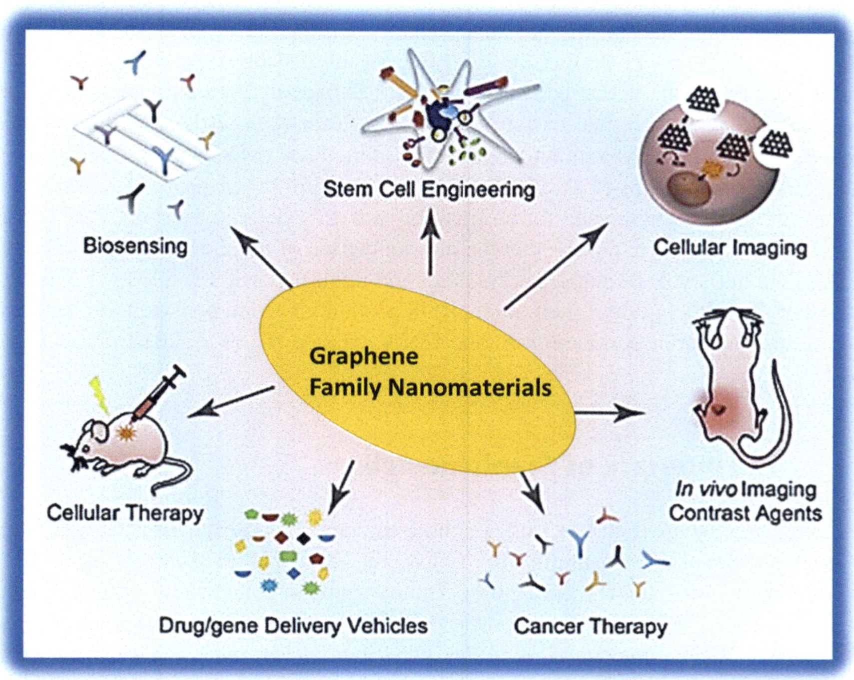

Fig. 2 General biomedical applications of graphene family nanomaterials, including biosensing, tissue regeneration, bioimaging, drug delivery, cancer therapy, and cellular therapy (Reproduced from Chimene et al. 2015, Copyright (2015) John Wiley and Sons)

nanotubes (Yang et al. 2013). The graphene layers intercalate in either a hexagonal or rhombohedral arrangement with weak van der Waals forces and are highly reactive with the edges than the center (Geim 2009). Single and multilayered graphene, graphene oxide (GO), and reduced GO (rGO) are collectively called graphene family nanomaterials (GFNs) (Lee et al. 2015a, b; Shin et al. 2016). It was reported that the number of layers of the material influences alteration in its physicochemical properties. In particular, the graphene stacked by more than 10 layers acts like graphite (Geim and Novoselov 2007). There are many conventional methods to produce graphene, including mechanical cleavage of graphite (Moon et al. 2020), electrochemical exfoliation (Vadivel et al. 2020), and chemical vapor deposition (Novoselov et al. 2004; Shang et al. 2019). Graphene oxide containing carboxylic, epoxide, and hydroxyl groups in its plane acts as an amphiphilic template to stabilize hydrophobic molecules with π-π interactions in a solution (Kim et al. 2010). When GO is subjected to reduction reaction by the chemical, thermal, or UV treatment, it is converted to reduced graphene oxide (Park et al. 2009; Jin et al. 2015).

GFNs are useful in neutralizing free radicals possessing substantial antioxidant properties (Qiu et al. 2014). They can interact with various biomacromolecules such as enzymes, proteins, peptides, and DNA and hence show remarkable properties in tissue regeneration (Sanchez et al. 2012). Some literature works showed that GFNs could differentiate and proliferate stem cells in neuronal (Hong et al. 2014), bone (Shin et al. 2015a, b, c; Park et al. 2016; Kim et al. 2017a, b), muscle (Shin et al. 2015a, b, c; Lee et al. 2016a, b; Shin et al. 2016; Shin et al. 2018a, b), vascular (Shin et al. 2017a, b), and cardiac tissue regeneration (Shin et al. 2018a, b) due to their excellent mechanical strength and electrical conductivity (Menaa et al. 2015; Kenry et al. 2018; Geetha Bai et al. 2019). The remarkable optical properties and large surface area render them used for bioimaging and drug delivery, respectively (Zhao et al. 2017). There were reports that the internalization of graphene into the cells can accumulate into cytoplasmatic, perinuclear, and nuclear portion and trigger depolarization of the mitochondrial membrane, ROS production, lipid peroxidation, caspase activation, chromatin condensation, and DNA damage (Li et al. 2012; Pinto et al. 2013).

3 The Interface of Graphene-Cells

Graphene was isolated at first from its three-dimensional parent material, graphite, by Andre Geim and Konstantin Novoselov, the 2010 Nobel laureates in Physics (Novoselov et al. 2004). Since then, many papers concerning graphene and graphene-related materials were produced with several biomedical applications. The studies of the interaction of GFNs with mammalian cells are of paramount importance in assessing the use of graphene in biomedical applications (Zhang et al. 2016). The GFNs and their derivatives contact the cell components, including plasma membrane and membrane organelles, and interfere with cellular metabolism. Scientific reports reveal that the internalization of graphene may cause completely different effects on different cell types. They can transport therapeutic agents intracellularly without affecting cellular functions. Sometimes, they can lead to cell intoxication involving other cellular mechanisms (Kostarelos et al. 2007; Shi et al. 2011; Zhao et al. 2011; Bianco 2013; Mao et al. 2014).

GFNs can be internalized into cells via four categories of endocytosis depending on cell types: caveolae-mediated endocytosis (Schulz et al. 2012), clathrin-mediated endocytosis (Mäger et al. 2012), phagocytosis (Akinc and Battaglia 2013), and micropinocytosis (Sato et al. 2009). The research group of Linares, in 2014, investigated the impact of eight different inhibitors that affect the endocytic mechanism after internalizing GO nanosheets into three different cell types, such as HepG2 hepatoma cells, RAW-264.7 macrophages, and Saos-2 osteoblasts, and found that macropinocytosis was a general mechanism (Linares et al. 2014). It was reported that the particle size (Mu et al. 2012), shape (Khim Chng et al. 2014; Kim et al. 2016), and surface chemistry (Gratton et al. 2008) determine the internalization of GFNs into cells. Computational studies indicated that graphene's entry is initiated at the corners and edge asperities, which are present at the irregular edges of fabricated materials (Li et al. 2013). Local piercing of graphene nanoparticles generated by these sharp protuberances initiates membrane propagation along the extended edge, thereby avoiding the high energy barrier. These salient findings provided a fundamental understanding of interactions at the interface of GFNs and biological systems.

After internalization into the cells, graphene is trapped within the endosome/lysosome system, the basement of intracellular interactions. In 2014, Li and coworkers reported that GO nano assemblies were mainly distributed in acidic lysosomes of mouse embryonic fibroblasts after their internalization into the cells (Li et al. 2014). It was discovered that GFNs could interact with the mitochondrial morphology and the functions, including the decrease in membrane potential, the impairment of ATP production, and the overproduction of ROS (Lammel et al. 2013; Zhou et al. 2014). Duch and coworkers informed that GO could donate electrons to complexes I and II of the electron transport chain and accelerate the generation of ROS (Duch et al. 2011). Jin and coworkers described that GO could penetrate A549 lung cancer cells and resided within the cytoplasm and nucleus (Jin et al. 2014). The cytoskeleton is also an essential cellular component that interacts with graphene-related nanomaterials. Matesanz and coworkers demonstrated that GO nanosheets were located on F-actin filaments of Saos-2 osteoblasts after their cell permeation and influenced cell cycle and cell death through a cytoskeleton-dependent manner (Matesanz et al. 2013). GFNs could interact with blood components also.

4 Description About Book Chapters

Firstly, this book provides an overview of the chemical composition and physical properties of graphene and its derivatives and their potential toxicity and biosafety. The book also covers the potentials of some graphene-based multidimensional formations, including particulates, coatings, matrices, patterns, foams, and 3D structures as new innovative options for stem cell fate modulation and tissue regeneration. Together with these potentials, the book subsequently focuses on the capability of MFGNs as agents for drug delivery, bioimaging, theranostics, and therapeutics, as well as their usefulness as potential platforms for nanobiosensors, biochips, and antibacterial agents. The book is divided into twelve in-depth chapters, each of which is organized to enrich the understanding of the functionality of GFNs and their varied biomedical applications.

Chapter "Principles and Biomedical Application of Graphene Family Nanomaterials" demonstrates the basic principles of each GFNs, a list of publications under the title of "biomedical application of graphene family nanomaterials," and an overview of biomedical applications. Chapter "Differential Toxicity of Graphene Family Nanomaterials Concerning Morphology" emphasizes the importance of studying differential toxicity of graphene family nanomaterials based on their morphology, including nanosheets, quantum dots, nanoplatelets, and nanoribbons. It reports the recent findings of cellular and animal level toxicity studies of GFNs and highlighted progress, challenges, and new toxicological approaches in the field. A detailed analysis of the role of GFNs in neurogenesis has been provided in chapter "Graphene-based Materials for Efficient Neurogenesis," monitoring the improvement of neural stem cells and differentiation into neural lineages. The basic principles of electrochemical biosensor devices and different generations of glucose sensors have been described in chapter "Functional Graphene Nanomaterials-based Hybrid Scaffolds for Osteogenesis and Chondrogenesis." Also, it presents the novel advances in electrochemical sensors using the GFNs. It is well known that glucose biosensing takes an approximately 70% share of the biosensor market as diabetic patients monitor their sugar levels several times a day (Hammond et al. 2016). The contribution of GFNs in skin wound healing and tissue regeneration has been specified in chapter "Role of Graphene Family Nanomaterials in Skin Wound Healing and Regeneration." Different 3D structures viz. electrospun nanofibrous mat, film, hydrogel, and sponge containing GFNs have been illustrated for in vivo skin tissue regeneration of excisional, incisional, diabetic, and infection type wound models (Mahmoudi et al. 2017; Shams et al. 2017). GFNs based drug delivery systems have been demonstrated with different preparation methods and their physicochemical properties in chapter "Graphene-based Nanomaterials as Drug Delivery Carriers." It discusses biodistribution and biosafety of such drug formulations also.

Biomedical imaging modalities using GFNs highlighting fluorescence imaging, Raman imaging, magnetic resonance imaging, and photoacoustic imaging have been described in chapter "Graphene-based Nanomaterials for Biomedical Imaging." This chapter addresses various challenges such as weak fluorescence intensity, quantum yield, broad emission band, and batch-to-batch variation, which can be caused by the top-down based method. Owing to high photostability, aqueous solubility, and low toxicity, GFNs have drawn remarkable interest in the field of nanomedicine as a superior theragnostic agent. Various theragnostic agents have been used for targeted delivery of antibiotics or chemotherapeutics to treat diseases like Alzheimer's disease, cancer, multidrug resistance diseases, and more (Chapter "Graphene: A

Promising Theranostic Agent"). Different GFNs based PTT assisted photodynamic, gene, chemo, and immunotherapeutic strategies have been established to improve cancer treatment in chapter "Graphene as Photothermal Therapeutic Agents." Further, it provides insights into novel GFNs based multifunctional PTT assisted combinatorial therapeutics with both imaging as well as a therapeutic activity. GFNs have unique physicochemical properties such as intrinsic near-infrared absorption and the ability to transduce the absorbed energy into heat, making their use in photothermal therapy.

Biosensors are the devices that convert the information of physicochemical or biological samples into readable signals (Hammond et al. 2016). Chapter "Graphene for Nanobiosensors and Nanobiochips" describes GFNs as the functional elements of biosensors and different GFNs based device formats such as disposable paper-based devices, lab-on-a-chips, and wearable biosensors. In chapter "Antibacterial Activity of Graphene-based Nanomaterials," the antibacterial activity of GFNs has been reviewed, highlighting the importance of size, morphology, and composites development. An increasing amount of research has been focused on evaluating the antibacterial effects and bactericidal mechanism of GFNs and their hybrid materials.

5 Concluding Remarks and Future Outlook

GFNs with different morphologies have been reported for various biomedical applications; however, only nanosheet nanomaterials have contributed more to skin tissue regeneration. GFNs can be used as drug molecules themselves as well as drug delivery carriers (McCallion et al. 2016; Priyadarsini et al. 2018). It is suggested that the composites of drug molecules with graphene-based nanomaterials improve biocompatibility and lead to effective drug release in the drug delivery field. The loading/conjugation of various chemotherapeutic drugs, PDT agents, metallic nanoparticles, organic dyes, and targeting ligands over GFNs would improve their multifunctional applications. Graphene-based theranostic agents with a high quantum yield can be synthesized if the purification step becomes more effortless. The mechanical property, biocompatibility, the osteogenic and chondrogenic properties have been found significantly increased when the desired polymers/nanomaterials/drugs are conjugated using appropriate scaffold fabrication methods (Lee et al. 2014; Lee et al. 2015a, b). Utilizing GFNs is an excellent approach to achieve non-invasive monitoring of stem cell behaviors and solve some limitations of potential stem cell therapies, including low neural differentiation and cellular toxicity. When the nanomaterials are dispersed within a three-dimensional scaffold, the resulting graphene-based materials take advantage of enhancing stem cell therapy. The photoluminescent properties of graphene-based nanomaterials can be enhanced through functionalization or cutting graphene sheets into smaller ones. Minimizing batch-to-batch variation by controlling the size distribution, preventing photoluminescent quenching, and improving the quantum yield would increase the performance of GFNs in the field of bioimaging.

A systematic biological study is required to evaluate the biochemical interaction between the nanoparticles and the macromolecules. The organ-specific toxicity of GFNs must be given uttermost importance. The lack of information about the clearance of GFNs from the body restricts their use in human clinical trials. The surface properties like zeta potential, functional groups, and polydispersion index should be discussed alone in detail when GFNs are stabilized with encapsulating agents for increasing their solubility and biosafety.

We hope that the researchers with interdisciplinary backgrounds will develop the application of GFNs by contemplating the challenges and the concerning suggestions. The pivotal roles of this book make it essential reading for scientists and engineers of all the biomedical research fields seeking to fully utilize novel GFNs-incorporated composite materials and develop various functional devices based on them. Besides its value as an introductory book for graduate students, it offers a well-organized reference guide for all individuals working in biomedical science and engineering.

Acknowledgments This research was supported by National Research Foundation of Korea (NRF) funded by the Ministry of Science (NRF-2021R1A2C2006013) and by Korea Evaluation Institute of Industrial Technology (KEIT) grant funded by the Ministry of Trade, Industry and Energy (MOTIE, Korea) (No. 20014399).

References

Aillon KL, Xie Y, El-Gendy N, Berkland CJ, Forrest ML (2009) Effects of nanomaterial physicochemical properties on in vivo toxicity. Adv Drug Deliv Rev 61(6):457–466

Akinc A, Battaglia G (2013) Exploiting endocytosis for nanomedicines. Cold Spring Harb Perspect Biol 5(11):a016980

Andreescu S, Ornatska M, Erlichman JS, Estevez A, Leiter JC (2012) Biomedical applications of metal oxide nanoparticles. In: Matijević E (ed) Fine Particles in Medicine and Pharmacy. Springer, Boston, MA, pp 57–100. https://doi.org/10.1007/978-1-4614-0379-1_3

Bianco A (2013) Graphene: safe or toxic? The two faces of the medal. Angew Chem Int Ed 52 (19):4986–4997

Butler SZ, Hollen SM, Cao L, Cui Y, Gupta JA, Gutiérrez HR, Heinz TF, Hong SS, Huang J, Ismach AF, Halperin EJ, Kuno M, Plashnitsa VV, Robinson RD, Ruoff RS, Salahuddin S, Shan J, Shi L, Spencer MG, Terrones M, Windl W, Goldberger JE (2013) Progress, challenges, and opportunities in two-dimensional materials beyond graphene. ACS Nano 7(4):2898–2926

Chen S, Duan J, Ran J, Jaroniec M, Qiao SZ (2013) N-doped graphene film-confined nickel nanoparticles as a highly efficient three-dimensional oxygen evolution electrocatalyst. Energy Environ Sci 6(12):3693–3699

Chhowalla M, Shin HS, Eda G, Li L-J, Loh KP, Zhang H (2013) The chemistry of two-dimensional layered transition metal dichalcogenide nanosheets. Nat Chem 5(4):263–275

Chimene D, Alge DL, Gaharwar AK (2015) Two-dimensional nanomaterials for biomedical applications: emerging trends and future prospects. Adv Mater 27(45):7261–7284

Dewapriya MAN, Rajapakse RKND, Meguid SA (2020) 2 - Mechanical properties of two-dimensional materials: atomistic modeling and future directions. In: Yang E-H, Datta D, Ding J, Hader G (eds) Synthesis, modeling, and characterization of 2D materials, and their heterostructures. Elsevier, Amsterdam, The Netherlands, pp 9–35. https://doi.org/10.1016/B978-0-12-818475-2.00002-7

Duch MC, Budinger GRS, Liang YT, Soberanes S, Urich D, Chiarella SE, Campochiaro LA, Gonzalez A, Chandel NS, Hersam MC, Mutlu GM (2011) Minimizing oxidation and stable nanoscale dispersion improves the biocompatibility of graphene in the lung. Nano Lett 11 (12):5201–5207

Geetha Bai R, Muthoosamy K, Manickam S, Hilal-Alnaqbi A (2019) Graphene-based 3D scaffolds in tissue engineering: fabrication, applications, and future scope in liver tissue engineering. Int J Nanomedicine 14:5753–5783

Geim AK (2009) Graphene: status and prospects. Science 324(5934):1530–1534

Geim AK, Novoselov KS (2007) The rise of graphene. Nat Mater 6(3):183–191

Gratton SEA, Ropp PA, Pohlhaus PD, Luft JC, Madden VJ, Napier ME, DeSimone JM (2008) The effect of particle design on cellular internalization pathways. Proc Natl Acad Sci 105 (33):11613–11618

Hammond JL, Formisano N, Estrela P, Carrara S, Tkac J (2016) Electrochemical biosensors and nanobiosensors. Essays Biochem 60(1):69–80

Hong SW, Lee JH, Kang SH, Hwang EY, Hwang Y-S, Lee MH, Han D-W, Park J-C (2014) Enhanced neural cell adhesion and neurite outgrowth on graphene-based biomimetic substrates. Biomed Res Int 2014:212149

Jia Y, Yi X, Li Z, Zhang L, Yu B, Zhang J, Wang X, Jia X (2020) Recent advance in biosensing applications based on two-dimensional transition metal oxide nanomaterials. Talanta 219: 121308

Jin C, Wang F, Tang Y, Zhang X, Wang J, Yang Y (2014) Distribution of graphene oxide and TiO_2-graphene oxide composite in A549 cells. Biol Trace Elem Res 159(1–3):393–398

Jin L, Lee JH, Jin OS, Shin YC, Kim MJ, Hong SW, Lee MH, Park JC, Han DW (2015) Stimulated osteogenic differentiation of human mesenchymal stem cells by reduced graphene oxide. J Nanosci Nanotechnol 15(10):7966–7970

Jones CF, Grainger DW (2009) In vitro assessments of nanomaterial toxicity. Adv Drug Deliv Rev 61(6):438–456

Kenry LWC, Loh KP, Lim CT (2018) When stem cells meet graphene: opportunities and challenges in regenerative medicine. Biomaterials 155:236–250

Khan K, Tareen AK, Aslam M, Wang R, Zhang Y, Mahmood A, Ouyang Z, Zhang H, Guo Z (2020) Recent developments in emerging two-dimensional materials and their applications. J Mater Chem C 8(2):387–440

Khatoon N, Chu MQ, Zhou CH (2020) Nanoclay-based drug delivery systems and their therapeutic potentials. J Mater Chem B 8(33):7335–7351

Khim Chng EL, Chua CK, Pumera M (2014) Graphene oxide nanoribbons exhibit significantly greater toxicity than graphene oxide nanoplatelets. Nanoscale 6(18):10792–10797

Kim F, Cote LJ, Huang J (2010) Graphene oxide: surface activity and two-dimensional assembly. Adv Mater 22(17):1954–1958

Kim SE, Kim MS, Shin YC, Eom SU, Lee JH, Shin D-M, Hong SW, Kim B, Park J-C, Shin BS, Lim D, Han D-W (2016) Cell migration according to shape of graphene oxide micropatterns. Micromachines 7(10):186

Kim J, Park S-J, Min D-H (2017a) Emerging approaches for graphene oxide biosensor. Anal Chem 89(1):232–248

Kim J-W, Shin YC, Lee J-J, Bae E-B, Jeon Y-C, Jeong C-M, Yun M-J, Lee S-H, Han D-W, Huh J-B (2017b) The effect of reduced graphene oxide-coated biphasic calcium phosphate bone graft material on osteogenesis. Int J Mol Sci 18(8):1725

Kostarelos K, Lacerda L, Pastorin G, Wu W, Wieckowski S, Luangsivilay J, Godefroy S, Pantarotto D, Briand J-P, Muller S, Prato M, Bianco A (2007) Cellular uptake of functionalized carbon nanotubes is independent of functional group and cell type. Nat Nanotechnol 2 (2):108–113

Lakshmanan R, Maulik N (2018) Graphene-based drug delivery systems in tissue engineering and nanomedicine. Can J Physiol Pharmacol 96(9):869–878

Lammel T, Boisseaux P, Fernández-Cruz ML, Navas JM (2013) Internalization and cytotoxicity of graphene oxide and carboxyl graphene nanoplatelets in the human hepatocellular carcinoma cell line Hep G2. Part Fibre Toxicol 10:27

Lee JH, Shin YC, Jin OS, Lee EJ, Han D-W, Kang SH, Hong SW, Ahn JY, Kim SH (2012) Cytotoxicity evaluations of pristine graphene and carbon nanotubes in fibroblastic cells. J Korean Phys Soc 61(6):873–877

Lee E, Lee J, Shin Y, Hwang D-G, Kim J, Jin OS, Jin L, Hong SW, Han D-W (2014) Graphene oxide-decorated PLGA/collagen hybrid fiber sheets for application to tissue engineering scaffolds. Biomater Res 18(1):18–24

Lee JH, Shin YC, Jin OS, Kang SH, Hwang Y-S, Park J-C, Hong SW, Han D-W (2015a) Reduced graphene oxide-coated hydroxyapatite composites stimulate spontaneous osteogenic differentiation of human mesenchymal stem cells. Nanoscale 7(27):11642–11651

Lee JH, Shin YC, Lee S-M, Jin OS, Kang SH, Hong SW, Jeong C-M, Huh JB, Han D-W (2015b) Enhanced osteogenesis by reduced graphene oxide/hydroxyapatite nanocomposites. Sci Rep 5 (1):18833

Lee J, Kim J, Kim S, Min DH (2016a) Biosensors based on graphene oxide and its biomedical application. Adv Drug Deliv Rev 105(Pt B):275–287

Lee JH, Lee Y, Shin YC, Kim MJ, Park JH, Hong SW, Kim B, Oh J-W, Park KD, Han D-W (2016b) In situ forming gelatin/graphene oxide hydrogels for facilitated C2C12 myoblast differentiation. Appl Spectrosc Rev 51(7–9):527–539

Li Y, Liu Y, Fu Y, Wei T, Le Guyader L, Gao G, Liu R-S, Chang Y-Z, Chen C (2012) The triggering of apoptosis in macrophages by pristine graphene through the MAPK and TGF-beta signaling pathways. Biomaterials 33(2):402–411

Li Y, Yuan H, von dem Bussche A, Creighton M, Hurt RH, Kane AB, Gao H (2013) Graphene microsheets enter cells through spontaneous membrane penetration at edge asperities and corner sites. PNAS 110(30):12295–12300

Li Y, Lu Z, Li Z, Nie G, Fang Y (2014) Cellular uptake and distribution of graphene oxide coated with layer-by-layer assembled polyelectrolytes. J Nanopart Res 16(5):2384

Linares J, Matesanz MC, Vila M, Feito MJ, Gonçalves G, Vallet-Regí M, Marques PAAP, Portolés MT (2014) Endocytic mechanisms of graphene oxide nanosheets in osteoblasts, hepatocytes and macrophages. ACS Appl Mater Interfaces 6(16):13697–13706

Liu S, Pan X, Liu H (2020) Two-dimensional nanomaterials for photothermal therapy. Angew Chem Int Ed Engl 59(15):5890–5900

Mäger I, Langel K, Lehto T, Eiríksdóttir E, Langel Ü (2012) The role of endocytosis on the uptake kinetics of luciferin-conjugated cell-penetrating peptides. Biochim Biophys Acta 1818 (3):502–511

Mahmoudi N, Eslahi N, Mehdipour A, Mohammadi M, Akbari M, Samadikuchaksaraei A, Simchi A (2017) Temporary skin grafts based on hybrid graphene oxide-natural biopolymer nanofibers as effective wound healing substitutes: pre-clinical and pathological studies in animal models. J Mater Sci Mater Med 28(5):73

Mao J, Guo R, Yan L-T (2014) Simulation and analysis of cellular internalization pathways and membrane perturbation for graphene nanosheets. Biomaterials 35(23):6069–6077

Matesanz M-C, Vila M, Feito M-J, Linares J, Gonçalves G, Vallet-Regi M, Marques PAAP, Portolés M-T (2013) The effects of graphene oxide nanosheets localized on F-actin filaments on cell-cycle alterations. Biomaterials 34(5):1562–1569

McCallion C, Burthem J, Rees-Unwin K, Golovanov A, Pluen A (2016) Graphene in therapeutics delivery: problems, solutions and future opportunities. Eur J Pharm Biopharm 104:235–250

Menaa F, Abdelghani A, Menaa B (2015) Graphene nanomaterials as biocompatible and conductive scaffolds for stem cells: impact for tissue engineering and regenerative medicine. J Tissue Eng Regen Med 9(12):1321–1338

Meng S, Zhang Y, Wang H, Wang L, Kong T, Zhang H, Meng S (2020) Recent advances on TMDCs for medical diagnosis. Biomaterials 269:120471

Moon JY, Kim M, Kim SI, Xu S, Choi JH, Whang D, Watanabe K, Taniguchi T, Park DS, Seo J, Cho SH, Son S-K, Lee J-H (2020) Layer-engineered large-area exfoliation of graphene. Sci Adv 6(44):eabc6601

Mu Q, Su G, Li L, Gilbertson BO, Yu LH, Zhang Q, Sun Y-P, Yan B (2012) Size-dependent cell uptake of protein-coated graphene oxide nanosheets. ACS Appl Mater Interfaces 4 (4):2259–2266

Mu H, Wang Z, Yuan J, Xiao S, Chen C, Chen Y, Chen Y, Song J, Wang Y, Xue Y, Zhang H, Bao Q (2015) Graphene-Bi_2Te_3 heterostructure as saturable absorber for short pulse generation. ACS Photonics 2(7):832–841

Novoselov KS, Geim AK, Morozov SV, Jiang D, Zhang Y, Dubonos SV, Grigorieva IV, Firsov AA (2004) Electric field effect in atomically thin carbon films. Science 306(5696):666–669

Palombo M, Deshmukh M, Myers D, Gao J, Szekely Z, Sinko PJ (2014) Pharmaceutical and toxicological properties of engineered nanomaterials for drug delivery. Annu Rev Pharmacol Toxicol 54:581–598

Park S, An J, Jung I, Piner RD, An SJ, Li X, Velamakanni A, Ruoff RS (2009) Colloidal suspensions of highly reduced graphene oxide in a wide variety of organic solvents. Nano Lett 9(4):1593–1597

Park KO, Lee JH, Park JH, Shin YC, Huh JB, Bae J-H, Kang SH, Hong SW, Kim B, Yang DJ, Han D-W, Yeum JH (2016) Graphene oxide-coated guided bone regeneration membranes with enhanced osteogenesis: spectroscopic analysis and animal study. Appl Spectrosc Rev 51 (7–9):540–551

Park SJ, Kim J, Kang S, Cha HJ, Shin H, Park J, Jang YS, Woo JS, Won C, Min DH (2020) Discovery of direct-acting antiviral agents with a graphene-based fluorescent nanosensor. Sci Adv 6(22):eaaz8201

Pattnaik S, Kumar Sutar M (2019) Graphene base nanocomposites: an overview. Mater Today Proc 18:5432–5437

Pinto AM, Gonçalves IC, Magalhães FD (2013) Graphene-based materials biocompatibility: a review. Colloids Surf B: Biointerfaces 111:188–202

Priyadarsini S, Mohanty S, Mukherjee S, Basu S, Mishra M (2018) Graphene and graphene oxide as nanomaterials for medicine and biology application. J Nanostructure Chem 8(2):123–137

Qiu Y, Wang Z, Owens ACE, Kulaots I, Chen Y, Kane AB, Hurt RH (2014) Antioxidant chemistry of graphene-based materials and its role in oxidation protection technology. Nanoscale 6 (20):11744–11755

Ryoo SR, Yim Y, Kim YK, Park IS, Na HK, Lee J, Jang H, Won C, Hong S, Kim SY, Jeon NL, Song JM, Min D-H (2018) High-throughput chemical screening to discover new modulators of microRNA expression in living cells by using graphene-based biosensor. Sci Rep 8(1):11413

Sanchez VC, Jachak A, Hurt RH, Kane AB (2012) Biological interactions of graphene-family nanomaterials: an interdisciplinary review. Chem Res Toxicol 25(1):15–34

Sato K, Nagai J, Mitsui N, Ryoko Y, Takano M (2009) Effects of endocytosis inhibitors on internalization of human IgG by Caco-2 human intestinal epithelial cells. Life Sci 85 (23):800–807

Schulz WL, Haj AK, Schiff LA (2012) Reovirus uses multiple endocytic pathways for cell entry. J Virol 86(23):12665–12675

Shams E, Yeganeh H, Naderi-Manesh H, Gharibi R, Mohammad Hassan Z (2017) Polyurethane/siloxane membranes containing graphene oxide nanoplatelets as antimicrobial wound dressings: in vitro and in vivo evaluations. J Mater Sci Mater Med 28(5):75–75

Shang L, Qi Y, Lu H, Pei H, Li Y, Qu L, Wu Z, Zhang W (2019) 7 - Graphene and graphene oxide for tissue engineering and regeneration. In: Cui W, Zhao X (eds) Theranostic bionanomaterials. Elsevier, Amsterdam, The Netherlands, pp 165–185. https://doi.org/10.1016/B978-0-12-815341-3.00007-9

Shi X, von dem Bussche A, Hurt RH, Kane AB, Gao H (2011) Cell entry of one-dimensional nanomaterials occurs by tip recognition and rotation. Nat Nanotechnol 6(11):714–719

Shin YC, Lee JH, Jin OS, Kang SH, Hong SW, Kim B, Park J-C, Han D-W (2015a) Synergistic effects of reduced graphene oxide and hydroxyapatite on osteogenic differentiation of MC3T3-E1 preosteoblasts. Carbon 95:1051–1060

Shin YC, Lee JH, Jin L, Kim MJ, Kim Y-J, Hyun JK, Jung T-G, Hong SW, Han D-W (2015b) Stimulated myoblast differentiation on graphene oxide-impregnated PLGA-collagen hybrid fibre matrices. J Nanobiotechnol 13(1):21

Shin YC, Lee JH, Kim MJ, Hong SW, Kim B, Hyun JK, Choi YS, Park J-C, Han D-W (2015c) Stimulating effect of graphene oxide on myogenesis of C2C12 myoblasts on RGD peptide-decorated PLGA nanofiber matrices. J Biol Eng 9(1):22

Shin SR, Li YC, Jang HL, Khoshakhlagh P, Akbari M, Nasajpour A, Zhang YS, Tamayol A, Khademhosseini A (2016) Graphene-based materials for tissue engineering. Adv Drug Deliv Rev 105(Pt B):255–274

Shin YC, Kim J, Kim SE, Song S-J, Hong SW, Oh J-W, Lee J, Park J-C, Hyon S-H, Han D-W (2017a) RGD peptide and graphene oxide co-functionalized PLGA nanofiber scaffolds for vascular tissue engineering. Regener Biomater 4(3):159–166

Shin YC, Song S-J, Hong SW, Jeong SJ, Chrzanowski W, Lee J-C, Han D-W (2017b) Multifaceted biomedical applications of functional graphene nanomaterials to coated substrates, patterned arrays and hybrid scaffolds. Nano 7(11):369

Shin YC, Kang SH, Lee JH, Kim B, Hong SW, Han D-W (2018a) Three-dimensional graphene oxide-coated polyurethane foams beneficial to myogenesis. J Biomater Sci Polym Ed 29 (7–9):762–774

Shin YC, Song SJ, Hong SW, Oh JW, Hwang YS, Choi YS, Han D-W (2018b) Graphene-functionalized biomimetic scaffolds for tissue regeneration. Adv Exp Med Biol 1064:73–89

Smith MJ, Brown JM, Zamboni WC, Walker NJ (2014) From immunotoxicity to nanotherapy: the effects of nanomaterials on the immune system. Toxicol Sci 138(2):249–255

Vadivel S, Tejangkura W, Sawangphruk M (2020) Graphite/graphene composites from the recovered spent Zn/carbon primary cell for the high-performance anode of lithium-ion batteries. ACS Omega 5(25):15240–15246

Wang X, Cheng L (2019) Multifunctional two-dimensional nanocomposites for photothermal-based combined cancer therapy. Nanoscale 11(34):15685–15708

Yang M-Q, Zhang N, Xu Y-J (2013) Synthesis of fullerene-, carbon nanotube-, and graphene-TiO_2 nanocomposite photocatalysts for selective oxidation: a comparative study. ACS Appl Mater Interfaces 5(3):1156–1164

Zeng M, Xiao Y, Liu J, Yang K, Fu L (2018) Exploring two-dimensional materials toward the next-generation circuits: from monomer design to assembly control. Chem Rev 118(13):6236–6296

Zhang B, Wei P, Zhou Z, Wei T (2016) Interactions of graphene with mammalian cells: molecular mechanisms and biomedical insights. Adv Drug Deliv Rev 105(Pt B):145–162

Zhang H, Fan T, Chen W, Li Y, Wang B (2020a) Recent advances of two-dimensional materials in smart drug delivery nano-systems. Bioact Mater 5(4):1071–1086

Zhang L, Meng Y, Pan G, Xia S (2020b) Experimental and theoretical investigations into the performance and mechanism of CO_2 capture by 3D and 2D ZnAl layered double hydroxides. Inorg Chem 59(23):17722–17731

Zhao F, Zhao Y, Liu Y, Chang X, Chen C, Zhao Y (2011) Cellular uptake, intracellular trafficking, and cytotoxicity of nanomaterials. Small 7(10):1322–1337

Zhao H, Ding R, Zhao X, Li Y, Qu L, Pei H, Yildirimer L, Wu Z, Zhang W (2017) Graphene-based nanomaterials for drug and/or gene delivery, bioimaging, and tissue engineering. Drug Discov Today 22(9):1302–1317

Zhou H, Zhang B, Zheng J, Yu M, Zhou T, Zhao K, Jia Y, Gao X, Chen C, Wei T (2014) The inhibition of migration and invasion of cancer cells by graphene via the impairment of mitochondrial respiration. Biomaterials 35(5):1597–1607

Zhou X, Sun H, Bai X (2020) Two-dimensional transition metal dichalcogenides: synthesis, biomedical applications and biosafety evaluation. Front Bioeng Biotechnol 8(236):1–13

The manufacturer's authorised representative in the EU is Springer Nature Customer Service Centre GmbH, Europaplatz 3, 69115 Heidelberg, Germany. If you have any concerns regarding our products, please contact ProductSafety@springernature.com

Printed and bound by CPI Group (UK) Ltd, Croydon, CR0 4YY
30/09/2026
02225248-0002